Lecture Notes in Computer Science 8741

Commenced Publication in 1973
Founding and Former Series Editors:
Gerhard Goos, Juris Hartmanis, and Jan van Leeuwen

Roman Kontchakov   Marie-Laure Mugnier (Eds.)

Web Reasoning and Rule Systems

8th International Conference, RR 2014
Athens, Greece, September 15-17, 2014
Proceedings

 Springer

Volume Editors

Roman Kontchakov
Birkbeck, University of London
Department of Computer Science
and Information Systems
Malet Street
London WC1E 7HX, UK
E-mail: roman@dcs.bbk.ac.uk

Marie-Laure Mugnier
LIRMM
161, rue ADA
34395 Montpellier Cedex 5, France
E-mail: mugnier@lirmm.fr

ISSN 0302-9743   e-ISSN 1611-3349
ISBN 978-3-319-11112-4   e-ISBN 978-3-319-11113-1
DOI 10.1007/978-3-319-11113-1
Springer Cham Heidelberg New York Dordrecht London

Library of Congress Control Number: 2014947223

LNCS Sublibrary: SL 3 – Information Systems and Application, incl. Internet/Web
and HCI

Typesetting: Camera-ready by author, data conversion by Scientific Publishing Services, Chennai, India

Printed on acid-free paper

Springer is part of Springer Science+Business Media (www.springer.com)

Preface

Web Reasoning aims to develop semantic-based techniques for exploiting and and making sense of data on the Web. Web data is distributed over numerous sources, which are dynamic, heterogenous, often incomplete, possibly contradictory and even unreliable. These features of web data require new methodologies and paradigms, adequate representation languages and practically efficient and robust algorithms. These challenging issues concern not only the Semantic Web but more generally modern information systems.

Ontologies are at the core of Web Reasoning. They are typically specified in languages based on description logics, rule-based formalisms, or combinations of the two. Recent developments in the field have built on close relationships with logic programming and databases, with a strong renewed interest for Datalog, the language of deductive databases. In this context, ontology-based data access, a paradigm of answering queries over data enriched with ontological knowledge, has emerged as a prominent direction. Ontology-based data integration and exchange have also attracted attention from both the academia and industry. The International Conference on Web Reasoning and Rule Systems (RR) is a major forum for discussion of these issues, and other issues relevant to Web Reasoning, and dissemination of the latest results in the field.

This volume contains the proceedings of the 8th RR conference, held from 15 to 17 September 2014 in Athens, Greece. The conference program featured 4 invited talks: keynotes by Frank van Harmelen and Markus Krötzsch, an industry talk by Stephan Grimm and a tutorial by Nasos Drosopoulos and Ilianna Kollia. The 9 full papers and 9 technical communications of this volume were included for presentation at the conference. The latter are shorter papers mainly describing preliminary and ongoing work, systems and applications, and new ideas of interest to the RR audience.

Accepted papers were selected out of 33 submissions, which included 19 full papers and 14 technical communications. Each submission received at least 3 reviews. After much discussion, 9 full papers and 7 technical communications were accepted, and further 3 full papers were accepted as technical communications, of which one was withdrawn. The conference also hosted a Doctoral Consortium with a number of poster presentations and 3 abstracts in these proceedings. As in recent years, the conference was co-located with the Reasoning Web Summer School (in the 10th edition), held just before RR in Athens.

We would like to thank the members of the Program Committee and the additional reviewers for their efforts to produce fair and thorough evaluation of the submitted papers, the Local Organization Committee headed by Manolis Koubarakis, the general chair Axel Polleres, the sponsorship chair Giorgos Stamou, the publicity chair Giorgos Stoilos, the Doctoral Consortium chair Francesco Ricca and of course the authors of the scientific papers and the invited speakers. Furthermore we are grateful to the sponsors for their generous support: NSF, Google, Artificial Intelligence Journal, Oracle, Optique, ICCS-NTUA, EETN, Inria and Siemens. Last, but not least, we thank the people behind EasyChair for providing resources and a marvelous conference management system.

September 2014   Roman Kontchakov
                                                         Marie-Laure Mugnier

Organization

General Chair

Axel Polleres — Vienna University of Economics and Business (WU), Austria

Program Chairs

Roman Kontchakov — Birkbeck, University of London, UK
Marie-Laure Mugnier — University of Montpellier 2, LIRMM/Inria, France

Local Organization Chair

Manolis Koubarakis — National and Kapodistrian University of Athens, Greece

Doctoral Consortium Chair

Francesco Ricca — University of Calabria, Italy

Sponsorship Chair

Giorgos Stamou — National Technical University of Athens, Greece

Publicity Chair

Giorgos Stoilos — National Technical University of Athens, Greece

Program Committee

José Júlio Alferes — Universidade Nova de Lisboa, Portugal
Darko Anicic — Siemens AG, Germany
Marcelo Arenas — Pontificia Universidad Católica de Chile, Chile
Jean-François Baget — Inria/LIRMM, France
Marcello Balduccini — Drexel University, USA

Additional Reviewers

Ahmetaj, Shqiponja
Fink, Michael
Klarman, Szymon
Mutharaju, Raghava
Alviano, Mario

Gutiérrez Basulto,
  Víctor
Kramdi, Seifeddine
Rullo, Antonino
Casini, Giovanni

Jung, Jean Christoph
Liang, Senlin
Ugarte, Martin

Doctoral Consortium Program Committee

Leopoldo Bertossi	Carleton University, Canada
Pedro Cabalar	University of Corunna, Spain
Wolfgang Faber	University of Huddersfield, UK
Joohyung Lee	Arizona State University, USA
Domenico Lembo	Sapienza Università di Roma, Italy
Marco Manna	University of Calabria, Italy
Marco Maratea	University of Genova, Italy
Alessandra Mileo	INSIGHT Centre for Data Analytics, NUI Galway, Ireland
Hans Tompits	Vienna University of Technology, Austria

Sponsors

Optique

SIEMENS

Semantic Technologies
in Selected Industrial Applications

Stephan Grimm

Siemens AG, Corporate Technology
Munich, Germany
stephan.grimm@siemens.com

Abstract. Semantic technology around knowledge representation and reasoning offers promising methods and tools for industrial applications. This talk will give an insight into selected projects where semantic technology has been successfully applied in innovative technology fields. It will illustrate that research on reasoning, rule-based systems and ontologies does have an impact in areas like power generation, industrial automation or health care, to name just a few.

Siemens is a leading industrial player in various innovative technology areas, such as power generation, industrial automation, traffic control or health care applications, to name just a few. The R&D department *Corporate Technology* is layered across the Siemens business units and is organized in various technology fields with the mission of transferring the latest research results into in-house business innovations.

Within the technology field of *Business Analytics and Monitoring* also semantic technologies are being researched on. They cover methods for the processing of highly structured data like ontologies, rule-based systems and deductive as well as abductive and inductive inference mechanisms, but also methods for data-driven interpretation, such as natural language processing and machine learning techniques. All those aspects of semantic technologies find an application in many innovative technology areas, the following being an incomplete list of examples drawn from past and ongoing research activities.

- Reasoning about ontologies in the light-weight description logic OWL EL is applied to the diagnosing of turbine sensor data in order to detect operational faults and to find their root causes, as reported in [6].
- Rule-based inference and complex event processing are applied in combination with ontologies for monitoring the operation of industrial manufacturing plants, as described in [2].
- A technology stack for CEP-style reasoning with the ETALIS framework [4] on the Gumstix[1] embedded controller is sketched in [5] for condition monitoring and diagnostics of technical devices.

---

[1] www.gumstix.com

– OWL DL reasoning is applied to the automated validation of plant engineering models to support plant engineers in finding misconceptions prior to
  building up industrial plants, as reported in [3].
– Semantic Media Wiki [7] is applied to the capturing and interactive visual-
  ization of knowledge about complex industrial plants in order to support the
  plant engineering phase, as e.g. reported in [1].
– An ontology of diseases and symptoms is used to infer likely diseases based
  on semantic annotations to clinical data helping clinicians to make diagnoses,
  as described in [8].

References

1. Abele, L., Grimm, S.: Knowledge-based Integration of Industrial Plant Models. In:
   Proceedings of the 39th Conference of the IEEE Industrial Electronics Society (2013)
2. Abele, L., Grimm, S., Zillner, S., Kleinsteuber, M.: An Ontology-Based Approach
   for Decentralized Monitoring and Diagnostics. In: IEEE International Conference
   on Industrial Informatics (2014)
3. Abele, L., Legat, C., Grimm, S., Müller, A.W.: Ontology-based Validation of Plant
   Models. In: IEEE International Conference on Industrial Informatics (2013)
4. Anicic, D., Fodor, P., Stühmer, R., Stojanovic, N.: Efficient Logic-Based Complex
   Event Processing and Reactivity Handling. In: International Conference on Computational Science and Engineering, CSE 2009, pp. 56–63 (2009)
5. Grimm, S., Hubauer, T., Runkler, T.A., Pachajoa, C., Rempe, F., Seravalli, M.,
   Neumann, P.: A CEP Technology Stack for Situation Recognition on the Gumstix
   Embedded Controller. In: GI-Jahrestagung, pp. 1925–1930 (2013)
6. Grimm, S., Watzke, M., Hubauer, T., Cescolini, F.: Embedded $\mathcal{EL}+$ Reasoning on
   Programmable Logic Controllers. In: Cudré-Mauroux, P., et al. (eds.) ISWC 2012,
   Part II. LNCS, vol. 7650, pp. 66–81. Springer, Heidelberg (2012)
7. Krötzsch, M., Vrandečić, D., Völkel, M.: Semantic MediaWiki. In: Cruz, I., Decker,
   S., Allemang, D., Preist, C., Schwabe, D., Mika, P., Uschold, M., Aroyo, L.M. (eds.)
   ISWC 2006. LNCS, vol. 4273, pp. 935–942. Springer, Heidelberg (2006)
8. Oberkampf, H., Zillner, S., Bauer, B., Hammon, M.: Interpreting Patient Data using
   Medical Background Knowledge. In: Proceedings of the 3rd International Confer-
   ence on Biomedical Ontology, ICBO 2012, Graz, Austria. KR-MED Series. CEUR-
   WS.org. (2012)

Table of Contents

Technical Communications

Posters

Doctoral Consortium

P ≠ P

Why Some Reasoning Problems Are More Tractable Than Others

Markus Krötzsch

Technische Universität Dresden, Germany

Abstract. Knowledge representation and reasoning leads to a wide range of computational problems, and it is of great interest to understand the difficulty of these problems. Today this question is mainly studied using computational complexity theory and algorithmic complexity analysis. For example, entailment in propositional Horn logic is P-complete and a specific algorithm is known that runs in linear time. Unfortunately, tight algorithmic complexity bounds are rare and often based on impractical algorithms (e.g., $O(n^{2.373})$ for transitive closure by matrix multiplication), whereas computational complexity results abound but are very coarse-grained (e.g., many P-complete problems cannot be solved in linear time).

In this invited paper, we therefore advocate another approach to gauging the difficulty of a computation: we reformulate computational problems as query answering problems, and then ask how powerful a query language is needed to solve these problems. This reduces reasoning problems to a computational model – query answering – that is supported by many efficient implementations. It is of immediate practical interest to know if a problem can be reduced to query answering in an existing database system. On the theoretical side, it allows us to distinguish problems in a more-fine grained manner than computational complexity without being specific to a particular algorithm. We provide several examples of this approach and discuss its merits and limitations.

1 Introduction

There are two main reasons for studying the complexity of computational problems. On the practical side, we want to know what is needed to solve the problem: how much time and memory will it take? On the theoretical side, we want to understand the relative difficulty of the problem as compared to others. For example, there is value in understanding that one problem is NP-complete while another is ExpTime-complete, even though we have no proof that the former is strictly easier than the latter.

The previous example also illustrates that computational complexity does often not provide insights about the worst-case complexity of algorithms implementing the problem. Such insights can rather be obtained by more detailed algorithmic analysis for specific problems. For example, it is known that the classical inference problem of computing the transitive closure of a binary relation can be solved in $O(n^{2.3729})$ [20].

Unfortunately, however, such detailed analysis is missing for most problems in knowledge representation and reasoning, and we have to be content with much coarser bounds obtained from some naive algorithm (such as the immediate $O(n^3)$ bound for transitive closure). Indeed, the efforts required for a more in-depth analysis are substantial: the study of transitive closure algorithms has a history of over four decades, and

R. Kontchakov and M.-L. Mugnier (Eds.): RR 2014, LNCS 8741, pp. 1–22, 2014.

it is still unknown if the conjectured optimum of $O(n^2)$ can be reached [20]. Moreover, the best known algorithms are of limited value in practice, since the underlying view of transitive closure as matrix multiplication is rarely feasible. Even the existing results thus provide little information about the actual "difficulty" of the problem.

Computational complexity is therefore often the preferred measure for the complexity of reasoning problems (and many other problems) [17]. By abstracting worst-case estimates from specific functions to general classes of functions, we obtain broad complexity classes. One strength of this approach is that such classes are usually stable both under "simple" transformations of the problem and under modifications of the underlying computational model (the Turing Machine). This suggests that they really capture an intrinsic aspect of a computational problem. Another strength of the theory is that many (though not all) complexity classes are comparable and form a linear order. Although we often do not know if the order is strict, this provides us with a one-dimensional scale on which to specify computational difficulty.

The price to pay for this nice and clean theory is that it often tells us very little about the "real" difficulty of a problem either. Complexity classes are very broad. The class of P-complete problems, e.g., includes problems that can be solved in linear time as well as problems that are inherently quadratic, cubic, or even worse. Moreover, important complexity classes are based on non-deterministic models of computation that do not correspond to existing hardware, which is one of the reasons why it is hard to explain why exactly an NP-complete problem should be easier to implement than a PSPACE-hard problem on a real (deterministic) computer.

Summing up, we have the choice between an arduous algorithmic analysis that provides detailed results for specific cases, and a more abstract complexity analysis that leads to more general but very coarse-grained classifications. In both cases, the practical significance of our insights will vary. The object of this paper is to explore a middle ground in between these two extremes.

Computation as Query Answering. We propose to view computational problems as query answering problems over a database, and to explore their "difficulty" by studying how powerful a query language is needed to solve them. Indeed, query languages come in a great variety – including simple *conjunctive queries*, *regular path queries* and their extensions [18,4,6], highly expressive recursive *Datalog* queries [1], and a range of expressive fragments of Datalog [2] and higher-order logic [19] – providing us with a large space for a fine-grained classification of problems.

Databases are essentially relational structures, i.e., (hyper)graphs, and many problems admit natural representations in this format. Indeed, it is well known that the class of constraint satisfaction problems can naturally be expressed in terms of query answering [9]. In general, such a view is particularly suggestive if the original problem is based on a *set* of axioms, constraints, production rules, etc., where the actual order does not affect the meaning of the problem. This is typical for logical theories. A database, viewed as a graph, is similarly unordered, while traditional computational models require us to impose an order on the input. It is known that this has a profound impact on the computational properties of the problem: query languages are more powerful if we provide a total order on the elements of a database [10].

We will adopt a fairly simple and direct view of computational problems as graphs (databases). After this step, some query languages are sufficiently expressive to solve the original problem, while others are not. The challenge is to make this characterisation as tight as possible, i.e., to identify the "exact" point at which a query language becomes too simple to express a problem. For such results to be meaningful, we need to restrict ourselves to a relevant class of query languages that defines the space in which we want to locate computational problems. Indeed, the space of *all* query languages is so fine-grained that each problem would fall into its own class, incomparable to the others.

Our analysis will thus always be relative to a choice of query languages that provide the desired level of detail and the connection to practical tools and scenarios. We give two examples for such a choice. The first is the space of Datalog queries where the arity of intensional predicates (those that occur in rule heads) is bounded by some constant. The simplest case is *monadic Datalog*, where only unary predicates can appear in rule heads. Each increase of this arity leads to a strictly more expressive query language [2], resulting in an infinite, linear hierarchy of languages. Our second example is the space of known navigational query languages. This space consists of specific languages proposed in several papers, and we can apply our approach to relate relevant practical tasks to these practical languages.

Contributions. The main contribution of this paper is to show that this approach is indeed feasible in practice. On the one hand, we find a variety of practical problems to which it is applicable. On the other hand, we show that it is indeed possible to obtain tight classifications of these problems relative to practical query languages. This requires a number of techniques for understanding the expressive power of query languages, which is an interesting side effect of this work. We argue that this view can provide fresh answers to the two original questions of computational complexity:

- It can provide relevant insights into the practical feasibility of a problem. Our chosen computational model – database query answering – is supported by many implementations. Determining whether or not a certain problem can be expressed in a practical query language can tell us something about its implementability that might be just as relevant as theoretical worst-case complexity.
- It allows us to compare the relative difficulty of problems in classes that are more fine-grained than those of complexity theory. By showing that one problem requires strictly more query power than another, we can compare problems within the framework of a relevant class of query languages. In particular, this gives us a yardstick for comparing problems within P.

In exchange for these benefits, we must give up some of the advantages of complexity theory and algorithmic analysis:

- The more fine-grained structure of query languages is tied to the fact that there are many different, mutually incomparable query languages. "Difficulty," when measured in these terms, becomes a multi-dimensional property. We can retain a one-dimensional view by restricting the freedoms we allow ourselves for choosing a query language.

- Although query answering is a practical task on which plenty of practical experience is available, it does not provide specific runtime bounds in the sense of algorithmic analysis. Even if we encode a problem using, e.g., conjunctive queries with transitive closure, we do not know which transitive closure algorithm is actually used.
- The results we obtain are not as universal or stable as those of complexity theory. Whether a problem can be solved by a query of a certain type depends on the specific representation of the problem. Syntactic structure is relevant. In compensation, we can often specify very precisely at which point a query language becomes too simple to express a problem – a similar precision can rarely be attained in complexity theory since the exact relationship between nearby complexity classes if often open.

Overall, we therefore believe that our approach is, not a replacement for, but a valuable addition to the available set of tools. Indeed, we will show a number of concrete examples where we gain specific insights that go beyond what other methods can tell us.

We use the problem of computing positive and negative entailments in propositional Horn logic as a simple example to exercise our approach. In Section 2, we introduce this example and state our main theorem: the computation of negative entailments in propositional Horn logic requires a strictly more expressive variant of Datalog than the computation of positive entailments. Towards a proof of this method, Section 3 gives formal definitions for *database*, *query*, and *retrieval problem*. We introduce Datalog and its proof trees in Section 4, derive useful characteristics of its expressivity in Section 5, and further refine this method to work with database encodings of retrieval problems in Section 6. This allows us to prove the main theorem in Section 7. In Section 8, we give an overview of several other results that have been obtained for Datalog using similar methods on various ontology languages. In Section 9, we briefly review recent query-based reasoning approaches for sub-polynomial reasoning problems, using navigational queries instead of Datalog. Section 10 sums up our results, discusses the relationship to adjacent areas, and gives some perspectives for the further development of this field.

2 Example: Propositional Horn Logic

For a concrete example, we will instantiate our framework for the case of reasoning with propositional Horn logic. Given a set of *propositional letters* **A**, the set of *propositional Horn logic rules* is defined by the following grammar:

$$\textbf{Body} ::= \top \mid \textbf{A} \mid \textbf{A} \wedge \textbf{A} \tag{1}$$

$$\textbf{Head} ::= \bot \mid \textbf{A} \tag{2}$$

$$\textbf{Rule} ::= \textbf{Body} \rightarrow \textbf{Head} \tag{3}$$

The constants $\top$ and $\bot$ represent *true* and *false*, respectively. We restrict to binary conjunctions in rule bodies here; all our results extend to the case of n-ary bodies (considered as nested binary conjunctions), but we prefer the simplest possible formulation here. A *propositional Horn logic theory* is a set of propositional Horn logic rules. Logical entailment is defined as usual.

Given a Horn logic theory T we can now ask whether T is satisfiable. This is the case whenever $\bot$ is *not* derived starting from $\top$. More generally, one can also ask whether T entails some proposition b or some propositional implication $a \rightarrow b$. All of these problems are easily reducible to one another, and it is a classic result that all of them can be solved in linear time [8].

In fact, a known linear-time algorithm computes the set of all propositions that T entails to be true [8]. Interestingly, however, the situation changes if we want to compute the set of all propositions that are entailed to be false: no linear time algorithm is known for this case.[1]

What explains this unexpected asymmetry? For any given pair of propositions a and b, we can decide whether $T \models a \rightarrow b$ in linear time. We could use this to decide $T \models a \rightarrow \bot$ for all propositions a, but this would require a linear number of linear checks, i.e., quadratic time overall. Complexity theory does not provide any help either: any of the above decision problems (those with a yes-or-no answer) is P-complete [7]. On the one hand, P does not distinguish between linear and quadratic algorithms; on the other hand, the problem we are interested in is not a decision problem in the first place.

Considering that we are interested in *querying* for a set of propositions, query languages do indeed suggest themselves as a computational formalism. Here we use Datalog as one of the most well-known recursive query languages [1]. However, query languages act on databases, while our problem is given as a set of rules.

Fortunately, there are natural ways to represent Horn logic theories as databases. For this example, we use a simple encoding based on binary relations b (binary body), u (unary body) and h (head), as well as unary relations t (true) and f (false). Each rule, each propositional letter, and each of the constants $\top$ and $\bot$ are represented by a vertex in our graph. A rule $\rho : c \wedge d \rightarrow e$ is encoded by relations $b(c, \rho)$, $b(d, \rho)$, and $h(\rho, e)$, and analogously for unary rules but with u instead of b to encode the single proposition in the body. The unary relations t and f contain exactly the vertices $\top$ and $\bot$, respectively. Note that we have one explicit vertex per rule to ensure that there is no confusion between the premises of multiple rules with the same head. A Horn logic theory is now translated to a graph by translating each rule individually. The graphs of several rules can touch in the same proposition vertices, while all the rule vertices are distinct.

For the remainder of this paper, we assume that all logical theories under consideration are consistent – this alleviates us from checking the special case that all entailments are valid due to inconsistency. Now it is easy to give a Datalog query for the set of all propositions that are entailed to be true, expressed by the head predicate T:

$$t(x) \rightarrow T(x) \tag{4}$$

$$T(x) \wedge u(x, v) \wedge h(v, z) \rightarrow T(z) \tag{5}$$

$$T(x) \wedge T(y) \wedge b(x, v) \wedge b(y, v) \wedge h(v, z) \rightarrow T(z) \tag{6}$$

---

[1] The question if this is unavoidable and why was put forward as an open problem at the recent Dagstuhl Seminar 14201 on *Horn formulas, directed hypergraphs, lattices and closure systems*.

It is easy to verify that T contains exactly those propositions that are inferred to be true. How does this compare to the problem of finding the false propositions? The following Datalog query computes this set in the predicate F:

$$\rightarrow I(w, w) \tag{7}$$

$$I(w, w) \wedge t(x) \rightarrow I(w, x) \tag{8}$$

$$I(w, x) \wedge u(x, v) \wedge h(v, z) \rightarrow I(w, z) \tag{9}$$

$$I(w, x) \wedge I(w, y) \wedge b(x, v) \wedge b(y, v) \wedge h(v, z) \rightarrow I(w, z) \tag{10}$$

$$I(w, x) \wedge f(x) \rightarrow F(w) \tag{11}$$

We use an auxiliary binary predicate I (implication) to compute implications between propositions. Rule (7) asserts that everything implies itself using a variable w that appears only in the head. Using such an *unsafe* rule here is not crucial: instead, one could also create several rules with non-empty bodies to find all proposition vertices.

Comparing the query in lines (4)–(6) with the query in lines (7)–(11), we can see that the only head predicate in the former is the unary T, whereas the latter also uses a binary head predicate I. It is known that Datalog with binary head predicates is strictly more expressive than monadic Datalog (where only unary head predicates are allowed) [2], so in this sense our encoding of the second problem takes up more query expressivity in this example. This might be a coincidence – maybe we just missed a simpler approach of computing false propositions – but it turns out that it is not:

Theorem 1. *To compute all propositions that are entailed to be false by a propositional Horn theory in Datalog, it is necessary to use head predicates of arity two or more.*

We prove this result in Section 7. The theorem confirms that computing false proposition is in a sense inherently more difficult than computing true propositions, when measuring difficulty relative to the expressive power of Datalog. Similar to traditional complexity classes, this insight does not provide us with specific runtime bounds: it neither shows that true propositions can be computed in linear time, nor that false propositions require at least quadratic time. Yet, it achieves what neither complexity theory nor algorithmic analysis have accomplished: to provably separate the two problems with respect to a general computational model.

Of course, the result is relative to the choice of Datalog as a model of computation. However, this covers a wide range of conceivable reasoning algorithms: almost every deterministic, polytime saturation procedure can be viewed as a syntactic variant of Datalog. Our result therefore suggests that every such procedure will face some inherent difficulties when trying to compute false propositions in linear time.

To give a proof for Theorem 1, we need to introduce various other techniques first. We begin by defining the framework of our investigation in a more rigorous way.

3 Databases, Queries, and Computational Problems

In this section, we give more precise definitions of the terms *database* and *query* that we used rather informally until now. Moreover, we specify the general type of computational problems that our approach is concerned with, and we give a canonical way of viewing these problems in terms of databases.

A database signature is a finite set Σ of relation symbols, each with a fixed arity ≥ 0. A database D over Σ consists of an *active domain* Δ^D and, for each relation symbol $r \in \Sigma$ of arity n, an n-ary relation $r^D \subseteq (\Delta^D)^n$. We require that Δ^D is countable (usually it is even finite). Other names for databases in this sense are *relational structure*, *(predicate-logic) interpretation*, and *directed hypergraph*. In particular, a database over a signature with only binary relations is the same as a directed graph with labelled edges.

In the most general sense, a query language is a decidable language (i.e., we can decide if something is a query or not) together with an interpretation that assigns to each query a function from databases to the *results* of the query over this database.

This is a very general definition. In this paper, we restrict *query* to mean *logic-based query under set semantics*. Accordingly, a query is a formula of second-order logic, where every second-order variable must be bound by a quantifier and first-order variables may occur bound or free. The number of free first-order variables is called the *arity* of the query. For simplicity, we will not consider queries that contain constants or function symbols. A query with arity 0 is called *Boolean*.

Let Q be a query of arity n, and let D be a database. A *solution* of Q is a mapping μ from free variables in Q to Δ^D, such that $D \models \mu(Q)$, i.e., the formula Q is satisfied by the database (interpretation) D when interpreting each free variable x as $\mu(x)$. In other words, we check if D is a model for "$\mu(Q)$" under the standard second-order logic semantics (the details are inessential here; we give the semantics for concrete cases later). The *result* $Q(D)$ of Q over D is the set of all solutions of Q over D. Boolean queries have at most one solution (the unique function μ with empty domain), i.e., they are either satisfied by the database or not. We tacitly identify query solutions μ with n-tuples $\langle \mu(x_1), \ldots, \mu(x_n) \rangle$, where $x_1, \ldots, x_n$ is the sequence of free variables ordered by their first occurrence in Q.

In traditional complexity theory, computational problems are often phrased as *word problems*: the input of the Turing machine is an arbitrary word over an input alphabet; the task is to decide whether this string is contained in the language or not. Input strings in this sense could be viewed as (linear) graphs, but this encoding would be impractical if we want query languages to return the actual answer to the problem. Instead, the graph we use should encode the logical relationships that are given in the input, rather than a serialization of these relations in a string. At the same time, the encoding of the problem must not make any significant changes to the formulation of the problem – in particular, it should not pre-compute any part of the solution.

Many problems in reasoning are naturally given as sets of formulae, where each formula has a well-defined term structure, i.e., formulae are defined as terms built using logical operators and signature symbols. A general framework to discuss such terms are (multi-sorted) term algebras. Instead of giving a full introduction to this field here, we give some examples.

Example 1. Formulae of propositional logic can be viewed as terms over a set of *operators*: we usually use binary operators $\wedge$, $\vee$, and $\rightarrow$; unary operator $\neg$; and sometimes also nullary operators $\top$ and $\bot$. Formulae (terms) are formed from these operators and a countably infinite set of propositional letters $\mathbf{A}$ (in algebraic terms, this set is called *generating set*) in the usual way.

Example 2. Terms of first-order logic are also terms in this sense. Here, the "operators" are given by a signature of function symbols with associated arity (we treat constants as functions of arity 0). Terms are constructed from a countably infinite set of variables $\mathbf{V}$ (the generators): every variable $x \in \mathbf{V}$ is a term; and for every n-ary function f and terms $t_1, \ldots, t_n$, the expression $f(t_1, \ldots, t_n)$ is a term.

Example 3. Formulae of first-order logic can also be viewed as terms using the well-known logical operators. In this case, the set of generators are the logical atoms, constructed using a first-order signature of predicate symbols and an underlying language of first-order terms. One can view such formulae as terms over a multi-sorted algebra, where we use multiple *sorts* (first-order terms, formulae, ...) and the arity of operators (function symbol, predicate, logical operator) is given as the signature of a function over sorts. For example, the operator $\wedge$ is of type formula $\times$ formula $\rightarrow$ formula while a ternary predicate is of type term $\times$ term $\times$ term $\rightarrow$ formula. If we introduce a sort variable, we can view quantifiers as binary operators variable $\times$ formula $\rightarrow$ formula.

As we are mainly interested in logic here, we use *formula* (rather than *term*) to refer to logical expressions. There is usually some choice on how to capture a logical language using (multi-sorted) operators and generating sets, but in all cases formulae have a natural representation as a labelled tree structure:

Definition 1. *Consider a set $\mathbf{L}$ of formulae over some signature of operators (which might include predicates and function symbols). For a formula $\varphi \in \mathbf{L}$, the set $\mathsf{ST}(\varphi)$ of subterms of φ is defined inductively as usual: $\varphi \in \mathsf{ST}(\varphi)$ and, if $\varphi = o(\varphi_1, \ldots, \varphi_n)$ for some n-ary operator o, then $\mathsf{ST}(\varphi_i) \subseteq \mathsf{ST}(\varphi)$ for each $i \in \{1, \ldots, n\}$. Moreover, φ is associated with a graph $\mathsf{G}(\varphi)$ defined as follows:*

- $\mathsf{G}(\varphi)$ *contains the vertex* $[\varphi]$,
- *if φ is of the form $o(\varphi_1, \ldots, \varphi_n)$ then $[\varphi]$ is labelled by t_o and, for each $i \in \{1, \ldots, n\}$, $\mathsf{G}(\varphi)$ contains an edge from $[\varphi]$ to $[\varphi_i]$ with label r_o^i, and $\mathsf{G}(\varphi_i) \subseteq \mathsf{G}(\varphi)$.*

We call a finite set $T \subseteq \mathbf{L}$ a theory. The subterms of T are defined by $\mathsf{ST}(T) := \bigcup_{\varphi \in T} \mathsf{ST}(\varphi)$. T is associated with a graph $\mathsf{G}(T) := \bigcup_{\varphi \in T} \mathsf{G}(\varphi)$ over the set of vertices $\{[\varphi] \mid \varphi \in \mathsf{ST}(T)\}$, i.e., vertices for the same subterm are identified in $\mathsf{G}(T)$.

Note that the number of vertices and edges of $\mathsf{G}(T)$ is linear in the size of T. The names we use to refer to vertices can also be of linear size, but they are only for our reference and should not be considered to be part of the database (the query languages we study do not "read" labels). Thus, the translation is linear overall.

Example 4. The propositional Horn rule $b \wedge c \rightarrow d$ is represented by the following graph structure $\mathsf{G}(b \wedge c \rightarrow d)$, where $t_\rightarrow$ and $t_\wedge$ are vertex labels:

$$[b \wedge c \rightarrow d] \xrightarrow{r_\rightarrow^1} [b \wedge c] \xrightarrow{r_\wedge^1} [b]$$
$$t_\rightarrow \qquad\qquad t_\wedge \qquad\qquad r_\wedge^2 \searrow [c]$$
$$r_\rightarrow^2 \searrow [d]$$

A propositional Horn theory is translated by taking the union of the graphs obtained by translating each of its rules. For example, the translation of the theory $\{a \rightarrow b, a \rightarrow c, b \wedge c \rightarrow d\}$ can be visualized as follows, where we omit the vertex labels:

$$[d] \xleftarrow{r_\rightarrow^2} [b \wedge c \rightarrow d] \xrightarrow{r_\rightarrow^1} [b \wedge c] \xrightarrow{r_\wedge^1} [b] \xleftarrow{r_\rightarrow^2} [a \rightarrow b] \xrightarrow{r_\rightarrow^1} [a]$$
$$\xrightarrow{r_\wedge^2} [c] \xleftarrow{r_\rightarrow^2} [a \rightarrow c] \quad r_\rightarrow^1$$

This provides us with a canonical representation of a broad class computational problems as graphs (and thus databases). Many problems are naturally given by sets of terms or formulae, in particular most logical reasoning tasks. Sometimes a slightly simpler encoding could be used too (e.g., the graphs in Example 4 are more "verbose" than those used for the intuitive introduction in Section 2), but results obtained for the canonical encoding usually carry over to such simplifications.

It is easy to formulate decision problems with respect to this encoding. However, our use of query languages allows us to generalise this to a broader class of *retrieval problems*, which will be the main notion of computational problem we study.

Definition 2. *Consider a set* **L** *of logical formulae. For* $n \geq 0$, *an* n-ary retrieval problem *is given by the following:*

- *a set* $\mathscr{T}$ *of theories of* **L**;
- *a mapping* Θ *from theories* $T \in \mathscr{T}$ *to subsets of* $\mathsf{ST}(T)^n$.

An n-ary retrieval problem *is* solved *by an* n-ary query Q *if, for all theories* $T \in \mathscr{T}$, *the result* $Q(\mathsf{G}(T))$ *of* Q *over* $\mathsf{G}(T)$ *is exactly the set* $\Theta(T)$ *(where we identify vertices* $[\varphi]$ *with subterms* φ).

Decision problems are obtained as retrieval problems of arity $n = 0$. Computing all true (respectively false) propositions in Horn logic is a unary retrieval problem. Also note that we do not require Q to decide whether the given database is actually of the form $\mathsf{G}(T)$ for some theory $T \in \mathscr{T}$: the query only needs to be correct for databases that actually encode instances of our computational problem.

4 Datalog with Bounded Arity

In this paper, we mainly consider Datalog as our query language [1] and obtain a hierarchy of increasing expressivity by varying the maximal arity of head predicates. In this section, we define the basic notions that we require to work with Datalog.

Let Σ be a database signature as in Section 3. A Datalog query over Σ is based on an extended set of relation (or predicate) symbols $\Sigma' \supseteq \Sigma$. Predicates in Σ are called *extensional* (or *extensional database*, EDB), and predicates in $\Sigma' \setminus \Sigma$ are called *intensional* (or *intensional database*, IDB). We also use a fixed, countably infinite set **V** of variables.

A *Datalog atom* is an expression $p(x_1, \ldots, x_n)$ where $p \in \Sigma'$ is of arity n, and $x_i \in \mathbf{V}$ for $i = 1, \ldots, n$. We do not consider Datalog queries with constant symbols here. An IDB (EDB) atom is one that uses an IDB (EDB) predicate. A *Datalog rule* is a formula of the form $B_1 \wedge \ldots \wedge B_\ell \rightarrow H$ where B_i and H are Datalog atoms, and H is an IDB

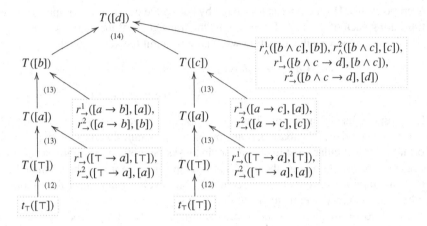

Fig. 1. Proof tree for Example 5 with leaf nodes combined in dotted boxes; labels (12)–(14) refer to the rule applied in each step

atom. The premise of a rule is also called its *body*, and the conclusion is called its *head*. A *Datalog query* $\langle P, g \rangle$ is a set of Datalog rules P with a *goal predicate* $g \in \Sigma' \setminus \Sigma$.

Under the logical perspective on queries of Section 3, a Datalog query corresponds to a second-order formula with IDB predicates representing second-order variables. For our purposes, however, it makes sense to define the semantics of Datalog via *proof trees*.

Consider a database D with active domain Δ^D. A *ground atom* for an n-ary (IDB or EDB) predicate p is an expression of the form $p(d_1, \ldots, d_n)$ where $d_1, \ldots, d_n \in \Delta^D$. A *variable assignment* μ for D is a function $\mathbf{V} \to \Delta^D$. The *ground instance* of an atom $p(x_1, \ldots, x_n)$ under μ is the ground atom $p(\mu(x_1), \ldots, \mu(x_n))$.

A *proof tree* for a Datalog query $\langle P, g \rangle$ over a database D is a structure $\langle N, E, \lambda \rangle$ where N is a finite set of nodes, $E \subseteq N \times N$ is a set of edges of a directed tree, and λ is a labelling function that assigns a ground atom to each node, such that the following holds for each node $n \in N$ with label $\lambda(n) = p(d_1, \ldots, d_k)$:

- if p is an EDB predicate, then n is a leaf node and $\langle d_1, \ldots, d_k \rangle \in p^D$;
- if p is an IDB predicate, then there is a rule $B_1 \wedge \ldots \wedge B_\ell \to H \in P$ and a variable assignment μ such that $\lambda(n) = \mu(H)$ and the set of child nodes $\{m \mid \langle n, m \rangle \in E\}$ is of the form $\{m_1, \ldots, m_\ell\}$ where $\lambda(m_i) = \mu(B_i)$ for each $i = 1, \ldots, \ell$.

A tuple $\langle d_1, \ldots, d_k \rangle$ is a solution of $\langle P, g \rangle$ over D if there is a proof tree for $\langle P, g \rangle$ over D with root label $g(d_1, \ldots, d_k)$.

Example 5. The Datalog query (4)–(6) of Section 2 can be reformulated for the canonical graphs of Section 3, as illustrated in Example 4. Note that $\top$ is a nullary operator.

$$t_\top(x) \to T(x) \tag{12}$$

$$T(x) \wedge r^1_\to(v, x) \wedge r^2_\to(v, z) \to T(z) \tag{13}$$

$$T(x) \wedge T(y) \wedge r^1_\wedge(w, x) \wedge r^2_\wedge(w, y) \wedge r^1_\to(v, w) \wedge r^2_\to(v, z) \to T(z) \tag{14}$$

Here, $r^1_\rightarrow$, $r^2_\rightarrow$, $r^1_\wedge$, $r^2_\wedge$, and $t_\top$ are EDB predicates from Σ, and T is the only additional IDB predicate in $\Sigma' \setminus \Sigma$. For example, rule (14) states that, whenever x and y are true, and there is a rule v of form $x \wedge y \rightarrow z$, then z is also true. Now consider the propositional Horn theory $\{\top \rightarrow a, a \rightarrow b, a \rightarrow c, b \wedge c \rightarrow d\}$, which is the same as in Example 4 but with an added implication $\top \rightarrow a$. The proof tree of $T([d])$ is shown in Fig. 1.

5 The Limits of Datalog Expressivity

To prove Theorem 1, we need to show that no monadic Datalog query can correctly compute the false propositions from a propositional Horn theory. To accomplish this, we first study the properties of retrieval problems that *can* be solved in monadic Datalog. We can then proceed to show that some of these properties are violated by (certain instances of) the logical entailment question that we are interested in.

Datalog has a number of general properties that we can try to exploit. Most obviously, Datalog is *deterministic*, or, in logical terms, it does not support disjunctive information. However, this feature is not immediately related to the arity of predicates. Another general property is that Datalog is *positive* in the sense that it only takes positive information into account during query evaluation: the absence of a structure cannot be detected in Datalog. In logical terms, this corresponds to a lack of negation; in model-theoretic terms, it corresponds to *closure of models under homomorphisms*. This feature is an important tool in our investigations:

Definition 3. *Consider databases D_1 and D_2 with active domains Δ^D_1 and Δ^D_2 and over the same database signature Σ. A function $\mu : \Delta^D_1 \rightarrow \Delta^D_2$ is a* homomorphism *from D_1 to D_2 if, for all n-ary relations $r \in \Sigma$ and elements $d_1, \ldots, d_n \in \Delta^D_1$, we have that $\langle d_1, \ldots, d_n \rangle \in r^{D_1}$ implies $\langle \mu(d_1), \ldots, \mu(d_n) \rangle \in r^{D_2}$.*

Consider a query Q over Σ. Q is closed under homomorphisms *if, for all databases D_1 and D_2 over Σ, and every homomorphism $\mu : D_1 \rightarrow D_2$, we have $\mu(Q(D_1)) \subseteq Q(D_2)$, where $\mu(Q(D_1))$ is the set obtained by applying μ to each element in the query result $Q(D_1)$. In particular, if Q is Boolean and D_1 is a model of Q, then D_2 is a model of Q.*

It is not hard to show that Datalog is closed under homomorphisms in this sense. The utility of this observation is that it allows us to restrict attention to a much more select class of databases. Intuitively speaking, if D_1 has a homomorphism to D_2, then D_1 is less specific than D_2. We are interested in cases where this simplification does not eliminate any query results. This is captured in the following notion.

Definition 4. *Consider a database signature Σ. A set of databases $\mathscr{D}$ is a* covering *of a query Q if, for all databases D over Σ, there is a database $D' \in \mathscr{D}$ and a homomorphism $\mu : D' \rightarrow D$, such that $Q(D) = \mu(Q(D'))$.*

Intuitively speaking, a covering of a query represents every way in which the query can return a result. In most cases, a minimal covering (i.e., a covering not properly containing another covering) does not exist. Nevertheless, coverings provide a powerful proof mechanism for showing the expressive limits of a (positive) query language. The general strategy is: (1) show that every query of the language admits a covering that consists only of databases with a specific structural property, and (2) then find a retrieval problem that does not admit a covering with this property.

For example, Datalog admits natural coverings based on proof trees. The labels of the leaves of a proof tree define a "minimal" database in which the root of the proof tree can be inferred. However, this sub-database can still be much more specific than needed to obtain a correct proof tree.

Example 6. The derivation of Example 5 is possible in any database that contains the leaf labels of the according proof tree, shown in dotted boxes in Fig. 1. The subtrees for $T([b])$ and $T([c])$ both rely on the same implication $\top \rightarrow a$. Intuitively speaking, this use of a in two different branches of the proof tree is not required by the Datalog query: if we replace one of the uses of a by a', e.g., by replacing the rule $a \rightarrow c$ with $\{a' \rightarrow c, \top \rightarrow a'\}$ then we are able to derive the same result. There is a homomorphism from the modified database to the original one, mapping both a' and a to a.

This idea of renaming elements that occur only in parallel branches of a proof tree can be formalised for arbitrary Datalog queries as follows.

Definition 5. *Consider a database D, a Datalog query $\langle P, g \rangle$, and a proof tree $t = \langle N, E, \lambda \rangle$ of $\langle P, g \rangle$ on D. The* interface *of a node $n \in N$ is the set of constants from the active domain Δ^D of D that occur in its label $\lambda(n)$.*

The diversification *of the proof tree t with root node n_r is constructed recursively:*

- *for all constants $d \in \Delta^D$ that are not in the interface of n_r, introduce a fresh constant d' that has not yet been used in t or in this construction yet, and replace every occurrence of d in labels of t by d';*
- *apply the diversification procedure recursively to all subtrees of t that have a non-leaf child node of n_r as their root.*

We tacitly assume that Δ^D contains all required new constants (or we extend it accordingly). Note that the replacement of constants may not be uniform throughout the tree, i.e., several fresh constants might be introduced to replace uses of a single constant d.

Example 7. Figure 2 shows the diversified version of the proof tree for Example 5 as shown in Fig. 1. Here, we use superscript numbers to distinguish several versions of renamed vertices, where the number indicates the node in which the constant was introduced. Note that we treat expressions like $[a \rightarrow b]$ as database constants, ignoring the internal structure that their name suggests. Likewise, a diversified name like $[a \rightarrow b]^2$ is just a constant symbol. The two tree nodes originally labelled $T([a])$ now refer to distinct vertices $[a]^2$ and $[a]^4$ in the graph. The set of leaf facts shown in dotted boxes forms a diversified database with the following structure (with vertex labels $t_\rightarrow$, $t_\wedge$, and $t_\top$ omitted):

$$[d] \xleftarrow{r_\rightarrow^2} [b \wedge c \rightarrow d]^1$$
$$\downarrow r_\rightarrow^1$$
$$[b \wedge c]^1 \xrightarrow{r_\wedge^1} [b]^1 \xleftarrow{r_\rightarrow^2} [a \rightarrow b]^2 \xrightarrow{r_\rightarrow^1} [a]^2 \xleftarrow{r_\rightarrow^2} [\top \rightarrow a]^3 \xrightarrow{r_\rightarrow^1} [\top]^3$$
$$\searrow r_\wedge^2 \quad [c]^1 \xleftarrow{r_\rightarrow^2} [a \rightarrow c]^4 \xrightarrow{r_\rightarrow^1} [a]^4 \xleftarrow{r_\rightarrow^2} [\top \rightarrow a]^5 \xrightarrow{r_\rightarrow^1} [\top]^5$$

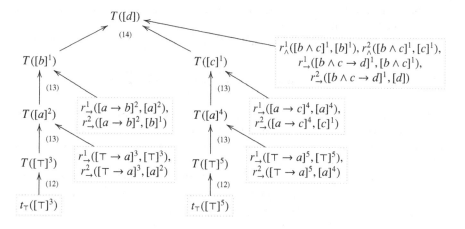

Fig. 2. Diversified version of the proof tree in Fig. 1

This database still allows for the fact $T([d])$ to be entailed, and it can be mapped by a homomorphism into the original database of Fig. 1 (partly illustrated in Example 4).

The previous example illustrates that diversified proof trees lead to diversified databases. The example considered only a single solution of the query; a more general formulation is as follows. For a Datalog query $\langle P, g \rangle$ and a database D, let $T(D, P, g)$ be the set of all proof trees for $\langle P, g \rangle$ over D. Let $\bar{T}(D, P, g)$ be a set containing one diversification for each tree in $T(D, P, g)$, where every fresh constant that was introduced during the diversification of a proof tree is distinct from all other constants throughout all diversified trees. Now the diversified database $D|_{\langle P,g \rangle}$ is defined as the union of all leaf node labels of proof trees in $\bar{T}(D, P, g)$.

Theorem 2. *Let $\mathscr{D}$ be the set of all databases over a signature Σ.[2] For every Datalog query $\langle P, g \rangle$ over Σ, the set $\{D|_{\langle P,g \rangle} \mid D \in \mathscr{D}\}$ is a covering for $\langle P, g \rangle$.*

This theorem can be used to show the limits of Datalog expressivity. Namely, the process of diversification is restricted by the size of the *interface* of a node in the proof tree, which in turn is closely related to the IDB arity. Therefore, a bound on the maximal IDB arity of a Datalog query affects the structure of diversifications. For example, diversifications of monadic Datalog queries are databases that can be decomposed into graphs that touch only in a single vertex, and which form a tree-like structure (technically, this is a so-called *tree decomposition* where we restrict the number of nodes that neighbouring bags may have in common). Given a query task that we conjecture to be unsolvable for monadic Datalog, we only need to find cases that do not admit such coverings. This occurs when diversification leads to databases for which strictly less answers should be returned.

For the case of retrieval problems in the sense of Definition 2, however, this approach is not quite enough yet. Indeed, a diversified database may fail to be a correct instance of

---

[2] This is a set since we assume that every database is defined over a countable active domain.

the original retrieval problem. For instance, the diversified database of Example 7 does not correspond to any database that can be obtained by translating a propositional Horn theory, since two distinct vertices $[\top]^3$ and $[\top]^5$ are labelled with the unary relation $t_\top$.

6   Finding Diversification Coverings for Retrieval Problems

To make diversification applicable to retrieval problems, we must perform renamings in a way that guarantees that the resulting database is still (part of) a correct translation of some problem instance. Intuitively speaking, reasoning problems (and all other problems over a set of terms or formulae) are based on certain sets of atomic symbols – propositions, constants, predicate names, etc. – of which we have an unlimited supply. These sets correspond to the *generators* in a term algebra, in contrast to the *operators* of the algebra. In logic, the generators are often provided by a *signature*, but we confine our use of this term to databases and queries, and we will speak of generators and operators instead to make the distinction.

Example 8. For propositional Horn rules as defined in (1)–(3), the set **A** of propositional letters is a set of generators, whereas the constants $\top$ and $\bot$ are nullary operators. Indeed, we can freely rename propositions to create new expressions with similar properties, but there is always just a single $\top$ and $\bot$.

Applied to the idea of diversification from Section 5, this means that we can easily rename a vertex $[a]$ to $[a]'$, since the latter can be understood as a vertex $[a']$ for a new proposition a'. However, we cannot rename $[\top]$ to $[\top]'$, since it is not possible to introduce another $\top$. Additional care is needed with vertices introduced for complex formulae, since they are in a one-to-one relationship with all of their subterms. For example, a vertex $[b \wedge c]$ must always have exactly two child vertices $[b]$ and $[c]$ – we cannot rename these children independently, or even introduce multiple children (a renamed version and the original version). These informal considerations motivate the following definition:

Definition 6. *Consider a retrieval problem $\langle \mathscr{T}, \Theta \rangle$ over a language $\mathbf{L}$ of formulae as in Definition 2. Let $T \in \mathscr{T}$ be an instance of this retrieval problem, let $\langle P, g \rangle$ be a Datalog query, and let $t = \langle N, E, \lambda \rangle$ be a proof tree of $\langle P, g \rangle$ on $\mathsf{G}(T)$. The $\mathbf{L}$-interface of a node $n \in N$ is the set of generators of $\mathbf{L}$ that occur in the label $\lambda(n)$.*

The $\mathbf{L}$-diversification of the proof tree t with root node n_r is constructed recursively:

- *for all generators $d \in \Delta^D$ that are not in the $\mathbf{L}$-interface of n_r, introduce a fresh generator a' that has not yet been used in t or in this construction yet, and replace every occurrence of a in labels of t by a';*
- *apply the diversification procedure recursively to all subtrees of t that have a non-leaf child node of n_r as their root.*

Note that we replace generator symbols that are part of vertex names of the form $[\varphi]$, leading to new vertices (constants), which we assume to be added to the active domain. As before, the replacement of constants may not be uniform throughout the tree. $\mathbf{L}$-diversifications of (sets of) databases can be defined as in Section 5.

Example 9. The **L**-diversification of the proof tree in Fig. 1 for **L** being the set of propositional Horn rules leads to the following diversified database:

$$[d] \xleftarrow{r^2_\rightarrow} [b^1 \wedge c^1 \to d]$$

$$\downarrow r^1_\rightarrow$$

$$[b^1 \wedge c^1] \xrightarrow{r^1_\wedge} [b^1] \xleftarrow{r^2_\rightarrow} [a^2 \to b^1] \xrightarrow{r^1_\rightarrow} [a^2] \xleftarrow{r^2_\rightarrow} [\top \to a^2] \xrightarrow{r^1_\rightarrow} [\top]$$

$$\searrow r^2_\wedge \quad [c^1] \xleftarrow{r^2_\rightarrow} [a^4 \to c^1] \xrightarrow{r^1_\rightarrow} [a^4] \xleftarrow{r^2_\rightarrow} [\top \to a^4] \quad \nearrow r^1_\rightarrow$$

This corresponds to the propositional Horn theory $\{\top \to a^2, \top \to a^4, a^2 \to b^1, a^4 \to c^1, b^1 \wedge c^1 \to d\}$, which does indeed entail that d is true while being more general than the original theory in the sense of Definition 3.

It is not hard to see that **L**-diversification is a restricted form of diversification that imposes additional restrictions for renaming vertices. Therefore, Theorem 2 still holds in this case. In addition, one can show that every **L**-diversified database is a subset of a database that is a correct encoding of some instance of the retrieval problem. For this to hold, we need to assume that the retrieval problem instances $\mathscr{T}$ contain all theories that can be obtained by renaming arbitrary occurrences of generator symbols, and by taking unions of theories in $\mathscr{T}$:

Theorem 3. *Consider a retrieval problem $\langle \mathscr{T}, \Theta \rangle$ over a language **L**, a theory $T \in \mathscr{T}$ and a Datalog query $\langle P, g \rangle$. Let $\mathsf{G}(T)|^{\mathbf{L}}_{\langle P,g \rangle}$ denote the **L**-diversification of $\mathsf{G}(T)$, i.e., the database that satisfies all ground instances that occur as labels of leaf nodes of **L**-diversified proof trees of $\langle P, g \rangle$ over $\mathsf{G}(T)$.*

If $\mathscr{T}$ is closed under unions of theories and replacements of generator symbols, then there is a theory $T|_{\langle P,g \rangle} \in \mathscr{T}$ such that $\mathsf{G}(T)|^{\mathbf{L}}_{\langle P,g \rangle} \subseteq \mathsf{G}(T|_{\langle P,g \rangle})$.

What this tells us is that **L**-diversification is a valid method for producing coverings of Datalog queries that satisfy the constraints of a certain problem encoding. It remains to use them to obtain a proof of Theorem 1.

7   Proof of Theorem 1

In this section, we bring together the tools prepared so far to complete the proof for the minimum arity of Datalog queries that solve the problem of retrieving all false propositions of a Horn logic theory.

First and foremost, we must find a suitable family of problematic cases for which an increased arity is necessary. Picking a single problem is not enough, since every finite structure can trivially be recognised by a single Datalog rule of sufficient size. We use Horn logic theories $(T_k)_{k \geq 1}$ of the form

$$T_k := \{b_1 \wedge a \to \bot, b_2 \to b_1, b_3 \to b_2, \ldots, b_k \to b_{k-1}, a \to b_k\}.$$

The structure of the graph $\mathsf{G}(T_k)$ of T_k is illustrated in Fig. 3. Clearly, each T_k entails a to be false (while all other propositions might be true or false).

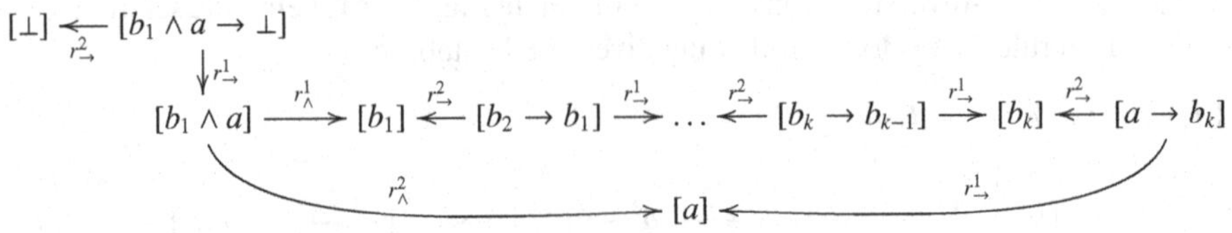

$$[\bot] \xleftarrow{r_\to^2} [b_1 \wedge a \to \bot]$$

$$\big\downarrow r_\to^1$$

$$[b_1 \wedge a] \xrightarrow{r_\to^1} [b_1] \xleftarrow{r_\to^2} [b_2 \to b_1] \xrightarrow{r_\to^1} \ldots \xleftarrow{r_\to^2} [b_k \to b_{k-1}] \xrightarrow{r_\to^1} [b_k] \xleftarrow{r_\to^2} [a \to b_k]$$

$$\xrightarrow[r_\wedge^2]{} [a] \xleftarrow{r_\to^1}$$

Fig. 3. Illustration of the databases obtained from the propositional Horn theory T_k

Now suppose for a contradiction that there is a monadic Datalog query $\langle P, g \rangle$ that solves the retrieval problem of computing false propositions from a propositional Horn theory. Thus, g has arity 1 and the answer of $\langle P, g \rangle$ on T_k is $\{[a]\}$. We show that this implies that $\langle P, g \rangle$ computes some wrong answers too. Indeed, by Theorems 2 and 3, for any T_k we can find an **L**-diversification T_k' for which $\langle P, g \rangle$ still returns $\{[a]\}$ as an answer. We claim that there is T_k for which this answer must be wrong, since the correct answer for any **L**-diversification of T_k would be $\{\}$.

To prove this, let m be the maximal number of EDB atoms in the body of some rule in P. We set k to be $\lceil m/2 \rceil + 2$, and consider a proof tree for the solution $[a]$ of $\langle P, g \rangle$ over $\mathsf{G}(T_k)$. The root of this tree must be labelled $g([a])$.

Observation 1. All of the edges in $\mathsf{G}(T_k)$ as sketched in Fig. 3 are essential for the entailment to hold: a substructure that lacks any of these edges could always be based on some different theory that does not entail a to be false. Therefore, we can assume that each atom of $\mathsf{G}(T_k)$ occurs as a label for a leaf node in the proof tree we consider.

Observation 2. We can assume that every rule body in P, when viewed as a graph with variables as vertices and body atoms as edges, is connected and contains the variable in the head of the rule. Indeed, if a rule would contain a connected component B in the body that does not include the variable in the head, then one of three cases would apply: (1) B does not match any database of the form $\mathsf{G}(T)$ for a propositional Horn theory T, and the whole rule can be removed; (2) B matches only graphs $\mathsf{G}(T)$ for theories T that are unsatisfiable, and the whole rule can be removed (since we restrict to satisfiable theories); (3) B matches a part of $\mathsf{G}(T)$ for some satisfiable theory T, and we can change our problem formulation to include a disjoint copy of T in T_k so as to make B redundant.

From Observations 1 and 2 we obtain that any proof tree for $g([a])$ on T_k must "scan" $\mathsf{G}(T_k)$ along connected subgraphs that overlap in at least one vertex. The number of edges of these graphs is bounded by the number of body atoms in the largest body P.

Now let G_r be the graph that consists of the EDB ground atom labels of direct children of the root of the proof tree, and let $\bar{G}_r$ be $\mathsf{G}(T_k) \setminus G_r$. By our choice of k, the number of body atoms of any rule is strictly smaller than $2k - 1$, so G_r contains less than $2k - 1$ edges. Thus, $\bar{G}_r$ contains at least 3 edges of the cyclic structure of length $2k + 2$ shown in Fig. 3. Note that the edges of $\bar{G}_r$ that occur in the cycle must form a chain.

By Observation 1, the edges of $\bar{G}_r$ occur as a label in the proof tree below some child node of the root. However, all nodes have an interface of size 1 in monadic Datalog, so only the generator symbols that occur together in a single vertex can be preserved in $\bar{G}_r$ after L-diversification as in Definition 6. It is easy to see, however, that there is no single vertex in G_r that contains enough generator symbols. For example, one of the minimal choices for $\bar{G}_r$ consists of the ground atoms $\{r_\rightarrow^1([b_i \rightarrow b_{i-1}], [b_i]), r_\rightarrow^2([b_{i+1} \rightarrow b_i], [b_i]), r_\rightarrow^1([b_{i+1} \rightarrow b_i], [b_{i+1}])\}$. In this case, G_r contains the vertices $[b_i \rightarrow b_{i-1}]$ and $[b_{i+1}]$ in the graph in its body, so it is possible that IDB child nodes in the proof tree have these vertices in their head. However, the graph that a subtree with $[b_i \rightarrow b_{i-1}]$ in its head matches after L-diversification has the form $\bar{G}_r' = \{r_\rightarrow^1([b_i \rightarrow b_{i-1}], [b_i]), r_\rightarrow^2([b_{i+1}' \rightarrow b_i], [b_i]), r_\rightarrow^1([b_{i+1}' \rightarrow b_i], [b_{i+1}'])\}$ where b_{i+1} was diversified to b_{i+1}'. Recall that G_r still uses the original node b_{i+1}, so the cycle does not connect in this place.

Analogously, a subtree with root vertex $[b_{i+1}]$ cannot recognize $\bar{G}_r$ either (in this case, it is b_{i-1} that can be diversified). Any chain of three edges in the cycle of Fig. 3 yields a minimal graph $\bar{G}_r$, so there are several more cases to consider; the arguments are exactly the same in each of them.

Thus, we find that the diversified graph $G(T_k)'$ does not contain a cycle any more. By Theorem 2, the answer of $\langle P, g \rangle$ on $G(T_k)'$ is still $\{[a]\}$. By Theorem 3 and Observation 1, however, there is a theory T_k' with $G(T_k)' \subseteq G(T_k')$ which does not entail a to be false. Hence, $\langle P, g \rangle$ provides incorrect results for T_k', which yields the required contradiction and finishes the proof.

This completes the proof of Theorem 1. It is interesting to note that the following modified theory $\hat{T}_k$ would not work for the proof: $\{b_1 \wedge a \rightarrow \perp, b_2 \wedge a \rightarrow b_1, \ldots, b_k \wedge a \rightarrow b_{k-1}, a \rightarrow b_k\}$. This modified theory still entails a to be false, but now every axiom contains the generator symbol a. This prevents L-diversification of a if the Datalog proof uses the nodes $[b_{i+1} \wedge a \rightarrow b_i]$ in all head atoms. Indeed, one can find a monadic Datalog query that correctly computes a to be false in all cases of this special form, and that never computes any wrong answers.

This example illustrates the crucial difference between our framework of studying retrieval problems and the traditional study of query language expressivity in database theory. Indeed, monadic Datalog cannot recognize graph structures as used in $\hat{T}_k$, yet it is possible in our setting by using the additional assumptions that the given database was obtained by translating a valid input.

8   Further Applications Based on Datalog Arity

Sections 2–7 laid out a general method of classifying the difficulty of reasoning problems based on the minimal arity of IDB predicates required to solve them. This particular instance of our approach is of special interest, due to the practical and theoretical relevance of Datalog. In this section, we give further examples where similar techniques have been used to classify more complicated reasoning tasks, and we relate this work to general works in descriptive complexity.

Many practical reasoning algorithms are based on the deterministic application of inference rules, and those that run in polynomial time can often be formulated in Datalog.

A prominent example is reasoning in lightweight ontology languages. The W3C Web Ontology Language defines three such languages: OWL EL, OWL RL, and OWL QL [16,15]. OWL RL was actually designed to support ontology-based query answering with rules, but OWL EL, too, is generally implemented in this fashion [3,13,12].

Relevant reasoning tasks for these languages are the computation of instances and subclasses, respectively, of class expressions in the ontology – both problems are P-complete for OWL EL and OWL RL alike. Yet, the problems are not equivalent when comparing the required Datalog expressivity, if we consider them as retrieval problems that compute pairs of related elements:

Theorem 4 ([13]). *For OWL EL ontologies*

– *retrieving all instances-of relationships requires Datalog of IDB arity at least 3;*
– *retrieving all subclass-of relationships requires Datalog of IDB arity at least 4.*

Theorem 5 ([14]). *For OWL RL ontologies retrieving all subclass-of relationships requires Datalog of IDB arity at least 4.*

Both theorems are interesting in that they provide practically relevant problems where IDB arities of 2 or even 3 are not enough. This illustrates that our approach is meaningful beyond the (possibly rather special) case of monadic Datalog. Instead, we obtain a real hierarchy of expressivity. Nevertheless, the proof techniques used to show these results follow the same pattern that we introduced for the simpler example given herein; in particular, diversifications of proof trees play an important role [13,14].

The practical significance of these results is two-fold. On the one hand, experience shows that problems of higher minimal arity are often more difficult to implement in practice. In case of OWL EL, the arity required for class subsumption drops to 3 if a certain feature, called *nominal classes*, is omitted. And indeed, this feature has long been unsupported by reasoners [11]. Nevertheless, it is important to keep in mind that neither complexity theory nor algorithmic complexity allow us to conclude that Datalog of higher IDB arities is necessarily harder to implement, so we can only refer to practical experiences here.

On the other hand, even if it is possible in theory that efficient algorithms are unaffected by minimal IDB arities, our results impose strong syntactic restrictions on how such algorithms can be expressed in rules. This is particularly relevant for the case of OWL RL, where reasoning is traditionally implemented by rules that operate on the RDF syntax of OWL. RDF describes a graph as a set of *triples*, which can be viewed as labelled binary edges or, alternatively, as unlabelled ternary hyperedges. In either case, however, this syntax does not provide us with the 4-ary predicates needed by Theorem 5. This asserts that it is impossible to describe OWL RL reasoning using RDF-based rules[3] as implemented in many practical systems.

Finally, it is worth noting that OWL reasoning is a very natural candidate for our approach, since the official RDF syntax of OWL is already in the shape of a database in our sense. In many ways, this syntax is very similar to the canonical syntax we introduced in Section 3. The only difference is that the presence of logical operators

---

[3] Under the standard assumption that such rules cannot add elements to the active domain of the database during reasoning.

of arbitrary arity in OWL necessitates the use of linked lists in the graph encoding. However, the use of only binary operators is a special case of this encoding, so the minimal IDB arities established here remain valid.

9   Reasoning with Navigational Query Languages

Datalog is an obvious choice for solving P-complete reasoning problems. For problems of lower complexity, however, it is more natural to consider query languages of lower data complexity. In particular, many *navigational query languages* – which view databases as a graph structure along which to navigate – are in NLogSpace for data complexity. Such languages are typically contained in linear, monadic Datalog. In a recent work, it was demonstrated how to implement this idea to solve OWL QL reasoning using the prominent SPARQL 1.1 query language [5]. Here, we give a brief overview of these results and relate them to our general framework.

Traditionally, OWL QL reasoning is often implemented by using *query rewriting*, where a reasoning task is transformed into a data access problem. This, however, is different from our setting since the ontological schema is already incorporated for computing the queries used to access the database. Bischoff et al. now introduced what they called *schema-agnostic query rewriting* where the ontology is stored in the database and not used for building the query [5]. This corresponds to our formalisation of reasoning as a retrieval problem, with the only difference that Bischoff et al. assume the standard RDF serialization of OWL ontologies rather than our canonical transformation.

SPARQL 1.1 is much weaker than Datalog, but it also supports a basic type of recursion in the form of regular expressions that can be used to specify patterns for paths in the RDF graph. This can be used for OWL QL reasoning. Bischoff et al. thus obtain fixed SPARQL 1.1 queries for retrieving all subclass-of, instance-of, and subproperty-of relationships. We omit the details here for lack of space. To provide an extremely simplified example: in a logic that only supports a binary relation subClassOf, it is possible to retrieve the entailed subclass-of relations with a single query x subClassOf* y, where * denotes the Kleene star (zero or more repetitions), and x and y are variables. It is straightforward to translate this toy example to SPARQL 1.1. Supporting all of OWL QL requires somewhat more work.

Schema-agnostic query rewriting certainly has some practical merit, allowing us to use SPARQL 1.1 database systems as OWL QL reasoners, but what does it tell us about the difficulty of the problem? For one thing, SPARQL 1.1 is already a fairly minimal navigational query language. More expressive options include nSPARQL [18], XPath [4], and other forms of nested path queries [6]. Indeed, it turns out that one feature of OWL QL – symmetric properties – cannot be supported in SPARQL 1.1 but in nSPARQL [5]. This is interesting since the feature does not otherwise add to the complexity of reasoning. In fact, one can express it easily using other features that cause no such problems for SPARQL 1.1.[4]

Nevertheless, the landscape of navigational query languages is less systematic than the neat hierarchy of Datalog of increasing IDB arity. Therefore, such results are more

---

[4] In detail, the RDF graph structure p rdfs:subPropertyOf _:b . _:b owl:inverseOf p can be matched as part of regular path queries, while p rdf:type owl:SymmetricProperty cannot.

interesting from a practical viewpoint (What is possible on existing graph database systems?) than from a theoretical one (Is one problem harder than the other in a principled way?). However, the further development of graph query languages may lead to a more uniform landscape that provides deeper insights.

10   Outlook and Open Problems

We have presented an approach of reformulating reasoning problems in terms of query answering problems, which, to the best of our knowledge, has not been phrased in this generality before. We argued that such a viewpoint presents several benefits: its practical value is to "implement" computing tasks in the languages that are supported by database management systems; its theoretical value is to connect the difficulty of these tasks to the rich landscape of query language expressivity.

A new result established herein showed that the computation of all positive entailments of propositional Horn logic is, in a concrete technical sense, easier than the computation of all negative entailments, at least when relying on deterministic rules of inference that can be captured in Datalog. Other results we cited showed how to implement ontological reasoning in Datalog and SPARQL 1.1, explained why reasoning in OWL EL seems to become "harder" when adding certain features, and showed that schema reasoning for OWL RL cannot be described using RDF-base rules [13,14,5]. The range of these results illustrates how the proposed approach can fill a gap in our current understanding of reasoning tasks, but many problems are still open.

Our proposal has close relationships to several other fields. The relative expressiveness of query languages is a traditional topic in database theory. However, as discussed in Section 7, related results cannot be transferred naively. When using queries to solve problems, we do not require queries to work on all databases, but only on those that actually encode an instance of the problem. This distinction is rarely important when using Turing machines for computation, but it is exposed by our more fine-grained approach.

Another related field is descriptive complexity theory, where one also seeks to understand the relationship between problems solved by a query language and problems solved by a certain class of Turing machines [10,9]. The big difference to our view is that the goal in descriptive complexity is to characterize existing complexity classes using query languages. To the contrary, we are most interested in query languages that are *not* equivalent to a complexity class in this sense, so as to discover more fine-grained distinctions between computational problems. Moreover, descriptive complexity focuses on decision problems on graphs, without considering translation (and expressivity relative to a problem encoding) or retrieval problems. Nevertheless, the deep results of descriptive complexity can provide important technical insights and methods for our field as well.

The obvious next step in this field is to apply these ideas to additional computational problems. Reasoning was shown to be a fruitful area of application, but sets of terms (which we called "theories") are also the input to problems in many other fields, such as formal languages, automata theory, and graph theory. It will be interesting to see if new insights in these fields can be obtained with query-based methods. In some cases, the first step is to recognize query-based approaches as being part of this general framework.

For example, Datalog has often been used as a *rule language* to solve computational problems, but it was rarely asked if simpler *query languages* could also solve the task.

A practical extension of these investigations is to explore the practical utility of these translations. Can we use existing database systems to perform complicated computations on large datasets for us? Empirical studies are needed to answer this.

Another general direction of research is to apply these ideas for the evaluation of query languages, thus turning around the original question. Indeed, problem reductions such as the ones we presented can motivate the need for a certain expressive feature in a query language. Again, this study has a practical side, since it may also guide the optimisation of query engines by providing meaningful benchmarks that are translated from other areas.

Finally, the theory we have sketched here is still in its development, and many questions remain open. Some of the methods we introduced are fairly general already, but their application in Section 7 was still rather pedestrian. It would be of great utility to flesh out more general properties, possibly graph-theoretic or algebraic, that can make it easier to see that a problem cannot be solved by means of certain queries. Another aspect that we ignored completely is the notion of problem *reductions*. Complexity theory uses many-to-one reductions to form larger complexity classes, but it is not clear which query languages can implement which kinds of reductions. Yet another possible extension would be to generalise the form of input problems we consider. While theories (as finite sets of terms) capture many problems, one could also consider more general formulations based on context-free grammars that describe problem instances.

Thus, overall, this paper is at best a starting point for further investigations in what will hopefully remain an exciting and fruitful field of study.

Acknowledgements. The author wishes to thank Sebastian Rudolph for extensive discussions on the subject and valuable feedback on this paper. Moreover, thanks are due to many colleagues whose input and collaborations have contributed to this paper, in particular Stefan Bischoff, Pierre Bourhis, Tomáš Masopust, Axel Polleres, and Michaël Thomazo. The author is grateful to Marie-Laure Mugnier and Roman Kontchakov for their kind invitation to RR 2014, where this work was presented. This work was supported by the DFG in project DIAMOND (Emmy Noether grant KR 4381/1-1).

References

1. Abiteboul, S., Hull, R., Vianu, V.: Foundations of Databases. Addison Wesley (1994)
2. Afrati, F.N., Cosmadakis, S.S.: Expressiveness of restricted recursive queries. In: Proc. 21st Symposium on Theory of Computing Conference (STOC 1989), pp. 113–126. ACM (1989)
3. Baader, F., Brandt, S., Lutz, C.: Pushing the $\mathcal{EL}$ envelope. In: Kaelbling, L., Saffiotti, A. (eds.) Proc. 19th Int. Joint Conf. on Artificial Intelligence (IJCAI 2005), pp. 364–369. Professional Book Center (2005)
4. Barceló, P., Libkin, L., Lin, A.W., Wood, P.T.: Expressive languages for path queries over graph-structured data. ACM Trans. Database Syst. 37(4), 31 (2012)
5. Bischoff, S., Krötzsch, M., Polleres, A., Rudolph, S.: Schema-agnostic query rewriting for SPARQL 1.1. In: Proc. 13th Int. Semantic Web Conf. (ISWC 2014). LNCS. Springer (to appear, 2014)

6. Bourhis, P., Krötzsch, M., Rudolph, S.: How to best nest regular path queries. In: Proc. 27th Int. Workshop on Description Logics (DL 2014). CEUR Workshop Proceedings, vol. 1193, pp. 404–415. CEUR-WS.org (2014)

7. Dantsin, E., Eiter, T., Gottlob, G., Voronkov, A.: Complexity and expressive power of logic programming. ACM Computing Surveys 33(3), 374–425 (2001)

8. Dowling, W.F., Gallier, J.H.: Linear-time algorithms for testing the satisfiability of propositional Horn formulae. J. Logic Programming 1(3), 267–284 (1984)

9. Feder, T., Vardi, M.: The computational structure of Monotone Monadic SNP and constraint satisfaction: A study through Datalog and group theory. SIAM Journal on Computing 28(1), 57–104 (1998)

10. Grohe, M.: From polynomial time queries to graph structure theory. Commun. ACM 54(6), 104–112 (2011)

11. Kazakov, Y., Krötzsch, M., Simančík, F.: Practical reasoning with nominals in the $\mathcal{EL}$ family of description logics. In: Brewka, G., Eiter, T., McIlraith, S.A. (eds.) Proc. 13th Int. Conf. on Principles of Knowledge Representation and Reasoning (KR 2012), pp. 264–274. AAAI Press (2012)

12. Kazakov, Y., Krötzsch, M., Simančík, F.: The incredible ELK: From polynomial procedures to efficient reasoning with $\mathcal{EL}$ ontologies. Journal of Automated Reasoning 53, 1–61 (2013)

13. Krötzsch, M.: Efficient rule-based inferencing for OWL EL. In: Walsh, T. (ed.) Proc. 22nd Int. Joint Conf. on Artificial Intelligence (IJCAI 2011), pp. 2668–2673. AAAI Press/IJCAI (2011)

14. Krötzsch, M.: The not-so-easy task of computing class subsumptions in OWL RL. In: Cudré-Mauroux, P., Heflin, J., Sirin, E., Tudorache, T., Euzenat, J., Hauswirth, M., Parreira, J.X., Hendler, J., Schreiber, G., Bernstein, A., Blomqvist, E. (eds.) ISWC 2012, Part I. LNCS, vol. 7649, pp. 279–294. Springer, Heidelberg (2012)

15. Krötzsch, M.: OWL 2 Profiles: An introduction to lightweight ontology languages. In: Eiter, T., Krennwallner, T. (eds.) Reasoning Web 2012. LNCS, vol. 7487, pp. 112–183. Springer, Heidelberg (2012), available at http://korrekt.org/page/OWL_2_Profiles

16. Motik, B., Cuenca Grau, B., Horrocks, I., Wu, Z., Fokoue, A., Lutz, C. (eds.): OWL 2 Web Ontology Language: Profiles. W3C Recommendation (October 27, 2009), http://www.w3.org/TR/owl2-profiles/

17. Papadimitriou, C.H.: Computational Complexity. Addison Wesley (1994)

18. Pérez, J., Arenas, M., Gutierrez, C.: nSPARQL: A navigational language for RDF. J. Web Semantics 8, 255–270 (2010)

19. Rudolph, S., Krötzsch, M.: Flag & check: Data access with monadically defined queries. In: Hull, R., Fan, W. (eds.) Proc. 32nd Symposium on Principles of Database Systems (PODS 2013), pp. 151–162. ACM (2013)

20. Williams, V.V.: Multiplying matrices faster than Coppersmith-Winograd. In: Karloff, H.J., Pitassi, T. (eds.) Proc. 44th Symposium on Theory of Computing Conference (STOC 2012), pp. 887–898. ACM (2012)

Web Reasoning for Cultural Heritage

Alexandros Chortaras, Nasos Drosopoulos, Ilianna Kollia, and Nikolaos Simou

National Technical University of Athens, Greece

Abstract. Cultural Heritage is the focus of a great and continually increasing number of R&D initiatives, aiming at efficiently managing and disseminating cultural resources on the Web. As more institutions make their collections available online and proceed to aggregate them in domain repositories, knowledge-based management and retrieval becomes a necessary evolution from simple syntactic data exchange. In the process of aggregating heterogeneous resources and publishing them for retrieval and creative reuse, networks such as Europeana and DPLA invest in technologies that achieve semantic data integration. The resulting repositories join the Linked Open Data cloud, allowing to link cultural heritage domain knowledge to existing datasets. Integration of diverse information is achieved through the use of formal ontologies, enabling reasoning services to offer powerful semantic search and navigation mechanisms.

Digital evolution of the Cultural Heritage field has grown rapidly in the last few years. Massive digitization and annotation activities have been taking place all over Europe and the United States. The strong involvement of companies, like Google, and the positive reaction of the European Union have led to a variety of converging actions towards digital cultural content generation from all possible sources, such as galleries, libraries, archives, museums and audiovisual archives [7]. The creation and evolution of Europeana,[1] as a unique point of access to European Cultural Heritage, has been one of the major achievements in this procedure.

The current state of the art in Cultural Heritage implements a model whereby many aggregators, content providers and projects feed their content into a national, thematic, or European portal, and this portal is then used by the end user to find cultural items. Typically, the content is described with the aid of standard sets of elements of information about resources (metadata schemas) that try to build an interoperability layer. Europeana has been developed to provide integrated access to objects from cultural heritage organizations, encompassing material from museums, libraries, archives and audiovisual archives. Several cross-domain, vertical or thematic aggregators have been deployed at regional, national and international level in order to reinforce this initiative by collecting and converting metadata about existing and newly digitized resources. Currently, more than 34 million cultural objects can be searched through the Europeana portal.

The Europeana Semantic Elements[2] (ESE) Model was an application profile used by Europeana to provide a generic set of terms that could be applied to heterogeneous materials allowing contributors to take advantage of their existing rich descriptions.

---

[1] http://www.europeana.eu

[2] http://www.europeana.eu/schemas/ese/

R. Kontchakov and M.-L. Mugnier (Eds.): RR 2014, LNCS 8741, pp. 23–28, 2014.

The latter constitute a knowledge base that is constantly growing and evolving, both by newly introduced annotations and digitization initiatives, as well as through the increased efforts and successful outcomes of the aggregators and the content providing organizations.

The new Europeana Data Model[3] (EDM) has been introduced as a data structure aiming to enable the linking of data and to connect and enrich descriptions in accordance with the Semantic Web developments. Its scope and main strength is the adoption of an open, cross-domain framework in order to accommodate the growing number of rich, community-oriented standards such as LIDO (Lightweight Information Describing Objects)[4] for museums, EAD (Encoded Archival Description)[5] for archives or METS (Metadata Encoding and Transmission Standard)[6] for libraries. Apart from its ability to support standards of high richness, EDM also enables source aggregation and data enrichment from a range of third party sources while clearly providing the provenance of all information.

Following ongoing efforts to investigate usage of the semantic layer as a means to improve user experience, we are facing the need to provide a more detailed semantic description of cultural content. Semantic description of cultural content, accessible through its metadata, would be of little use, if users were not in a position to pose their queries in terms of a rich integrated ontological knowledge. Currently this is performed through a data storage schema, which highly limits the aim of the query. *Semantic query answering* refers to the finding of answers to queries posed by users, based not only on string matching over data that are stored in databases, but also on the implicit meaning that can be found by reasoning based on the detailed *domain terminological knowledge*. In this way, content metadata can be terminologically described, semantically connected and used in conjunction with other useful, possibly complementary content and information, independently published on the web. A semantically integrated cultural heritage knowledge, facilitating access to cultural content is, therefore, achieved. The key is to semantically connect metadata with ontological domain knowledge through appropriate mappings. It is important to notice that the requirement of sophisticated query answering is even more demanding for experienced users (professionals, researchers, educators) in a specific cultural context.

The architecture of the system we have developed for metadata aggregation and semantic enrichment [5] is depicted in Figure 1. Cultural content providers (museums, libraries, archives) and aggregators wish to make their content visible to Europeana. This is performed by ingesting (usually a subset of) their content metadata descriptions to the Europeana portal. This is a rather difficult task, mainly due to the heterogeneity of the metadata storage schemas (from both technological and conceptual point of view) that need to be transformed to the EDM form. Using our system, the *Metadata Ingestion* module provides users with the ability to map and transform their data to EDM elements through a friendly graphical interface that also offers useful metadata manipulation functions. The result of this module is an EDM representation of the cultural

---

[3] http://www.europeana.eu/schemas/edm/

[4] http://www.lido-schema.org

[5] http://www.loc.gov/ead/

[6] http://www.loc.gov/standards/mets/

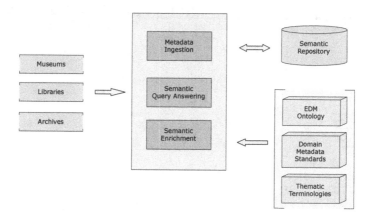

Fig. 1. The architecture of our metadata aggregation and semantic enrichment system

content metadata. Moreover, through the *Semantic Enrichment* module, the transformed metadata are serialized as RDF triples [6] and stored in the *Semantic Repository*.

The metadata elements are represented in the semantic repository as descriptions of individuals, i.e., connections of individuals with entities of the *terminological knowledge*. This knowledge is an ontological representation of EDM (the *EDM Ontology*), that is connected, on the one hand, to *Domain Metadata Standards* (Dublin Core, LIDO, CIDOC CRM[7], etc.) sharing terminology with them and providing the general description of 'Who?', 'What?', 'When?' and 'Where?' for every digital object and, on the other hand, to more specific terminological axioms providing details about species, categories, properties, interrelations, etc. (e.g., brooches are made of copper or gold). The latter knowledge (the *Thematic Ontologies*) is developed by the providers and aggregators and can be used both for semantic enrichment of content metadata, and for reasoning in the *Semantic Query Answering* module. Thus, it provides the user with the ability to build complex queries in terms of the above terminology and access cultural content effectively.

In the following we describe in more detail the main modules of the system.

Metadata Schema Mapping. The process of metadata mapping formalizes the notion of 'crosswalk' by hiding technical details and permitting semantic equivalences to emerge as the centrepiece. This module is based on the Metadata Interoperability Services tool (MINT) that has been successfully used in many Europeana aggregation projects.[8] It involves a graphical, web-based environment where interoperability is achieved by letting users create mappings between an input and a target schema. In Figure 2 a snapshot of the mapping editor of the system is shown where a LIDO schema is mapped to an EDM schema. On the left-hand side of the mapping editor one can find the interactive tree, which represents the snapshot of the XML schema that the user is

---

[7] http://www.cidoc-crm.org
[8] http://mint.image.ece.ntua.gr/redmine/projects/mint/wiki

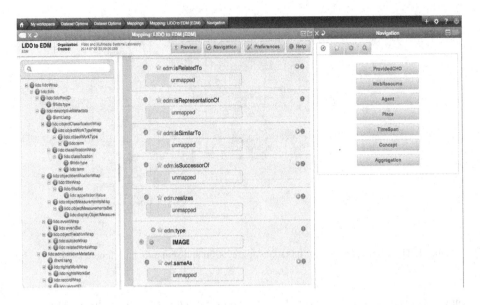

Fig. 2. Screenshot of the mapping editor

using as input for the mapping process. Moreover, the interface provides the user with groups of high level elements that constitute separate semantic entities of the target schema. These are presented on the right-hand side as buttons that are then used to access the set of corresponding sub-elements. This set is visualized on the middle part of the screen as a tree structure of embedded boxes, representing the internal structure of the complex element. The user is able to interact with this structure by clicking to collapse and expand every embedded box that represents an element along with all relevant information (attributes, annotations) defined in the XML schema document. To perform an actual mapping between the input and the target schema, the user can simply drag a source element and drop it on the respective target in the middle. User's mapping actions are expressed through Extensible Stylesheet Language Transformations (XSLT) that is a language used for transforming XML documents into other XML documents. XSLT stylesheets are stored and can be applied to any user data, can be exported and published as a well-defined, machine understandable crosswalk and shared with other users to act as template for their mapping needs.

Using the mapping tool the provided metadata are transformed to instances of the EDM ontology in RDF/XML serialization. An example output EDM RDF preview of a record is shown in Figure 3.

Semantic Enrichment and Query Answering. The transformation of the data of content providers to data in terms of the EDM ontology results in a set of RDF triples that are more like an attribute-value set for each object. Since the EDM ontology is a general ontology referring to metadata descriptions of each object, the usage of thematic ontologies for different domains is necessary in order to add semantically processable information to each object. For example, the information that an object is of type vase

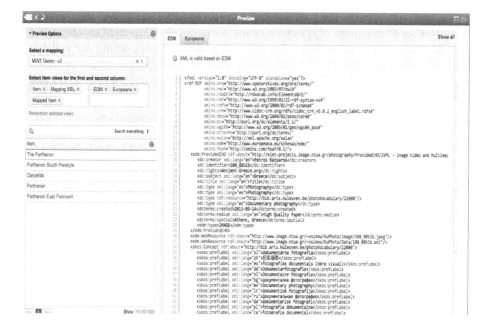

Fig. 3. EDM RDF preview

may not be adequate for a specific application; one may be interested in the specific type of vase, or, in absence of such information, in the characteristics that a vase should have in order to be classified to a specific type. First, thematic ontologies are created in collaboration with field experts. These ontologies include individuals which represent objects, classes which define sets of objects and object properties defining relationships between objects. Then the data values filling the attributes of the EDM-RDF instances are transformed to individuals of the thematic ontologies. These individuals are then grouped together to form classes as imposed by the thematic ontologies. The transformation of the data values to individuals is performed from a technical point of view by mapping the data values to IRIs (International Resource Identifiers). After this transformation the data is stored in a semantic repository, from where they can be extracted through queries.

The semantic query answering module of the system is based on the combination of two other query answering systems, namely Rapid and OWL-BGP. Rapid is an efficient query rewriting system for the OWL 2 QL and OWL 2 EL fragments [8][1]. According to query rewriting, a query issued over a terminology and a set of instance data stored in a data repository can be answered by first rewriting the query w.r.t. the terminological knowledge and then answering the rewritten query over the data alone (i.e., without taking the terminological knowledge into account). OWL-BGP[9] is a framework for efficiently answering SPARQL queries issued over expressive OWL 2 ontologies using the OWL 2 Direct Semantics entailment regime of SPARQL [4][3]. Currently, the HermiT reasoner [2] is used for performing inferences in OWL-BGP, but other reasoners could

---

[9] https://code.google.com/p/owl-bgp/

also be used. The input to the combined system is a SPARQL query and the output is the answer set of mappings of query variables to entities of the queried ontology. For example, let us assume that we have i) a terminological knowledge stating that metallic rings (:MetallicRing) are rings (:Ring), ii) the objects :ring1, :ring2, :ring3, of which :ring1 and :ring2 are metallic rings and are produced in :greece and :ring3 is a ring and iii) the following SPARQL query asking for rings and the place they are produced in:

SELECT $?x, ?y$ WHERE { $?x$ rdf:type :Ring. $?x$:producedIn $?y$. }

where $?x, ?y$ are variables. Note that the empty prefix is used for an imaginary example namespace. The answers to this query are the mappings: $?x \mapsto$:ring1, $?y \mapsto$:greece and $?x \mapsto$:ring2, $?y \mapsto$:greece. Note that :ring1 and :ring2 are rings since they are metallic rings and from the terminological knowledge we know that metallic rings are rings. Even though :ring3 is a ring, the place of production of it is not explicitly stated, nor can it be inferred from the given knowledge, hence, :ring3 is not returned as an answer mapping for $?x$.

This tutorial presents the state of the art in digital cultural heritage, focusing on interoperable content metadata aggregation and management, on thematic knowledge generation and usage for semantic search and semantic query answering over large volumes of content. The metadata schema mapping module is being used for content ingestion in the Europeana ecosystem, while the semantic enrichment and query answering modules can further improve the provided cultural heritage services in fields such as research, education, tourism.

Acknowledgements. This work has been partially funded by the projects Europeana Fashion (CIP-ICT-PSP.2011.2.1), AthenaPlus (CIP-ICT-PSP.2012.2.1) and Europeana Space (CIP-ICT-PSP-2013-7).

References

1. Chortaras, A., Trivela, D., Stamou, G.: Optimized query rewriting for OWL 2 QL. In: Bjørner, N., Sofronie-Stokkermans, V. (eds.) CADE 2011. LNCS, vol. 6803, pp. 192–206. Springer, Heidelberg (2011)
2. Glimm, B., Horrocks, I., Motik, B., Stoilos, G., Wang, Z.: HermiT: An OWL 2 Reasoner. Journal of Automated Reasoning (2014) (accepted for publication)
3. Kollia, I., Glimm, B.: Cost Based Query Ordering over OWL Ontologies. In: Cudré-Mauroux, P., Heflin, J., Sirin, E., Tudorache, T., Euzenat, J., Hauswirth, M., Parreira, J.X., Hendler, J., Schreiber, G., Bernstein, A., Blomqvist, E. (eds.) ISWC 2012, Part I. LNCS, vol. 7649, pp. 231–246. Springer, Heidelberg (2012)
4. Kollia, I., Glimm, B.: Optimizing SPARQL Query Answering over OWL Ontologies. Journal of Artificial Intelligence Research (JAIR) 48, 253–303 (2013)
5. Kollia, I., Tzouvaras, V., Drosopoulos, N., Stamou, G.B.: A Systemic Approach for Effective Semantic Access to Cultural Content. Semantic Web Journal 3(1), 65–83 (2012)
6. Manola, F., Miller, E. (eds.): Resource Description Framework (RDF): Primer. W3C Recommendation (February 10, 2004), http://www.w3.org/TR/rdf-primer/ 10 February
7. Simou, N., Evain, J., Drosopoulos, A., Tzouvaras, V.: Linked European Television Heritage. Semantic Web Journal (2013)
8. Trivela, D., Stoilos, G., Chortaras, A., Stamou, G.B.: Optimising resolution-based rewriting algorithms for dl ontologies. In: Proceedings of the 26th International Workshop on Description Logics, pp. 464–476 (2013)

Planning with Transaction Logic

Reza Basseda[1], Michael Kifer[1], and Anthony J. Bonner[2]

[1] Stony Brook University, USA
{rbasseda,kifer}@cs.stonybrook.edu
[2] University of Toronto, Canada
bonner@cs.toronto.edu

Abstract. Automated planning has been the subject of intensive research and is at the core of several areas of AI, including intelligent agents and robotics. In this paper, we argue that Transaction Logic is a natural specification language for planning algorithms, which enables one to see further afield and thus discover better and more general solutions than using one-of-a-kind formalisms. Specifically, we take the well-known *STRIPS* planning strategy and show that Transaction Logic lets one specify the *STRIPS* planning algorithm easily and concisely, and to prove its completeness. Moreover, extensions to allow indirect effects and to support action ramifications come almost for free. Finally, the compact and clear logical formulation of the algorithm made possible by this logic is conducive to fruitful experimentation. To illustrate this, we show that a rather simple modification of the *STRIPS* planning strategy is also complete and yields speedups of orders of magnitude.

1 Introduction

The classical problem of automated planning is at the core of several important areas, such as robotics, intelligent information systems, and multi-agent systems, and it has been preoccupying AI researchers for over forty years.

In this paper, we argue that a general logical theory, specifically *Transaction Logic* (or $\mathcal{TR}$) [10, 9, 8], provides multiple advantages for specifying, generalizing, and solving planning problems. Transaction Logic is an extension of classical logic with dedicated support for specifying and reasoning about actions, including sequential and parallel execution, atomicity of transactions, and more. To illustrate the point, we take the classical *STRIPS* planning problem [12, 19] and show that both the *STRIPS* framework and the associated planning algorithm easily and naturally lend themselves to compact representation in Transaction Logic.

We emphasize that this paper is *not* about *STRIPS* or about inventing new planning strategies (although we do—as a side effect). It is rather about the advantages of $\mathcal{TR}$ as a general tool for specifying a wide range of planning frameworks and strategies— *STRIPS* is just an illustration. One can likewise apply $\mathcal{TR}$ to HTN-like planning systems [20] and to various enhancements of *STRIPS*, like *RSTRIPS* and *ABSTRIPS* [21].

Our contention is that precisely because *STRIPS* is cast here as a purely logical problem in a suitable general logic, a number of otherwise non-trivial extensions become low-hanging fruits and we get them almost for free. In particular, *STRIPS* planning

R. Kontchakov and M.-L. Mugnier (Eds.): RR 2014, LNCS 8741, pp. 29–44, 2014.
© Springer International Publishing Switzerland 2014

can be naturally extended with intensional rules, which endows the framework with support for ramification [14] (i.e., with indirect effects of actions), and we show that the resulting planning algorithm is complete. Then, after inspecting the logic rules that simulate the *STRIPS* algorithm, we observe that more restrictive rules can be derived, which intuitively should involve a smaller search space. The new rules lead to a different *STRIPS*-like algorithm, which we call *fast STRIPS*, or *fSTRIPS*. We show that *fSTRIPS* is also a complete planning algorithm, and our experiments indicate that it can be orders of magnitude faster than the original.

A number of deductive planning frameworks have been proposed over the years [2–4, 16, 11, 15, 26, 27], but only a few of these approaches support any kind of action ramification. Most importantly, our work differs in the following respects.

- Many of the above approaches invent one-of-a-kind theories that are suitable only for the particular problem at hand. We, on the other hand, use a general logic, which integrates well with most other approaches to knowledge representation.
- These works typically first demonstrate how they can capture *STRIPS*-like actions and then rely on a theorem prover of some sort to find plans. This type of planning is called *naive*, as it has to contend with extremely large search spaces. In contrast, we capture not merely *STRIPS* actions—they are part of the basic functionality in $\mathcal{TR}$—but also the actual optimized *planning strategies* (*STRIPS*, HTN, etc.), which utilize heuristics to reduce the search space. That is, we first compactly express these heuristics as $\mathcal{TR}$ rules and *then* use the $\mathcal{TR}$ theorem prover to find plans. The effect is that the theorem prover, in fact, executes those specialized and more efficient algorithms.
- The clear and compact form used to represent the planning heuristics is suggestive of various optimizations, which lead to new and more efficient algorithms. We illustrate this with the example of discovery of *fSTRIPS*.

We are also unaware of anything similar to our results in the literature on the situation calculus or other first-order logic based methodologies (cf. [19, 22]).

Finally, several aspects found in the planning literature, like parallelization of plans and loops [18, 17, 24], are orthogonal to the results presented here and provide a natural direction for further extensions of our work.

This paper is organized as follows. Section 2 introduces the *STRIPS* planning framework and its extension in support of action ramification. Section 3 provides the necessary background on Transaction Logic in order to make this paper self-contained. Section 4 casts the extended *STRIPS* as a problem of posing a transactional query in Transaction Logic and shows that executing this query using the $\mathcal{TR}$'s proof theory makes for a sound and complete *STRIPS* planning algorithm. Section 5 introduces the *fSTRIPS* algorithm, which is also cast as a transactional query in $\mathcal{TR}$, and shows that this is also a complete planning strategy. In Section 6, we present our experiments that show that *fSTRIPS* can be orders of magnitude better than *STRIPS* both in time and space. Section 7 concludes the paper.

2   Extended *STRIPS*-Style Planning

In this section we first remind the reader a number of standard concepts in logic and then introduce the *STRIPS* planning problem.

We assume denumerable sets of variables $\mathcal{V}$, constants $\mathcal{C}$, and predicate symbols $\mathcal{P}$— all three sets being pairwise disjoint. The set of predicates, $\mathcal{P}$, is further partitioned into *extensional* ($\mathcal{P}_{ext}$) and *intensional* ($\mathcal{P}_{int}$) predicates. In *STRIPS*, actions update the state of a system by adding or deleting statements about predicates. In the original *STRIPS*, all predicates were extensional, and the addition of intensional predicates is a major enhancement, which allows us to deal with the so-called *ramification problem* [14], i.e., with indirect consequences of actions.

Atomic formulas (or just ***atoms***) have the form $p(t_1, ..., t_n)$, were $p \in \mathcal{P}$ and each t_i is either a constant or a variable. Extending the logical signature with function symbols is straightforward in our framework, but we avoid doing this here in order to save space.

An atom is ***extensional*** if $p \in \mathcal{P}_{ext}$ and ***intensional*** if $p \in \mathcal{P}_{int}$. A ***literal*** is either an atom or a negated extensional atom of the form $\neg p(t_1, ..., t_n)$. Negated intensional atoms are not allowed. (It is not too hard to extend our framework and the results to allow negated intensional atoms, but we refrain from doing so due to space limitations).

Extensional predicates represent database facts: they can be directly manipulated (inserted or deleted) by actions. Intensional predicate symbols are used for atomic statements defined by *rules*—they are *not* affected by actions directly. Instead, actions make extensional facts true or false and this indirectly affects the dependent intensional atoms. These indirect effects are known as action *ramifications* in the literature.

A ***fact*** is a ***ground*** (i.e., variable-free) extensional atom. A set $\mathbf{S}$ of literals is ***consistent*** if there is no atom, *atm*, such that both *atm* and $\neg atm$ are in $\mathbf{S}$.

A ***rule*** is a statement of the form *head* $\leftarrow$ *body* where *head* is an intensional atom and body is a conjunction of literals. A ***ground instance*** of a rule, R, is any rule obtained from R by a substitution of variables with constants from $\mathcal{C}$ such that different occurrences of the same variable are always substituted with the same constant. Given a set $\mathbf{S}$ of literals and a ground rule of the form $atm \leftarrow \ell_1 \wedge \cdots \wedge \ell_m$, the rule is *true* in $\mathbf{S}$ if either $atm \in \mathbf{S}$ or $\{\ell_1, ..., \ell_m\} \not\subseteq \mathbf{S}$. A (possibly non-ground) rule is *true* in $\mathbf{S}$ if all of its ground instances are true in $\mathbf{S}$.

Definition 1 (State). *Given a set $\mathbb{R}$ of rules, a **state** is a consistent set $\mathbf{S} = \mathbf{S}_{ext} \cup \mathbf{S}_{int}$ of literals such that*

1. *For each fact atm, either $atm \in \mathbf{S}_{ext}$ or $\neg atm \in \mathbf{S}_{ext}$.*
2. *Every rule of $\mathbb{R}$ is true in $\mathbf{S}$.*   □

Definition 2 (*STRIPS* Action). *A STRIPS **action** is a triple of the form $\alpha = \langle p_\alpha(X_1, ..., X_n), Pre_\alpha, E_\alpha \rangle$, where*

- *$p_\alpha(X_1, ..., X_n)$ is an intensional atom in which $X_1, ..., X_n$ are variables and $p_\alpha \in \mathcal{P}_{int}$ is a predicate that is reserved to represent the action α and can be used for no other purpose;*
- *Pre_α, called the **precondition** of α, is a set of literals that may include extensional as well as intensional literals;*

- E_α, called the **effect** of α, is a consistent set that may contain extensional literals only;
- The variables in Pre_α and E_α must occur in $\{X_1, ..., X_n\}$.[1]   □

Note that the literals in Pre_α can be both extensional and intensional, while the literals in E_α can be extensional only.

Definition 3 (Execution of a *STRIPS* Action). *A STRIPS action α is **executable** in a state $\mathbf{S}$ if there is a substitution $\theta : \mathcal{V} \longrightarrow \mathcal{C}$ such that $\theta(Pre_\alpha) \subseteq \mathbf{S}$. A **result of the execution** (with respect to θ) is the state $\mathbf{S}'$ such that $\mathbf{S}' = (\mathbf{S} \setminus \neg\theta(E_\alpha)) \cup \theta(E_\alpha)$, where $\neg E = \{\neg\ell \mid \ell \in E\}$. In other words, $\mathbf{S}'$ is $\mathbf{S}$ with all effects of $\theta(\alpha)$ applied.*   □

Note that $\mathbf{S}'$ is well-defined since E_α is consistent. Observe also that, if α has variables, the result of an execution, $\mathbf{S}'$, may depend on the chosen substitution θ.

The following simple example illustrates the above definition. We follow the standard logic programming convention whereby lowercase symbols represent constants and predicate symbols. The uppercase symbols denote variables that are implicitly universally quantified outside of the rules.

Example 1. Consider a world consisting of just two blocks and the action $pickup = \langle pickup(X, Y), \{clear(X)\}, \{\neg on(X, Y), clear(Y)\}\rangle$. Consider also the state $\mathbf{S} = \{clear(a), \neg clear(b), on(a, b), \neg on(b, a)\}$. Then the result of the execution of $pickup$ at state $\mathbf{S}$ with respect to the substitution $\{X \to a, Y \to b\}$ is $\mathbf{S}' = \{clear(a), clear(b), \neg on(a, b), \neg on(b, a)\}$. It is also easy to see that $pickup$ cannot be executed at $\mathbf{S}$ with respect to any substitution of the form $\{X \to b, Y \to ...\}$.   □

Definition 4 (Planning Problem). *A **planning problem** $\langle \mathbb{R}, \mathbb{A}, G, \mathbf{S}\rangle$ consists of a set of rules $\mathbb{R}$, a set of STRIPS actions $\mathbb{A}$, a set of literals G, called the **goal** of the planning problem, and an **initial state** $\mathbf{S}$. A sequence of actions $\sigma = \alpha_1, ..., \alpha_n$ is a **planning solution** (or simply a **plan**) for the planning problem if:*

- *$\alpha_1, ..., \alpha_n \in \mathbb{A}$; and*
- *there is a sequence of states $\mathbf{S}_0, \mathbf{S}_1, ..., \mathbf{S}_n$ such that*
  - *$\mathbf{S} = \mathbf{S}_0$ and $G \subseteq \mathbf{S}_n$ (i.e., G is satisfied in the final state);*
  - *for each $0 < i \leq n$, α_i is executable in state $\mathbf{S}_{i-1}$ and the result of that execution (for some substitution) is the state $\mathbf{S}_i$.*

In this case we will also say that $\mathbf{S}_0, \mathbf{S}_1, ..., \mathbf{S}_n$ is an execution of σ.   □

3   Overview of Transaction Logic

To make this paper self-contained, we provide a brief introduction to the parts of Transaction Logic that are needed for the understanding of this paper. For further details, the reader is referred to [7, 9, 10, 5, 8].

---

[1] Requiring the variables of Pre_α to occur in $\{X_1, ..., X_n\}$ is not essential for us: we can easily extend our framework and consider the extra variables to be existentially quantified.

$\mathcal{TR}$ is a faithful extension of the first-order predicate calculus and so all of that syntax carries over. In this paper, we focus on rules, however, so we will be dealing exclusively with that subset of the syntax from now on. The most important new connective that Transaction Logic brings in is the **serial conjunction**, denoted $\otimes$. It is a binary associative connective, like the classical conjunction, but it is not commutative. Informally, the formula $\phi \otimes \psi$ is understood as a composite action that denotes an *execution* of ϕ followed by an execution of ψ. The **concurrent conjunction** connective, $\phi \| \psi$, is associative *and commutative*. Informally, it says that ϕ and ψ can execute in an *interleaved* fashion. For instance, $(\alpha_1 \otimes \alpha_2) \| (\beta_1 \otimes \beta_2)$ can execute as $\alpha_1, \beta_1, \alpha_2, \beta_2$, or as $\alpha_1, \beta_1, \beta_2, \alpha_2$, or as $\alpha_1, \alpha_2, \beta_1, \beta_2$, while $(\alpha_1 \otimes \alpha_2) \otimes (\beta_1 \otimes \beta_2)$ can execute only as $\alpha_1, \alpha_2, \beta_1, \beta_2$. When ϕ and ψ are regular first-order formulas, both $\phi \otimes \psi$ and $\phi \| \psi$ reduce to the usual first-order conjunction, $\phi \wedge \psi$. The logic also has other connectives but they are beyond the scope of this paper.

In addition, $\mathcal{TR}$ has a general, extensible mechanism of **elementary updates** or elementary **actions,** which have the important effect of taking the infamous *frame problem* out of many considerations in this logic (see [9, 10, 7, 23, 6]). Here we will use only the following two types of elementary actions, which are specifically designed on complete *STRIPS* states (Definition 1): $+p(t_1, \ldots, t_n)$ and $-p(t_1, \ldots, t_n)$, where p denotes an *extensional* predicate symbol of appropriate arity and $t_1, \ldots, t_n$ are terms.

Given a state $\mathbf{S}$ and a *ground* elementary action $\alpha = +p(a_1, \ldots, a_n)$, an execution of α at state $\mathbf{S}$ deletes the literal $\neg p(a_1, \ldots, a_n)$ and adds the literal $p(a_1, \ldots, a_n)$. Similarly, executing $-p(a_1, \ldots, a_n)$ results in a state that is exactly like $\mathbf{S}$, but $p(a_1, \ldots, a_n)$ is deleted and $\neg p(a_1, \ldots, a_n)$ added. In some cases (e.g., if $p(a_1, \ldots, a_n) \in \mathbf{S}$), the action $+p(a_1, \ldots, a_n)$ has no effect, and similarly for $-p(a_1, \ldots, a_n)$.

A **serial rule** is a statement of the form

$$h \leftarrow b_1 \otimes b_2 \otimes \ldots \otimes b_n. \tag{1}$$

where h is an atomic formula and $b_1, \ldots, b_n$ are literals or elementary actions. The informal meaning of such a rule is that h is a complex action and one way to execute h is to execute b_1 then b_2, etc., and finally to execute b_n.

Thus, we now have regular first-order as well as serial-Horn rules. For simplicity (thought this is not required by $\mathcal{TR}$), we assume that the sets of intentional predicates that can appear in the heads of regular rules and those in the heads of serial rules are disjoint. Thus, we now have the following types of atomic statements:

- *Extensional atoms.*
- *Intentional atoms*: The atoms that appear in the heads of regular rules. These two categories of atoms populate database states and will be collectively called *fluents*. We will now allow any kind of fluent to be negated in the body of a serial rule of the form (1).
- *Elementary actions*: $+p$, $-p$, where p is an extensional atom.
- *Complex actions*: These are the atoms that appear at the head of the serial rules. Complex and elementary actions will be collectively called **actions**.

As remarked earlier, for fluents $f \otimes g$ is equivalent to $f \wedge g$ and we will often write $f \wedge g$ for fluents even if they occur in the bodies of serial rules. Note that a serial rule all of whose body literals are fluents is essentially a regular rule, since all the $\otimes$-connectives can be replaced with $\wedge$. Therefore, one can view the regular rules as a special case of serial rules.

The following example illustrates the above concepts.

$$move(X, Y) \leftarrow (on(X, Z) \wedge clear(X) \wedge clear(Y) \wedge \neg tooHeavy(X))$$
$$\otimes - on(X, Z) \otimes +on(X, Y) \otimes -clear(Y).$$
$$tooHeavy(X) \leftarrow weight(X, W) \wedge limit(L) \wedge W < L.$$
$$? - move(blk1, blk15) \otimes move(SomeBlk, blk1).$$

Here *on, clear, tooHeavy, weight*, and *limit* are fluents and the rest of atoms represent actions. The predicate *tooHeavy* is an intentional fluent, while *on, clear*, and *weight* are extensional fluents. The actions $+on(...)$, $-clear(...)$, and $-on(...)$ are elementary and the intentional predicate *move* represents a complex action. This example illustrates several features of Transaction Logic. The first rule is a serial rule defining of a complex action of moving a block from one place to another. The second rule defines the intensional fluent *tooHeavy*, which is used in the definition of *move* (under the scope of default negation). As the second rule does not include any action, it is a regular rule.

The last statement above is a *request to execute* a composite action, which is analogous to a query in logic programming. The request is to move block *blk1* from its current position to the top of *blk15* and then find some other block and move it on top of *blk1*. Traditional logic programming offers no logical semantics for updates, so if after placing *blk1* on top of *blk15* the second operation ($move(SomeBlk, blk1)$) fails (say, all available blocks are too heavy), the effects of the first operation will persist and the underlying database becomes corrupted. In contrast, Transaction Logic gives update operators the logical semantics of an *atomic database transaction*. This means that if any part of the transaction fails, the effect is as if nothing was done at all. For example, if the second action in our example fails, all actions are "backtracked over" and the underlying database state remains unchanged.

This semantics is given in purely model-theoretic terms and here we will only give an informal overview. The truth of any action in $\mathcal{TR}$ is determined over sequences of states—**execution paths**—which makes it possible to think of truth assignments in $\mathcal{TR}$'s models as executions. If an action, ϕ, defined by a set of serial rules, $\mathbb{P}$, evaluates to true over a sequence of states $\mathbf{D}_0, \ldots, \mathbf{D}_n$, we say that it can *execute* at state $\mathbf{D}_0$ by passing through the states $\mathbf{D}_1, ..., \mathbf{D}_{n-1}$, ending in the final state $\mathbf{D}_n$. This is captured by the notion of **executional entailment**, which is written as follows:

$$\mathbb{P}, \mathbf{D}_0 \ldots \mathbf{D}_n \models \phi \tag{2}$$

The next example further illustrates $\mathcal{TR}$ by showing a definition of a recursive action.

Example 2 (Pyramid building). The following rules define a complex operation of stacking blocks to build a pyramid. It uses some of the already familiar fluents and actions from the previous example. In addition, it defines the actions *pickup, putdown,* and a recursive action *stack.*

$$stack(0, AnyBlock) \leftarrow .$$
$$stack(N, X) \leftarrow N > 0 \otimes move(Y, X) \otimes stack(N - 1, Y) \otimes on(Y, X).$$
$$move(X, Y) \leftarrow X \neq Y \otimes pickup(X) \otimes putdown(X, Y).$$
$$pickup(X) \leftarrow clear(X) \otimes on(X, Y) \otimes -on(X, Y) \otimes +clear(Y). \tag{3}$$
$$pickup(X) \leftarrow clear(X) \otimes on(X, table) \otimes -on(X, table).$$
$$putdown(X, Y) \leftarrow clear(Y) \otimes \neg on(X, Z1) \otimes \neg on(Z2, X) \otimes$$
$$-clear(Y) \otimes +on(X, Y).$$

The first rule says that stacking zero blocks on top of X is a no-op. The second rule says that, for bigger pyramids, stacking N blocks on top of X involves moving some other block, Y, on X and then stacking $N - 1$ blocks on Y. To make sure that the planner did not remove Y from X while building the pyramid on Y, we are verifying that $on(Y, X)$ continues to hold at the end. The remaining rules are self-explanatory. □

Several inference systems for serial-Horn $\mathcal{TR}$ are described in [7]—all analogous to the well-known SLD resolution proof strategy for Horn clauses plus some $\mathcal{TR}$-specific inference rules and axioms. The aim of these inference systems is to prove statements of the form $\mathbb{P}, \mathbf{D} \cdots \vdash \phi$, called **sequents**. Here $\mathbb{P}$ is a set of serial rules and ϕ is a *serial goal*, i.e., a formula that has the form of a body of a serial rule, such as (1). A proof of a sequent of this form is interpreted as a proof that action ϕ defined by the rules in $\mathbb{P}$ can be successfully executed starting at state $\mathbf{D}$.

An inference succeeds if it finds an **execution** for the transaction ϕ, i.e., a sequence of database states $\mathbf{D}_1, \ldots, \mathbf{D}_n$ such that $\mathbb{P}, \mathbf{D}\,\mathbf{D}_1 \ldots \mathbf{D}_n \vDash \phi$. Here we will use the following inference system, which we present in a simplified form—only the version for ground facts and rules. The inference rules can be read either top-to-bottom (if *top* is proved then *bottom* is proved) or bottom-to-top (to prove *bottom* first prove *top*).

Definition 5 ($\mathcal{TR}$ **inference System**). *Let $\mathbb{P}$ be a set of rules (serial or regular) and $\mathbf{D}$, $\mathbf{D}_1, \mathbf{D}_2$ denote states.*

– *Axiom:* $\mathbb{P}, \mathbf{D} \cdots \vdash ()$, *where* () *is an empty clause (which is true at every state).*
– *Inference Rules*
  1. *Applying transaction definition: Suppose $t \leftarrow body$ is a rule in $\mathbb{P}$.*

  $$\frac{\mathbb{P}, \mathbf{D} \cdots \vdash body \otimes rest}{\mathbb{P}, \mathbf{D} \cdots \vdash t \otimes rest} \tag{4}$$

  2. *Querying the database: If $\mathbf{D} \models t$ then*

  $$\frac{\mathbb{P}, \mathbf{D} \cdots \vdash rest}{\mathbb{P}, \mathbf{D} \cdots \vdash t \otimes rest} \tag{5}$$

  3. *Performing elementary updates: If the elementary update t changes the state $\mathbf{D}_1$ into the state $\mathbf{D}_2$ then*

  $$\frac{\mathbb{P}, \mathbf{D}_2 \cdots \vdash rest}{\mathbb{P}, \mathbf{D}_1 \cdots \vdash t \otimes rest} \tag{6}$$

4. *Concurrency: If* ϕ_i, $i = 1, ..., n$ *are serial conjunctions then*

$$\frac{\mathbb{P}, \mathbf{D} \cdots \vdash \phi_1 \| ... \| \phi_j \| ... \| \phi_n}{\mathbb{P}, \mathbf{D}' \cdots \vdash \phi_1 \| ... \| \phi'_j \| ... \| \phi_n} \tag{7}$$

for any j, $1 \leq j \leq n$, *where* $\mathbf{D}'$ *is obtained from* $\mathbf{D}$ *and* ϕ'_j *from* ϕ_j *as in either of the inference rules (4-6) above.*

A ***proof*** of a sequent, seq_n, is a series of sequents, seq_1, seq_2, $\dots$, seq_{n-1}, seq_n, where each seq_i is either an axiom-sequent or is derived from earlier sequents by one of the above inference rules. This inference system has been proven sound and complete with respect to the model theory of $\mathcal{TR}$ [7]. This means that if ϕ is a serial goal, the executional entailment $\mathbb{P}, \mathbf{D}\,\mathbf{D}_1 \dots \mathbf{D}_n \models \phi$ holds if and only if there is a proof of $\mathbb{P}, \mathbf{D} \cdots \vdash \phi$ over the execution path $\mathbf{D}, \mathbf{D}_1, \dots, \mathbf{D}_n$, i.e., $\mathbf{D}_1, \dots, \mathbf{D}_n$ is the sequence of intermediate states that appear in the proof and $\mathbf{D}$ is the initial state. In this case, we will also say that such a proof proves the statement $\mathbb{P}, \mathbf{D}\,\mathbf{D}_1 \dots \mathbf{D}_n \vdash \phi$.

4   The $\mathcal{TR}$-*STRIPS* Planner

The informal idea of using $\mathcal{TR}$ as a planning formalism and an encoding of *STRIPS* as a set of $\mathcal{TR}$ rules first appeared in an unpublished report [7]. The encoding was incomplete and it did not include ramification and intensional predicates. We extend the original method with intentional predicates, make it complete, and formulate and prove the completeness of the resulting planner.

Definition 6 (Enforcement Operator). *Let G be a set of extensional literals. We define $Enf(G) = \{+p \mid p \in G\} \cup \{-p \mid \neg p \in G\}$. In other words, $Enf(G)$ is the set of elementary updates that makes G true.* □

Next we introduce a natural correspondence between *STRIPS* actions and $\mathcal{TR}$ rules.

Definition 7 (Actions as $\mathcal{TR}$ Rules). *Let $\alpha = \langle p_\alpha(\overline{X}), Pre_\alpha, E_\alpha \rangle$ be a STRIPS action. We define its **corresponding** $\mathcal{TR}$ rule, $tr(\alpha)$, to be a rule of the form*

$$p_\alpha(\overline{X}) \leftarrow (\wedge_{\ell \in Pre_\alpha} \ell) \otimes (\otimes_{u \in Enf(E_\alpha)} u). \tag{8}$$

Note that in (8) the actual order of action execution in the last component, $\otimes_{u \in Enf(E_\alpha)} u$, is immaterial, since all such executions happen to lead to the same state.

We now define a set of $\mathcal{TR}$ clauses that simulate the well-known *STRIPS* planning algorithm and extend this algorithm to handle intentional predicates and rules. The reader familiar with the *STRIPS* planner should not fail to notice that, in essence, these rules are a natural (and much more concise and general) verbalization of the classical *STRIPS* algorithm [12]. However—importantly—unlike the original *STRIPS*, these rules constitute a *complete* planner when evaluated with the $\mathcal{TR}$ proof theory.

Definition 8 ($\mathcal{TR}$ Planning Rules). *Let $\Pi = \langle \mathbb{R}, \mathbb{A}, G, \mathbf{S} \rangle$ be a STRIPS planning problem (see Definition 4). We define a set of $\mathcal{TR}$ rules, $\mathbb{P}(\Pi)$, which provides a sound and complete solution to the STRIPS planning problem. $\mathbb{P}(\Pi)$ has three disjoint parts, $\mathbb{P}_\mathbb{R}$, $\mathbb{P}_\mathbb{A}$, and $\mathbb{P}_G$, described below.*

- *The $\mathbb{P}_{\mathbb{R}}$ part: for each rule $p(\overline{X}) \leftarrow p_1(\overline{X}_1) \wedge \cdots \wedge p_k(\overline{X}_n)$ in $\mathbb{R}$, $\mathbb{P}_{\mathbb{R}}$ has a rule of the form*

$$achieve\_p(\overline{X}) \leftarrow \|_{i=1}^{n} achieve\_p_i(\overline{X}_i). \tag{9}$$

Rule (9) is an extension to the classical STRIPS planning algorithm and is intended to capture intentional predicates and ramification of actions; it is the only major aspect of our $\mathcal{TR}$-based rendering of STRIPS that was not present in the original in one way or another.

- *The part $\mathbb{P}_{\mathbb{A}} = \mathbb{P}_{actions} \cup \mathbb{P}_{atoms} \cup \mathbb{P}_{achieves}$ is constructed out of the actions in $\mathbb{A}$ as follows:*
  - *$\mathbb{P}_{actions}$: for each $\alpha \in \mathbb{A}$, $\mathbb{P}_{actions}$ has a rule of the form*

$$p_\alpha(\overline{X}) \leftarrow (\wedge_{\ell \in Pre_\alpha} \ell) \otimes (\otimes_{u \in Enf(E_\alpha)} u). \tag{10}$$

  This is the $\mathcal{TR}$ rule that corresponds to the action α, introduced in Definition 7.
  - *$\mathbb{P}_{atoms} = \mathbb{P}_{achieved} \cup \mathbb{P}_{enforced}$ has two disjoint parts as follows:*
    - *$\mathbb{P}_{achieved}$: for each extensional predicate $p \in \mathcal{P}_{ext}$, $\mathbb{P}_{achieved}$ has the rules*

$$achieve\_p(\overline{X}) \leftarrow p(\overline{X}).$$
$$achieve\_not\_p(\overline{X}) \leftarrow \neg p(\overline{X}). \tag{11}$$

    These rules say that if an extensional literal is true in a state then that literal has already been achieved as a goal.
    - *$\mathbb{P}_{enforced}$: for each action $\alpha = \langle p_\alpha(\overline{X}), Pre_\alpha, E_\alpha \rangle$ in $\mathbb{A}$ and each $e(\overline{Y}) \in E_\alpha$, $\mathbb{P}_{enforced}$ has the following rule:*

$$achieve\_e(\overline{Y}) \leftarrow execute\_p_\alpha(\overline{X}). \tag{12}$$

    This rule says that one way to achieve a goal that occurs in the effects of an action is to execute that action.
  - *$\mathbb{P}_{achieves}$: for each action $\alpha = \langle p_\alpha(\overline{X}), Pre_\alpha, E_\alpha \rangle$ in $\mathbb{A}$, $\mathbb{P}_{achieves}$ has the following rule:*

$$execute\_p_\alpha(\overline{X}) \leftarrow (\|_{\ell \in Pre_\alpha} achieve\_\ell) \otimes p_\alpha(\overline{X}). \tag{13}$$

  This means that to execute an action, one must first achieve the precondition of the action and then perform the state changes prescribed by the action.
- *$\mathbb{P}_G$: Let $G = \{g_1, ..., g_k\}$. Then $\mathbb{P}_G$ has a rule of the form:*

$$achieve_G \leftarrow (\|_{g_i=1}^{k} achieve\_g_i) \otimes (\wedge_{i=1}^{k} g_i). \tag{14}$$

Given a set $\mathbb{R}$ of rules, a set $\mathbb{A}$ of *STRIPS* actions, an initial state **S**, and a goal G, Definition 8 gives a set of $\mathcal{TR}$ rules that specify a planning strategy for that problem. To find a solution for that planning problem, one simply needs to place the request

$$? - achieve_G. \tag{15}$$

at a desired initial state and use the $\mathcal{TR}$'s inference system of Section 3 to find a proof. The inference system in question is sound and complete for *serial clauses*, and the rules in Definition 8 satisfy that requirement.

Example 3 (Planning rules for register exchange). Consider the classical problem of swapping two registers in a computer from [21]. The reason this problem is interesting is because it is the simplest problem where the original *STRIPS* is incomplete. Example 4 explains why and how our complete $\mathcal{TR}$-based planner handles the issue.

Consider two memory registers, x and y, with initial contents a and b, respectively. The goal is to find a plan to exchange the contents of these registers with the help of an auxiliary register, z. Let the extensional predicate $value(Reg, Val)$ represent the content of a register. Then the initial state of the system is $\{value(x, a), value(y, b)\}$. Suppose the only available action is $copy = \langle copy(Src, Dest, V), \{value(Src, V)\}, \{\neg value(Dest, V), value(Dest, V)\}\rangle$, which copies the value V of the source register, Src, to the destination register $Dest$. The old value of $Dest$ is erased and the value of Src is written over. The planning goal is $G = \{value(x, b), value(y, a)\}$. Per Definition 8, the planning rules for this problem are as follows.

Due to case (10):

$$copy(Src, Dest, V) \leftarrow value(Src, V) \otimes \\ -value(Dest, \_) \otimes +value(Dest, V). \qquad (16)$$

Due to (11), (12), and (13):

$$achieve\_value(R, V) \leftarrow value(R, V). \\ achieve\_not\_value(R, V) \leftarrow \neg value(R, V). \qquad (17)$$

$$achieve\_value(Dest, V) \leftarrow execute\_copy(Src, Dest, V). \qquad (18)$$

$$execute\_copy(Src, Dest, V) \leftarrow achieve\_value(Src, V) \otimes \\ copy(Src, Dest, V). \qquad (19)$$

Due to (14):

$$achieve_G \leftarrow (achieve\_value(x, b) \parallel achieve\_value(y, a)) \\ \otimes (value(x, b) \wedge value(y, a)). \qquad (20)$$

Case (9) of Definition 8 does not contribute rules in this example because the planning problem does not involve intensional fluents. □

As mentioned before, a solution plan for a *STRIPS* planning problem is a sequence of actions leading to a state that satisfies the planning goal. Such a sequence can be extracted by picking out the atoms of the form p_α from a successful derivation branch generated by the $\mathcal{TR}$ inference system. Since each p_α uniquely corresponds to a *STRIPS* action, this provides us with the requisite sequence of actions that constitutes a plan.

Suppose $seq_0, \ldots, seq_m$ is a deduction by the $\mathcal{TR}$ inference system. Let $i_1, \ldots, i_n$ be exactly those indexes in that deduction where the inference rule (4) was applied to some sequent using a rule of the form $tr(\alpha_{i_r})$ introduced in Definition 7. We will call $\alpha_{i_1}, \ldots, \alpha_{i_n}$ the **pivoting sequence of actions**. The corresponding **pivoting sequence of states** $\mathbf{D}_{i_1}, \ldots, \mathbf{D}_{i_n}$ is a sequence where each $\mathbf{D}_{i_r}$, $1 \le r \le n$, is the state at which α_{i_r} is applied. We will prove that the pivoting sequence of actions is a solution to the planning problem.

All theorems in this section assume that $\Pi = \langle \mathbb{R}, \mathbb{A}, G, \mathbf{D}_0 \rangle$ is a *STRIPS* planning problem and that $\mathbb{P}(\Pi)$ is the corresponding set of $\mathcal{TR}$ planning rules as in Definition 8.

Theorem 1 (Soundness of $\mathcal{TR}$ Planning). *Any pivoting sequence of actions in the derivation of* $\mathbb{P}(\Pi), \mathbf{D}_0 \ldots \mathbf{D}_m \vdash achieve_G$ *is a solution plan.*[2]

Completeness of a planning strategy means that, for any *STRIPS* planning problem, if there is a solution, the planner will find at least one plan. Completeness of $\mathcal{TR}$ planning is established by induction on the length of the plans.

Theorem 2 (Completeness of $\mathcal{TR}$ Planning). *If there is a plan that achieves the goal G from the initial state* $\mathbf{D}_0$ *then the* $\mathcal{TR}$*-based STRIPS planner will find a plan.*

Theorem 2 establishes the completeness of the planner that is comprised of the $\mathcal{TR}$ proof theory and the rules that express the original *STRIPS* strategy.

Recall that the *classical STRIPS* planner described in [12, 21] was incomplete. The next example illustrates the reason for this incompleteness and contrasts the situation to the $\mathcal{TR}$-based planner.

Example 4 (Register exchange, continued). Consider the register exchange problem of Example 3. The original *STRIPS* planner fails to find a plan if, in the initial state, the auxiliary register z has the value t distinct from a and b [21]. We will now illustrate how the $\mathcal{TR}$ based planner deals with this case. Let $\mathbb{P}$ be the set of $\mathcal{TR}$ rules (16-19) that constitute the planner for the $\mathcal{TR}$-based planner for this problem. Given the planning goal $G = \{value(x,b), value(y,a)\}$ and the initial state $\mathbf{D}_0$, where $\{value(x,a), value(y,b)\} \subseteq \mathbf{D}_0$, we will show how the $\mathcal{TR}$ inference system constructs a derivation (and thus a plan) for the sequent $\mathbb{P}, \mathbf{D}_0 \cdots \mathbf{D}_n \vdash achieve_G$ for some $\mathbf{D}_n$ such that $\{value(x,b), value(y,a)\} \subseteq \mathbf{D}_n$.

Consider the sequent $\mathbb{P}, \mathbf{D}_0 \cdots \vdash achieve_G$ that corresponds to the query (15). Applying the inference rule (4) to that sequent using the rule (20), we get:

$$\mathbb{P}, \mathbf{D}_0 \cdots \vdash (achieve\_value(x,b) \| achieve\_value(y,a))$$
$$\otimes (value(x,b) \wedge value(y,a))$$

Applying the inference rule (4) twice to the resulting sequent using the rules (18–19) with appropriate substitutions result in:

$$\mathbb{P}, \mathbf{D}_0 \cdots \vdash ((achieve\_value(z,b) \otimes copy(z,x,b)) \| achieve\_value(y,a))$$
$$\otimes (value(x,b) \wedge value(y,a))$$

Applying the inference rule (4) once more and again using the rules (18–19) we get:

$$\mathbb{P}, \mathbf{D}_0 \cdots \vdash ((achieve\_value(y,b) \otimes copy(y,z,b) \otimes copy(z,x,b))$$
$$\| achieve\_value(y,a)) \otimes (value(x,b) \wedge value(y,a))$$

One more application of the inference rule (4) but this time in conjunction with (17) yields:

$$\mathbb{P}, \mathbf{D}_0 \cdots \vdash ((value(y,b) \otimes copy(y,z,b) \otimes copy(z,x,b))$$
$$\| achieve\_value(y,a)) \otimes (value(x,b) \wedge value(y,a))$$

---

[2] Sequents of the form $\mathbb{P}(\Pi), \mathbf{D}_0 \ldots \mathbf{D}_m \vdash \ldots$ were defined at the very end of Section 3.

Since $value(y, b) \in \mathbf{D}_0$, we can eliminate it by the inference rule (5). Then we can replace the first *copy* using its definition (16) due to the inference rule (4).

$$\mathbb{P}, \mathbf{D}_0 \cdots \vdash ((-value(z, \_) \otimes +value(z, b) \otimes copy(z, x, b)) \\ \| achieve\_value(y, a)) \otimes (value(x, b) \wedge value(y, a))$$

Applying the inference rule (6) twice to the primitive updates at the front first yields

$$\mathbb{P}, \mathbf{D}_1 \cdots \vdash ((+value(z, b) \otimes copy(z, x, b)) \\ \| achieve\_value(y, a)) \otimes (value(x, b) \wedge value(y, a))$$

and then

$$\mathbb{P}, \mathbf{D}_2 \cdots \vdash (copy(z, x, b) \| achieve\_value(y, a)) \otimes (value(x, b) \wedge value(y, a))$$

where $\mathbf{D}_1$ is $\mathbf{D}_0$ with $value(z, t)$ (where t denotes the old value of z) deleted and $\mathbf{D}_2$ is $\mathbf{D}_1$ with $value(z, b)$ added.

Now we can use the inference rule (7) to explore the subgoal $achieve\_value(y, a)$. Namely, we can expand this subgoal with the inference rule (4) twice, first using (18–19) and then using (17), obtaining

$$\mathbb{P}, \mathbf{D}_2 \cdots \vdash (copy(z, x, b) \| (value(x, a) \otimes copy(x, y, a))) \\ \otimes (value(x, b) \wedge value(y, a))$$

Since $value(x, a)$ is true in $\mathbf{D}_2$, it can be removed. Finally, the two *copy*'s can be replaced by their definition (16) and then the remaining $+value(\ldots)$ and $-value(\ldots)$ can be executed using the inference rule (6). This will advance the database (via three intermediate states) to state $\mathbf{D}_6$ containing $\{value(x, b), value(y, a), value(z, b)\}$ in which both $value(x, b)$ and $value(y, a)$ are true. Therefore, the inference rule (5) can be used to derive the $\mathcal{TR}$ axiom $\mathbb{P}, \mathbf{D}_6 \cdots \vdash ()$, thus concluding the proof. The pivoting sequence of actions in this proof is $\langle copy(y, z, b), copy(x, y, a), copy(z, x, b) \rangle$, which constitutes the desired plan. $\qquad\square$

5   The *fSTRIPS* Planner

In this section, we introduce *fSTRIPS* — a modification of the previously introduced *STRIPS* transform, which represents to a new planning strategy, which we call *fast STRIPS*. We show that although the new strategy explores a smaller search space, it is still sound and complete. Section 6 shows that *fSTRIPS* can be orders of magnitude faster than *STRIPS*.

Definition 9 ($\mathcal{TR}$ **Planning Rules for *fSTRIPS***)**. *Let* $\Pi = \langle \mathbb{R}, \mathbb{A}, G, \mathbf{S} \rangle$ *be a STRIPS planning problem as in Definition 4 and* $\mathbb{P}(\Pi)$ *is as in Definition 8. We define* $\mathbb{P}_f(\Pi)$ *to be exactly as* $\mathbb{P}(\Pi)$ *except for the* $\mathbb{P}_{enforced}$ *part. For* $\mathbb{P}_f(\Pi)$*, we redefine* $\mathbb{P}^f_{enforced}$ *(the replacement of* $\mathbb{P}_{enforced}$*) as follows:*

For each action $\alpha = \langle p_\alpha(\overline{X}), Pre_\alpha, E_\alpha \rangle$ *in* $\mathbb{A}$ *and each* $e(\overline{Y}) \in E_\alpha$*,* $\mathbb{P}^f_{enforced}$ *has the following rule:*

$$achieve\_e(\overline{Y}) \leftarrow \neg e(\overline{Y}) \otimes execute\_p_\alpha(\overline{X}). \tag{21}$$

This rule says that an action, α, should be attempted only if it helps to achieve the currently pursued, unsatisfied goal.   □

The other key aspect of *fSTRIPS* is that it uses a modified (general, unrelated to planning) proof theory for $\mathcal{TR}$, which relies on *tabling*, a technique analogous to [25]. This theory was introduced in [13] and was shown to be sound and complete. Here we use it for two reasons. First, it terminates if the number of base fluents is finite. Second, it has the property that it will not attempt to construct plans that have extraneous loops and thus will not attempt to large and unnecessary parts of the search space.

To construct a plan, as before, we can extract a pivoting sequence of actions and show that the new pivoting sequence of actions is still a solution plan.

Similarly to Section 4, we assume till the end of this section that $\Pi = \langle \mathbb{R}, \mathbb{A}, G, \mathbf{D}_0 \rangle$ is a *STRIPS* planning problem, that $\mathbb{P}(\Pi)$ is the set of planning rules in Definition 8, and that $\mathbb{P}_f(\Pi)$ is the set of planning rules as specified in Definition 9.

Theorem 3 (Soundness of *fSTRIPS*). *Any pivoting sequence of actions in the derivation of $\mathbb{P}_f(\Pi), \mathbf{D}_0 \dots \mathbf{D}_m \vdash achieve_G$ is a solution plan.*

Theorem 4 (Completeness of *fSTRIPS*). *If there is a plan to achieve the goal G from an initial state, $\mathbf{D}_0$, then $\mathcal{TR}$ will find a plan.*

Theorem 5 (*fSTRIPS* Finds no More Plans than *STRIPS*). *Any plan found by the fSTRIPS planner will also be found by the STRIPS planner.*

In other words, the *STRIPS* strategy may generate more plans than *fSTRIPS*. The plans that are not generated by *fSTRIPS* are those that contain actions whose effects were not immediately required at the time of the action selection. This has the effect of ignoring longer plans when shorter plans are already found. The upshot of all this is that *STRIPS* has a larger search space to explore, and this explains the inferior performance of *STRIPS* compared to *fSTRIPS*, as the experiments in the next section show.

6 Experiments

In this section we briefly report on our experiments that compare *STRIPS* and *fSTRIPS*. The test environment was a tabled $\mathcal{TR}$ interpreter [13] implemented in XSB and running on Intel®Xeon(R) CPU E5-1650 0 @ 3.20GHz 12 CPU and 64GB memory running on Mint Linux 14 64-bit.

The actual test cases are taken from [1] and represent so called *State Modifying Policies*. A typical use of such a policy is to determine if a particular access request (say, to play digital contents) should be granted. The first test case, a *Movie Store*, is shown in Example 5. The second test case, a *Health Care Authorization* example, is too large to be included here and can be found at `http://ewl.cewit.stonybrook.edu/planning/` along with the first test case and all the necessary items needed to reproduce the results.

Example 5 (State Modifying Policy for a Movie Store). The following represents a policy where users can buy movies online, try them, and sell them, if not satisfied.

$$
\begin{aligned}
buy(X, M) &\leftarrow \neg bought(\_, M) \otimes +bought(X, M) \\
play1(X, M) &\leftarrow bought(X, M) \otimes \neg played1(X, M) \otimes +played1(X, M) \\
keep(X, M) &\leftarrow bought(X, M) \otimes \neg played1(X, M) \otimes +played1(X, M) \\
&\quad \otimes + happy(X, M) \\
play2(X, M) &\leftarrow played1(X, M) \otimes \neg played2(X, M) \otimes +played2(X, M) \\
play3(X, M) &\leftarrow played2(X, M) \otimes \neg played3(X, M) \otimes +played3(X, M) \\
sell(X, M) &\leftarrow played1(X, M) \otimes \neg played3(X, M) \otimes \neg happy(X, M) \\
&\quad \otimes + sold(X, M) \otimes -bought(X, M)
\end{aligned}
\tag{22}
$$

The first rule describes an action of a user, X, buying a movie, M. The action is possible only if the movie has not already been purchased by somebody. The second rule says that, to play a movie for the first time, the user must buy it first and not have played it before. The third rule deals with the case when the user is happy and decides to keep the movie. The remaining rules are self-explanatory.   □

A reachability query in a state modifying policy is a specification of a *target state* (usually an undesirable state), and the administrator typically wants to check if such a state is reachable by a sequence of actions. The target state specification consists of a set of literals, and the reachability query is naturally expressed as a planning problem. For instance, in Example 5, the second rule can be seen as a *STRIPS* action whose precondition is $\{bought(X, M) \otimes \neg played1(X, M)\}$ and the effect is $\{+played1(X, M)\}$. The initial and the target states in this example are sets of facts that describe the movies that have been bought, sold, and played by various customers.

Table 1. Results for different goal sizes (number of literals in the goals). The initial state is fixed and has 6 extensional atoms.

Size of goal	Movie Store				Size of goal	Health Care			
	STRIPS		*fSTRIPS*			*STRIPS*		*fSTRIPS*	
	CPU	Mem	CPU	Mem		CPU	Mem.	CPU	Mem.
6	0.0160	1095	0.0080	503	3	10.0240	246520	0.0400	2011
9	0.2760	14936	0.1360	6713	4	32.9540	774824	0.2040	8647
12	9.4120	409293	5.8840	184726	5	46.1380	1060321	0.3080	13622

Table 2. Results for different sizes (number of facts) in initial states. The planning goal is fixed: 6 extensional literals in the "movie store" case and 3 extensional literals in the "health care" case.

Size of initial state	Movie Store				Size of initial state	Health Care			
	STRIPS		*fSTRIPS*			*STRIPS*		*fSTRIPS*	
	CPU	Mem	CPU	Mem		CPU	Mem.	CPU	Mem.
20	9.2560	409293	5.8800	184726	3	0.148	5875	0.012	718
30	9.2600	409293	5.7440	184726	6	10.076	246519	0.04	2011
40	9.2520	409293	5.8000	184726	9	689.3750	9791808	0.124	5443
50	9.4120	409293	5.8840	184726	12	>1000	N/A	0.348	14832
60	9.3720	409293	5.8240	184726	18	>1000	N/A	0.94	38810

The main difference between the two test cases is that the Health Care example has many actions and intensional rules, while the movie store case has only six actions and no intensional predicates. As seen from Tables 1 and 2, for the relatively simple Movie Store example, *fSTRIPS* is about twice more efficient both in time and space.[3] However, in the more complex Health Care example, *fSTRIPS* is at least two orders of magnitude better both time-wise and space-wise. While in the Movie Store example the statistics for the two strategies seem to grow at the same rate, in the Health Care case, the *fSTRIPS* time appears to grow linearly, while the time for *STRIPS* grows quadratically.

7 Conclusion

This paper has demonstrated that the use of Transaction Logic accrues significant benefits in the area of planing. That is, the message is the benefits of $\mathcal{TR}$, not any particular planning heuristic. As an illustration, we have shown that sophisticated planning strategies, such as *STRIPS*, can be naturally represented in $\mathcal{TR}$ and that the use of this powerful logic opens up new possibilities for generalizations and devising new, more efficient algorithms. For instance, we have shown that once the *STRIPS* algorithm is cast as a set of rules in $\mathcal{TR}$, the framework can be extended, almost for free, to support such advanced aspects as action ramification, i.e., indirect effects of actions. Furthermore, by tweaking these rules just slightly, we obtained a new, much more efficient planner, which we dubbed *fSTRIPS* (fast *STRIPS*). These non-trivial insights were acquired merely due to the use of $\mathcal{TR}$ and not much else. The same technique can be used to cast even more advanced strategies such as *RSTRIPS*, *ABSTRIPS* [21], and HTN [20] as $\mathcal{TR}$ rules, and the *fSTRIPS* optimization straightforwardly applies to the first two.

There are several promising directions to continue this work. One is to investigate other planning strategies and, hopefully, accrue similar benefits. Other possible directions include non-linear plans and plans with loops [18, 17, 24]. For instance non-linear plans could be represented using Concurrent Transaction Logic [8], while loops are easily representable using recursive actions in $\mathcal{TR}$.

Acknowledgments. This work was supported, in part, by the NSF grant 0964196. We also thank the reviewers for valuable comments.

References

1. Becker, M.Y., Nanz, S.: A logic for state-modifying authorization policies. ACM Trans. Inf. Syst. Secur. 13(3), 20:1–20:28 (2010)
2. Bibel, W.: A deductive solution for plan generation. New Generation Computing 4(2), 115–132 (1986)
3. Bibel, W.: A deductive solution for plan generation. In: Schmidt, J.W., Thanos, C. (eds.) Foundations of Knowledge Base Management. Topics in Information Systems, pp. 453–473. Springer, Heidelberg (1989)
4. Bibel, W., del Cerro, L.F., Fronhöfer, B., Herzig, A.: Plan generation by linear proofs: On semantics. In: Metzing, D. (ed.) 13th German Workshop on Artificial Intelligence, Informatik-Fachberichte, GWAI 1989, vol. 216, pp. 49–62. Springer, Heidelberg (1989)

---

[3] Time is measured in seconds and memory in kilobytes.

5. Bonner, A., Kifer, M.: Transaction logic programming. In: Int'l Conference on Logic Programming, pp. 257–282. MIT Press, Budapest (1993)
6. Bonner, A., Kifer, M.: Applications of transaction logic to knowledge representation. In: Gabbay, D.M., Ohlbach, H.J. (eds.) ICTL 1994. LNCS, vol. 827, pp. 67–81. Springer, Heidelberg (1994)
7. Bonner, A., Kifer, M.: Transaction logic programming (or a logic of declarative and procedural knowledge). Tech. Rep. CSRI-323, University of Toronto (November 1995), http://www.cs.toronto.edu/~bonner/transaction-logic.html
8. Bonner, A., Kifer, M.: Concurrency and communication in transaction logic. In: Joint Int'l Conference and Symposium on Logic Programming, pp. 142–156. MIT Press, Bonn (1996)
9. Bonner, A., Kifer, M.: A logic for programming database transactions. In: Chomicki, J., Saake, G. (eds.) Logics for Databases and Information Systems, ch. 5, pp. 117–166. Kluwer Academic Publishers (March 1998)
10. Bonner, A.J., Kifer, M.: An overview of transaction logic. Theoretical Computer Science 133 (1994)
11. Cresswell, S., Smaill, A., Richardson, J.: Deductive synthesis of recursive plans in linear logic. In: Biundo, S., Fox, M. (eds.) ECP 1999. LNCS, vol. 1809, pp. 252–264. Springer, Heidelberg (2000)
12. Fikes, R.E., Nilsson, N.J.: STRIPS: A new approach to the application of theorem proving to problem solving. Artificial Intelligence 2(3-4), 189–208 (1971)
13. Fodor, P., Kifer, M.: Tabling for transaction logic. In: Proceedings of the 12th International ACM SIGPLAN Symposium on Principles and Practice of Declarative Programming, PPDP 2010, pp. 199–208. ACM, New York (2010)
14. Giunchiglia, E., Lifschitz, V.: Dependent fluents. In: Proceedings of International Joint Conference on Artificial Intelligence (IJCAI), pp. 1964–1969 (1995)
15. Guglielmi, A.: Concurrency and plan generation in a logic programming language with a sequential operator. In: Hentenryck, P.V. (ed.) ICLP, pp. 240–254. MIT Press (1994)
16. Hölldobler, S., Schneeberger, J.: A new deductive approach to planning. New Generation Computing 8(3), 225–244 (1990)
17. Kahramanogullari, O.: Towards planning as concurrency. In: Hamza, M.H. (ed.) Artificial Intelligence and Applications, pp. 387–393. IASTED/ACTA Press (2005)
18. Kahramanogullari, O.: On linear logic planning and concurrency. Information and Computation 207(11), 1229–1258 (2009); Special Issue: Martín-Vide, C., Otto, F., Fernau, H. (eds.): LATA 2008. LNCS, vol. 5196. Springer, Heidelberg (2008)
19. Lifschitz, V.: On the semantics of strips. In: Georgeff, M. (ed.) Lansky, Amy (eds, pp. 1–9. Morgan Kaufmann, San Mateo (1987)
20. Nau, D., Ghallab, M., Traverso, P.: Automated Planning: Theory & Practice. Morgan Kaufmann Publishers Inc., San Francisco (2004)
21. Nilsson, N.: Principles of Artificial Intelligence. Tioga Publ. Co., Paolo Alto (1980)
22. Reiter, R.: Knowledge in Action: Logical Foundations for Describing and Implementing Dynamical Systems. MIT Press, Cambridge (2001)
23. Rezk, M., Kifer, M.: Transaction logic with partially defined actions. J. Data Semantics 1(2), 99–131 (2012)
24. Srivastava, S., Immerman, N., Zilberstein, S., Zhang, T.: Directed search for generalized plans using classical planners. In: Proceedings of the 21st International Conference on Automated Planning and Scheduling (ICAPS 2011). AAAI (June 2011)
25. Swift, T., Warren, D.: Xsb: Extending the power of prolog using tabling. Theory and Practice of Logic Programming (2011)
26. Thielscher, M.: Computing ramifications by postprocessing. In: IJCAI, pp. 1994–2000. Morgan Kaufmann (1995)
27. Thielscher, M.: Ramification and causality. Artificial Intelligence 89(1-2), 317–364 (1997)

A Generalization of Approximation Fixpoint Theory and Application

Yi Bi[1], Jia-Huai You[2,*], and Zhiyong Feng[1]

[1] School of Computer Science and Technology, Tianjin University, Tianjin, China
[2] Department of Computing Science, University of Alberta, Edmonton, Canada
you@cs.ualberta.ca

Abstract. The approximation fixpoint theory (AFT) provides an algebraic framework for the study of fixpoints of operators on bilattices, and has been useful in dealing with semantics issues for various types of logic programs. The theory in the current form, however, only deals with consistent pairs on a bilattice, and it thus does not apply to situations where inconsistency may be part of a fixpoint construction. This is the case for FOL-programs, where a rule set and a first-order theory are tightly integrated. In this paper, we develop an extended theory of AFT that treats consistent as well as inconsistent pairs on a bilattice. We then apply the extended theory to FOL-programs and explore various possibilities on semantics. This leads to novel formulations of approximating operators, and new well-founded semantics and characterizations of answer sets for FOL-programs. The work reported here shows how consistent approximations may be extended to capture wider classes of logic programs whose semantics can be treated uniformly.

1 Introduction

AFT, also known as the theory of consistent approximations, is a powerful framework for the study of semantics of various types of logic programs [6,13]. Under this theory, the semantics of a logic program is defined by respective fixpoints closely related to an approximating operator on a bilattice. The approach is highly general as it only depends on mild conditions on approximating operators. The *well-founded fixpoint* of an approximating operator defines a well-founded semantics (WFS) and *exact stable fixpoints* define an answer set semantics. As different approximating operators may represent different intuitions, AFT provides an elegant way to treat semantics uniformly and allows to explore alternative semantics by choosing different approximating operators. We can understand the properties of a semantics even without a concrete approximating operator. For example, the least fixpoint approximates all other fixpoints and, mathematically this property holds for any approximating operator.

However, current AFT is not applicable to the types of logic programs where inconsistency needs to be treated explicitly. FOL-programs fall into this category. An FOL-program is a combined knowledge base $KB = (L, \Pi)$, where L is a theory of a decidable fragment of first-order logic and Π a set of rules possibly containing first-order

---

[*] Corresponding author.

R. Kontchakov and M.-L. Mugnier (Eds.): RR 2014, LNCS 8741, pp. 45–59, 2014.
© Springer International Publishing Switzerland 2014

formulas. Recent literature has shown extensive interests in combining ASP with fragments of classical logic, such as description logics (DLs) (see, e.g., [4,5,7,10,12,14,15]). In this way, logic programs can access external knowledge bases and are able to reason with them. In this paper, we use FOL-program as an umbrella term for approaches that allow first-order formulas to appear in rules (the so-called *tight* integration), for generality. The main interest of this paper is a generalization of AFT, which is motivated primarily by the need to allow operators to be defined on consistent as well as inconsistent pairs on a bilattice, and the usefulness of a uniform treatment of semantics for these programs. In general, we are interested in semantics for FOL-programs with the following features:

- The class of all FOL-programs are supported;
- Combined reasoning with closed world as well as open world is supported; and
- There is structural evidence that the former approximates the latter.

Under the first feature, we shall allow an atom with its predicate shared in the first-order theory L to appear in a rule head. This results in two-way flow of information between the knowledge base L and rule set Π, and enables inference within each component automatically. For example, assume L contains a formula that says students are entitled to educational discount, $\forall x\ St(x) \supset EdDiscount(x)$. Using the notation of DL, we would write $St \sqsubseteq EdDiscount$. Suppose in an application anyone who is not employed full time but registered for an evening class is given the benefit of a student. We can write a rule

$$St(X) \leftarrow EveningClass(X), not\ HasJob(X).$$

Thus, that such a person enjoys educational discount can be inferred directly from the underlying knowledge base L.

To support all FOL-programs, we need to treat inconsistencies explicitly. For example, consider an FOL-program, $KB = (L, \Pi)$, where $L = \{\forall x A(x) \supset C(x), \neg C(a)\}$ and $\Pi = \{A(a) \leftarrow not\ B(a); B(a) \leftarrow B(a)\}$. Let the Herbrand base be $\{A(a), B(a)\}$. In an attempt to compute the well-founded semantics of KB by an iterative process, we begin with the empty set; then, since $B(a)$ is false by closed world reasoning, we derive $A(a)$, resulting in an inconsistency with L. This reasoning process suggests that during an iterative process a consistent set of literals may be mapped to an inconsistent one and, in general, whether inconsistencies arise or not is not known *a priori* without actually performing the computation. The current AFT is not applicable here since we can only define approximating operators on consistent pairs on a bilattice.

The well-founded semantics has been defined for some subclasses of FOL-programs. The closest that one can find is the work of [10], which relies on syntactic restrictions so that the least fixpoint is computed over consistent sets of literals. To ensure that the construction is well-defined, it is assumed that DL axioms must be, or can be converted to, *tuple generating dependencies* (which are essentially Horn rules) plus constraints. Thus, the approach cannot be lifted to handle first-order formulas in general.

Combined reasoning is often needed in the real world. For example, we may write a rule

$$PrescribeTo(X, Q) \leftarrow Effective(X, Z), Contract(Q, Z), \neg isAllergicTo(Q, X)$$

to describe that an antibiotic is prescribed to a patient who contracted a bacterium, if the antibiotic against that bacterium is effective and patient is not allergic to it. Though *Effective* can be reasoned with under the closed world assumption, it is preferred to judge whether a patient is not allergic to an antibiotic under the open world assumption, e.g., it holds if it can be proved classically. This is in contrast with closed world reasoning whereas one may infer nonallergic due to lack of evidence for allergy.

For answer set semantics, solutions to this problem have been proposed in [14,15,17]. Essentially, with an appropriate separation of two kinds of predicates, we can verify whether a guess is indeed an answer set. However, we are not aware of any work in the context of the WFS that addresses this problem. As we will see in this paper, this is a nontrivial issue, and different possible logical treatments exist.

In the past, the general practice is to define WFS and ASP semantics separately, sometimes under different logic frameworks, study their properties, and determine which WFS approximates which answer set semantics. The last issue is important as the former is often practically easier to compute than that of the latter, and may be applied as an approximation, or used as constraint propagation in the computation of the latter.

As motivated above, in this paper we present the theoretical work on extending AFT and apply it to FOL-programs. The core of the paper is preceded by some definitions and followed by related work and remarks. The full report of this work with proofs can be found in [3].

2 Approximation Fixpoint Theory Revisited

2.1 Background

We assume familiarity with Knaster-Tarski fixpoint theory [18]. Briefly, a *lattice* $\langle \mathcal{L}, \leq \rangle$ is a *poset* in which every two elements have a least upper bound (lub) and a greatest lower bound (glb). A *chain* in a poset is a linearly ordered subset of $\mathcal{L}$. A poset $\langle \mathcal{L}, \leq \rangle$ is *chain-complete* if it contains a least element $\perp$ and if every chain $C \subseteq \mathcal{L}$ has a least upper bound in $\mathcal{L}$.

A complete lattice is chain-complete, but the converse does not hold in general. However, as shown in [6], a monotone operator on a chain-complete poset possesses fixpoints and a least fixpoint. We denote the least fixpoint of an operator $\mathcal{O}$ by $lfp(\mathcal{O})$.

Given a complete lattice $\langle \mathcal{L}, \leq \rangle$, $\langle \mathcal{L}^2, \leq, \leq_p \rangle$ denotes the induced (product) bilattice, where $\leq_p$ is called the *precision order* and defined as: for all $x, y, x', y' \in \mathcal{L}$, $(x, y) \leq_p (x', y')$ if $x \leq x'$ and $y' \leq y$. The $\leq_p$ ordering is a complete lattice ordering on $\mathcal{L}^2$. Below, we may refer to a lattice $\langle \mathcal{L}, \leq \rangle$ by $\mathcal{L}$ and the induced bilattice by $\mathcal{L}^2$.

We say that a pair $(x, y) \in \mathcal{L}^2$ is *consistent* if $x \leq y$, *inconsistent* otherwise, and *exact* if $x = y$. We denote the set of all consistent pairs by $\mathcal{L}^c$. A consistent pair $(x, y) \in \mathcal{L}^c$ defines an *interval*, denoted by $[x, y]$, which is defined by the set $\{z \mid x \leq z \leq y\}$. A consistent pair (x, y) in $\mathcal{L}^c$ can be seen as an *approximation* of every $z \in \mathcal{L}$ such that $z \in [x, y]$. In this sense, the precision order $\leq_p$ corresponds to the precision of approximation, while an exact pair approximates the only element in it.

2.2   An Extended Theory of Approximations

For a generalization beyond $\mathcal{L}^c$, two issues must be addressed. The first is on the notion of approximating operator. As analyzed below, if we adopt the original definition, inconsistencies will be left out. The other issue is how to make the original algebraic manipulation based on *stable revision operator* work in the new context.

An *approximating operator* $\mathcal{A}$ is a $\leq_p$-monotone operator on $\mathcal{L}^2$ that approximates an operator $\mathcal{O}$ on $\mathcal{L}$. In the original theory, it is required that $\mathcal{A}(x, x) = (\mathcal{O}(x), \mathcal{O}(x))$, for all $x \in \mathcal{L}$; i.e., $\mathcal{A}$ extends $\mathcal{O}$ on all exact pairs. If we want to allow a transition from a consistent pair to an inconsistent one, this condition is too strong. This is because an exact pair in general is not a maximal element in $\mathcal{L}^2$. For any exact pair (z, z), there are possibly inconsistent pairs (x, y) such that $(z, z) \leq_p (x, y)$. To see this, suppose $(x, y) \in \mathcal{L}^c$ and consider any $z \in [x, y]$. Since $(x, y) \leq_p (z, z)$, by the $\leq_p$-monotonicity of $\mathcal{A}$, we have $\mathcal{A}(x, y) \leq_p \mathcal{A}(z, z)$. If we further require $\mathcal{A}(z, z) = (\mathcal{O}(z), \mathcal{O}(z))$, then $\mathcal{Q}(z) \in [x, y]$. This is exactly what should happen if $\mathcal{A}(z, z)$ is consistent, as in this case we want (x, y) to approximate all fixpoints of $\mathcal{O}$ in $[x, y]$. But if $\mathcal{A}(z, z)$ is inconsistent, it is possible that $\mathcal{Q}(z)$ lies outside of $[x, y]$. This analysis suggests the following definition.

Definition 1. *Let $\mathcal{O}$ be an operator on $\mathcal{L}$ of a complete lattice $\langle \mathcal{L}, \leq \rangle$. We say that $\mathcal{A} : \mathcal{L}^2 \to \mathcal{L}^2$ is an* approximating operator *of $\mathcal{O}$ iff the following conditions are satisfied:*

– *For all $x \in \mathcal{L}$, if $\mathcal{A}(x, x)$ is consistent then $\mathcal{A}(x, x) = (\mathcal{O}(x), \mathcal{O}(x))$.*
– *$\mathcal{A}$ is $\leq_p$-monotone.*

Example 1. To see why the consistency condition "$\mathcal{A}(x, x)$ is consistent" in the definition is critical, consider a complete lattice where $\mathcal{L} = \{\bot, \top\}$ and $\leq$ is defined as usual. Let $\mathcal{O}$ be the identify function on $\mathcal{L}$. Then we have two fixpoints, $\mathcal{O}(\bot) = \bot$ and $\mathcal{O}(\top) = \top$. Let $\mathcal{A}$ be an identity function on $\mathcal{L}^2$ everywhere except $\mathcal{A}(\top, \top) = (\top, \bot)$. Thus, $\mathcal{A}(\top, \top)$ is inconsistent. It is easy to check that $\mathcal{A}$ is $\leq_p$-monotone. Since $\mathcal{A}(\bot, \bot) = (\mathcal{Q}(\bot), \mathcal{Q}(\bot))$, and $(\bot, \bot)$ is the only exact pair such that $\mathcal{A}(\bot, \bot)$ is consistent, $\mathcal{A}$ is an approximating operator of $\mathcal{O}$. But note that $\mathcal{A}(\top, \top) \neq (\mathcal{Q}(\top), \mathcal{Q}(\top))$, even though $\mathcal{O}(\top) = \top$. The fixpoint $\top$ of $\mathcal{O}$ is not captured by the operator $\mathcal{A}$ because $\mathcal{A}(\top, \top)$ is inconsistent. If we do not strengthen the definition by the consistency condition, mappings like the operator $\mathcal{A}$ above will be ruled out as approximating operators, which means we will fail to accommodate inconsistencies as we set out to do.[1]

Since $\mathcal{A}$ is a $\leq_p$-monotone operator on $\mathcal{L}^2$, its least fixpoint exists, which is called the *Kripke-Kleene fixpoint* of $\mathcal{A}$. Note that this fixpoint may be inconsistent, in which case it may not approximate any fixpoint of $\mathcal{O}$. It however does approximate every fixpoint x of $\mathcal{O}$ when $\mathcal{A}(x, x)$ is consistent.

Our main interest in this paper is in what are called the *well-founded* and *stable fixpoints* of $\mathcal{A}$. For this purpose, let us denote by $\mathcal{A}^1$ and $\mathcal{A}^2$ the projection of an approximating operator $\mathcal{A} : \mathcal{L}^2 \to \mathcal{L}^2$ on its first and second components, respectively, i.e., $\mathcal{A}^1(\cdot, v)$ is $\mathcal{A}$ with v fixed, and $\mathcal{A}^2(u, \cdot)$ is $\mathcal{A}$ with u fixed.

---

[1] This example specifies a system in which states are represented by a pair of factors - high and low. Here, all states are *stable* except the one in which both factors are high. This state may be transmitted to an "inconsistent state" with the first factor high and the second low. This state is the only inconsistent one, and it itself is stable.

In AFT, a key algebraic manipulation is performed by what is called a *stable revision operator*, which we will denote by $St_{\mathcal{A}}$. A pair (u, v) can be viewed as an approximation to any exact pair in the interval $[u, v]$, where u is *a lower estimate* and v *an upper estimate*. The pair, generated by $St_{\mathcal{A}}(u, v)$, consists of a new lower estimate and a new upper estimate, which may be computed by iterative processes, $x_0 = \bot, x_1 = \mathcal{A}^1(x_0, v), ..., x_{\alpha+1} = \mathcal{A}^1(x_\alpha, v), ...$ and $y_0 = u, y_1 = \mathcal{A}^2(u, y_0), ..., y_{\alpha+1} = \mathcal{A}^2(u, y_\alpha), ...$, respectively. It is clear that the operator $\mathcal{A}^1(\cdot, v)$ is defined on $\mathcal{L}$, and if $\mathcal{A}$ is $\leq_p$-monotone on $\mathcal{L}^2$ then $\mathcal{A}^1(\cdot, v)$ is monotone (i.e., $\leq$-monotone) on $\mathcal{L}$. As $\mathcal{L}$ is a complete lattice, the first iterative process above computes $lfp(\mathcal{A}^1(\cdot, v))$. However, for the second iterative process to compute the least fixpoint of $\mathcal{A}^2(u, \cdot)$, we need to ensure that $\mathcal{A}^2(u, \cdot)$ is an operator on the complete lattice $[u, \top]$.

In [6], the authors identify a subset of $\mathcal{L}^c$ with a desirable property, called $\mathcal{A}$-*reliability*. An element $(u, v) \in \mathcal{L}^c$ is said to be $\mathcal{A}$-*reliable* if $(u, v) \leq_p \mathcal{A}(u, v)$. With this property, it is proved that $\mathcal{A}^2(u, \cdot)$ is an operator on the complete lattice $[u, \top]$; thus because $\mathcal{A}^2(u, \cdot)$ is monotone, the new upper estimate can be computed as the least fixpoint of $\mathcal{A}^2(u, \cdot)$.

Now let us generalize the notion of $\mathcal{A}$-reliability to $\mathcal{L}^2$. That is, from now on let us call an element $(u, v) \in \mathcal{L}^2$ $\mathcal{A}$-*reliable* if $(u, v) \leq_p \mathcal{A}(u, v)$. Again, the current theory is not strong enough. For example, in Example 1, all pairs $(u, v) \in \mathcal{L}^2$ are $\mathcal{A}$-reliable, but $\mathcal{A}^2(u, \cdot)$ is not guaranteed to be an operator on $[u, \top]$, e.g., when (u, v) is $(\top, \top)$ (which is consistent), we have $\mathcal{A}^2(\top, \top) = \bot$, which is outside the interval $[\top, \top]$.

As we see above, that (u, v) is consistent does not guarantee that $\mathcal{A}(u, z)$, for any $z \in [u, v]$, is also consistent. Interestingly, for $\mathcal{A}^2(u, \cdot)$ to be an operator on $[u, \top]$, given that (u, v) is consistent, it is sufficient that $\mathcal{A}(u, u)$ is consistent.

Lemma 1. *Let $\langle \mathcal{L}, \leq \rangle$ be a complete lattice and $\mathcal{A} : \mathcal{L}^2 \to \mathcal{L}^2$ an approximating operator of an operator $\mathcal{O}$ on $\mathcal{L}$. If a consistent pair $(u, v) \in \mathcal{L}^2$ is $\mathcal{A}$-reliable and $\mathcal{A}(u, u)$ is consistent, then for every $x \in [u, \top]$, $\mathcal{A}^2(u, x) \in [u, \top]$.*

By Lemma 1 and the fact that $\mathcal{A}$ is $\leq_p$-monotone, when both (u, v) and $\mathcal{A}(u, u)$ are consistent, $\mathcal{A}^2(u, \cdot)$ is a monotone operator on $[u, \top]$. Thus the least fixpoint of $\mathcal{A}^2(u, \cdot)$ exists. Together with the fact that $\mathcal{A}^1(\cdot, v)$ is a monotone operator on $\mathcal{L}$, the following notion of *stable revision operator* is well-defined.

Definition 2. *Let $\mathcal{O}$ be an operator on $\mathcal{L}$ and $\mathcal{A}$ be an approximating operator of $\mathcal{O}$. Define the* stable revision operator, $St_{\mathcal{A}} : \mathcal{L}^r \to \mathcal{L}^2$, *where $\mathcal{L}^r$ is the set of $\mathcal{A}$-reliable pairs in $\mathcal{L}^2$, as follows:*

$$St_{\mathcal{A}}(u, v) = \begin{cases} (lfp(\mathcal{A}^1(\cdot, v)), \mathcal{A}^2(u, v)) & \text{if } (u, v) \text{ or } \mathcal{A}(u, u) \text{ is inconsistent} \\ (lfp(\mathcal{A}^1(\cdot, v)), lfp(\mathcal{A}^2(u, \cdot))) & \text{otherwise} \end{cases}$$

Intuitively, the definition says that in case that either (u, v) or $\mathcal{A}(u, u)$ is inconsistent we will compute $\mathcal{A}^2(u, v)$ to improve the upper bound, because in this case $\mathcal{A}^2(u, \cdot)$ is not guaranteed to be an operator on $[u, \top]$, and thus $lfp(\mathcal{A}^2(u, \cdot))$ may be ill-defined.

Note that if (u, v) is inconsistent because of $v < u$, by $\mathcal{A}$-reliability, $\mathcal{A}(u, v)$ possesses an (equal or) higher degree of inconsistency. This represents a "transition" from an inconsistent state to a possibly deeper inconsistent one. Also, $St_{\mathcal{A}}(u, v)$ may be inconsistent even when both (u, v) and $\mathcal{A}(u, u)$ are consistent, because the range of $lfp(\mathcal{A}^1(\cdot, v))$ is $\mathcal{L}$ and it is possible that $lfp(\mathcal{A}^1(\cdot, v)) \nleq lfp(\mathcal{A}^2(u, \cdot))$. This is rather interesting as it represents a "transition" from a consistent state to an inconsistent one.

The computation of $lfp(\mathcal{A}^1(\cdot, v))$ starts from the least element $\bot \in \mathcal{L}$, and in general there is no guarantee that this least fixpoint improves u. Thus we need another desirable property for the stable revision operator to behave as expected. We call an element $(u, v) \in \mathcal{L}^2$ $\mathcal{A}$-*prudent* if $u \leq lfp(\mathcal{A}^1(\cdot, v))$. Let us denote by $\mathcal{L}^{rp}$ the set of $\mathcal{A}$-reliable and $\mathcal{A}$-prudent pairs in $\mathcal{L}^2$.

By $\mathcal{A}$-reliability, a pair $(u, v) \in \mathcal{L}^{rp}$ is revised by operator $\mathcal{A}$ into a more precise approximation, or a more inconsistent state. The next lemma states that this is the case for the stable revision operator for all $\mathcal{A}$-reliable and $\mathcal{A}$-prudent pairs in $\mathcal{L}^2$, and any revised pair in $\mathcal{L}^{rp}$ can be "revised even more" continuously.

Lemma 2. *For every pair* $(u, v) \in \mathcal{L}^{rp}$, $(u, v) \leq_p St_{\mathcal{A}}(u, v)$, *and* $St_{\mathcal{A}}(u, v)$ *is also* $\mathcal{A}$*-reliable and* $\mathcal{A}$*-prudent.*

Let C be a chain in $\mathcal{L}^{rp}$. Denote by C^1 and C^2, respectively, the projection of C on its first and second elements. It is clear that $(lub(C^1), glb(C^2)) = lub(C)$. By adopting a proof of [6] (the proof of Proposition 3.10), it can be shown that for any chain C of $\mathcal{L}^{rp}$, $(lub(C^1), glb(C^2))$ is $\mathcal{A}$-reliable and $\mathcal{A}$-prudent. It follows from Lemma 2 that the operator $St_{\mathcal{A}}$ is defined on $\mathcal{L}^{rp}$. We therefore have

Theorem 1. *The structure* $\langle \mathcal{L}^{rp}, \leq_p \rangle$ *is a chain-complete poset that contains the least element* $(\bot, \top)$, *and the stable revision operator* $St_{\mathcal{A}}$ *is* $\leq_p$*-monotone on* $\mathcal{L}^{rp}$.

Definition 3. *Let* $\langle \mathcal{L}, \leq \rangle$ *be a complete lattice,* $\mathcal{A}$ *an approximating operator of some operator* $\mathcal{O}$ *on* $\mathcal{L}$, *and* $St_{\mathcal{A}}$ *the stable revision operator on* $\mathcal{L}^{rp}$. *The least fixpoint of* $St_{\mathcal{A}}$ *is called the* well-founded fixpoint *of* $\mathcal{A}$, *the fixpoints of* $St_{\mathcal{A}}$ *are called the* stable fixpoints *of* $\mathcal{A}$.

The extended theory is a nontrivial generalization of the original one, because the set of $\mathcal{A}$-reliable and $\mathcal{A}$-prudent pairs in $\mathcal{L}^2$ in general contains inconsistent pairs.

Finally, to compare the precisions of two approximating operators $\mathcal{A}$ and $\mathcal{B}$, we define: $\mathcal{A}$ is *more precise than* $\mathcal{B}$ if for all pairs $(u, v) \in \mathcal{L}^2$, $\mathcal{B}(u, v) \leq_p \mathcal{A}(u, v)$. Intuitively, that $\mathcal{A}$ is more precise than $\mathcal{B}$ implies that the stable fixpoints of $\mathcal{A}$ approximate the respective fixpoints of $\mathcal{O}$ more closely than those of $\mathcal{B}$. In particular, the well-founded fixpoint of $\mathcal{A}$ approximates all fixpoints of $\mathcal{O}$ and this approximation is as good as the well-founded fixpoint of $\mathcal{B}$, and probably better.

3   Application to FOL-Programs

3.1   Language and Notation

We assume a language of a decidable fragment of first-order logic, denoted $\mathcal{L}_{\Sigma}$, where $\Sigma = \langle F^n; P^n \rangle$, called a signature, and F^n and P^n are disjoint countable sets of n-ary function and n-ary predicate symbols, respectively. Constants are 0-ary functions. *Terms* are variables, constants, or functions in the form $f(t_1, ..., t_n)$, where each t_i is a term and $f \in F^n$. *First-order Formulas*, or just *formulas*, are defined as usual, so are the notions of *satisfaction*, *model*, and *entailment*.

Let Φ_P be a finite set of predicate symbols and Φ_C a nonempty finite set of constants such that $\Phi_C \subseteq F^n$. An *atom* is of the form $P(t_1, ..., t_n)$ where $P \in \Phi_P$ and each t_i is either a constant from Φ_C or a variable. A *negated atom* is of the form $\neg A$ where A is an atom. We do not assume any other restriction on the vocabularies, that is, Φ_P and P^n may have predicate symbols in common.

An *FOL-program* is a combined knowledge base $KB = (L, \Pi)$, where L is a first-order theory of $\mathcal{L}_\Sigma$ and Π a *rule base*, which is a finite collection of rules of the form

$$H \leftarrow A_1, \ldots, A_m, not\ B_1, \ldots, not\ B_n \tag{1}$$

where H is an atom, and A_i and B_i are atoms or formulas. By abuse of terminology, each A_i is called a *positive literal* and each $not\ B_i$ is called a *negative literal*. If B in $not\ B$ is a formula, we also call $not\ B$ a *negative formula*.

For any rule r, we denote by $head(r)$ the head of the rule and $body(r)$ its body, and we define $pos(r) = \{A_1, ..., A_m\}$ and $neg(r) = \{B_1, ..., B_n\}$.

A *ground instance* of a rule r in Π is obtained by replacing every free variable with a constant from Φ_C. In this paper, we assume that a rule base Π is already grounded if not said otherwise. When we refer to an atom/literal/formula, by default we mean it is a ground one.

Note that the body of a ground rule may contain arbitrary first-order sentences, which are first-order formulas with no free variables.

Given an FOL-program $KB = (L, \Pi)$, the *Herbrand base* of Π, denoted HB_Π, is the set of all ground atoms $P(t_1, ..., t_n)$, where $P \in \Phi_P$ occurs in KB and $t_i \in \Phi_C$.

We denote by Ω the set of all predicate symbols appearing in HB_Π such that $\Omega \subseteq P^n$. For distinction, we call atoms whose predicate symbols are not in Ω *ordinary*, and all the other formulas *FOL-formulas*. If $L = \emptyset$ and Π only contains rules of the form (1) where all H, A_i and B_j are ordinary atoms, then KB is called a *normal logic program*.

Any subset $I \subseteq HB_\Pi$ is called an *interpretation* of Π. It is also called a *total* interpretation or a *2-valued* interpretation. If I is an interpretation, we define $\bar{I} = HB_\Pi \setminus I$.

Let Q be a set of atoms. We define $\neg.Q = \{\neg A \mid A \in Q\}$. For a set of atoms and negated atoms S, we define $S^+ = \{A \mid A \in S\}$, $S^- = \{A \mid \neg A \in S\}$, and $S|_\Omega = \{A \in S \mid pred(A) \in \Omega\}$, where $pred(A)$ is the predicate symbol of A. Let $Lit_\Pi = HB_\Pi \cup \neg HB_\Pi$. A subset $S \subseteq Lit_\Pi$ is consistent if $S^+ \cap S^- = \emptyset$. For a first-order theory L, we say that $S \subseteq Lit_\Pi$ is *consistent with* L if the first-order theory $L \cup S|_\Omega$ is consistent (i.e., the theory is *satisfiable*). Note that when we say S is consistent with L, both S and L must be consistent. Similarly, a (2-valued) interpretation I is consistent with L if $L \cup I|_\Omega \cup \neg.\bar{I}|_\Omega$ is consistent.

Definition 4. *Let $KB = (L, \Pi)$ be an FOL-program and $I \subseteq HB_\Pi$ an interpretation. Define the satisfaction relation under L, denoted $\models_L$, as follows (the definition extends to conjunctions of literals):*

1. *For any ordinary atom $A \in HB_\Pi$, $I \models_L A$ if $A \in I$ and $I \models not\ A$ if $A \notin I$.*
2. *For any FOL-formula A, $I \models_L A$ if $L \cup I|_\Omega \cup \neg.\bar{I}|_\Omega \models A$, and $I \models_L not\ A$ if $I \not\models_L A$.*

Let $KB = (L, \Pi)$ be an FOL-program. For any $r \in \Pi$ and $I \subseteq HB_\Pi$, $I \models_L r$ if $I \not\models_L body(r)$ or $I \models_L head(r)$. I is a *model* of KB if I is consistent with L and I satisfies all rules in Π.

Example 2. To illustrate the flexibility provided by the parameter Ω, suppose we have a program $KB = (L, \Pi)$ where Π contains a rule that says any unemployed with disability receives financial assistance:

$$Assist(X) \leftarrow Disabled(X), not\ Employed(X)$$

Assume $\Omega = \Phi_P = \{Assist, Employed\}$ and $\Phi_C = \{a\}$. Then, the Herbrand base is $HB_\Pi = \{Assist(a), Employed(a)\}$. As answer sets and WFS are based on HB_Π, we say that the predicates appearing in HB_Π are interpreted under the closed world assumption. In particular, $Employed$ is interpreted under the closed world assumption, and as such, $not\ Employed(a)$ can be established for lack of evidence of employment. On the other hand, $Disabled(a)$ is not in HB_Π and thus its truth requires a direct, classic proof; in this case we say that $Disabled$ is interpreted under the open world assumption.

3.2   Semantics of FOL-Programs under Extended AFT

We first extend the standard immediate consequence operator to FOL-programs. This is the operator to be approximated, i.e., it is a concrete instance of operator $\mathcal{O}$ in Def. 1.

Let $KB = (L, \Pi)$ be an FOL-program. We define an operator on the complete lattice $\langle 2^{HB_\Pi}, \subseteq \rangle$, $\mathcal{K}_{KB}: 2^{HB_\Pi} \to 2^{HB_\Pi}$ as follows: for any $I \in 2^{HB_\Pi}$,

$$\mathcal{K}_{KB}(I) = \{head(r) \mid r \in \Pi, I \models_L body(r)\} \cup \{A \in HB_\Pi|_\Omega \mid I \models_L A\} \qquad (2)$$

This operator is essentially the standard immediate consequence operator augmented by *direct positive consequences*. Notice that elements in 2^{HB_Π} are 2-valued interpretations, which are inherently weak in representing inconsistency - if HB_Π is a fixpoint of $\mathcal{K}_{KB}$, in general it is not indicative of whether this is the result of inconsistency, or all the atoms in HB_Π are consistently derived. In the extended AFT, it is inconsistent pairs that tie up this loose end, by explicitly representing negative information.

According to extended AFT, given the lattice $\langle 2^{HB_\Pi}, \subseteq \rangle$, the induced bilattice is $\langle (2^{HB_\Pi})^2, \subseteq_p \rangle$. A pair (I, J) in $(2^{HB_\Pi})^2$ is consistent if $I \subseteq J$, which represents a *partial interpretation* $I \cup \neg.\bar{J}$ (also called *a 3-valued interpretation*), The pair (I, J) is inconsistent if $I \not\subseteq J$. In the sequel, we will use a consistent pair (I, J) as well as the corresponding set $I \cup \neg.\bar{J}$ to denote a partial interpretation.

Intuitively, a partial interpretation (I, J) means that the atoms in I are true, those in J are *potentially true*, hence those in $\bar{J}$ are false as they are not even potentially true.

Given a partial interpretation (I, J), (I', J') is said to be *a consistent extension of* (I, J) if $I \subseteq I' \subseteq J' \subseteq J$.

Below, we define two entailment relations under partial interpretations, the first of which is the standard entailment relation based on Kleene's 3-valued logic and the other is defined in terms of the first that differs only on requiring a check for FOL-formulas by all consistent extensions.

Definition 5. *Let $KB = (L, \Pi)$ be an FOL-program, (I, J) a 3-valued interpretation, and ϕ a literal. We define two entailment relations below, which extends naturally to conjunction of literals.*

- $(I, J) \models_L \phi$ *iff*
    - *if ϕ is an ordinary atom A then $A \in I$, and if ϕ is a negative ordinary literal not A then $A \in \bar{J}$;*
    - *if ϕ is an FOL-formula then $L \cup I|_\Omega \cup \neg.\bar{J}|_\Omega \models \phi$, and if ϕ is a negative formula not A, then $L \cup I|_\Omega \cup \neg.\bar{J}|_\Omega \not\models A$.*
- $(I, J) \Vdash_L \phi$ *iff*
    - $(I, J) \models_L \phi$, *and in addition,*
    - *if ϕ is an FOL-formula then $(I', J') \models_L \phi$ for every consistent extension (I', J') of (I, J), and if ϕ is a negative formula not A, then $(I', J') \not\models_L A$ for every consistent extension (I', J') of (I, J).*

In the extended AFT, semantics is defined by approximating operators, which map a pair (I, J) to a pair (I', J'). Roughly speaking, to guarantee stability, we want to add those atoms to I' that are not only true under (I, J), but also true in any interpretation that extends (I, J).[2]

3.3 Semantics by Approximating Operators

We define three closely related approximating operators on $(2^{HB_\Pi})^2$, with approximation precisions in increasing order, but with some costs for higher precision.

In the following, given an FOL-program $KB = (L, \Pi)$, a function $\mathcal{F}_{KB} : (2^{HB_\Pi})^2 \rightarrow (2^{HB_\Pi})^2$ is called *progressive* if $(I, J) \subseteq_p \mathcal{F}_{KB}(I, J)$.[3]

Approximating Operator Φ_{KB}: Standard Semantics

Definition 6. *(Operator Φ_{KB}) Let $KB = (L, \Pi)$ be an FOL-program, $(I, J) \in (2^{HB_\Pi})^2$, and $\mathcal{F}_{KB}$ a progressive function. The operators Φ_{KB}^1 and Φ_{KB}^2 are defined as: If (I, J) is inconsistent with L, then $\Phi_{KB}(I, J)$ is inconsistent and $\Phi_{KB}(I, J) = \mathcal{F}_{KB}(I, J)$; otherwise, for all $H \in HB_\Pi$,*

- $H \in \Phi_{KB}^1(I, J)$ *iff one of the following holds*
    - *(a)* $(I, J) \models_L H$.
    - *(b)* $\exists r \in \Pi$ *with* $head(r) = H$ *s.t.* $(I, J) \Vdash_L body(r)$.
- $H \in \Phi_{KB}^2(I, J)$ *iff* $(I, J) \not\models_L \neg H$ *and one of the following holds*
    - *(a)* $\exists(I', J')$ *with* $I \subseteq I' \subseteq J' \subseteq J$ *s.t.* $(I', J') \models_L H$.
    - *(b)* $\exists r \in \Pi$ *with* $head(r) = H$ *s.t. for every* $\phi \in body(r)$, *there is a consistent extension* (I', J') *of* (I, J) *such that* $(I', J') \models_L \phi$.

---

[2] Since the entailment relation for first-order theory is monotonic, the real effect is only on negative FOL-formulas.

[3] Notice that Theorem 1 only applies to $\mathcal{A}$-reliable and $\mathcal{A}$-prudent pairs. To be progressive is to be $\mathcal{A}$-reliable. As the function $\mathcal{F}_{KB}$ is applied only once, $\mathcal{A}$-prudence becomes unnecessary.

If (I, J) is inconsistent with L, the resulting pair must be inconsistent and is determined by $\mathcal{F}_{KB}(I, J)$. We leave this as a parameter to ensure flexibility and generality; e.g., if we insist that inconsistency be treated as in classical logic, $\mathcal{F}_{KB}(I, J)$ should lead to full triviality, i.e., $\mathcal{F}_{KB}(I, J) = (HB_\Pi, \emptyset)$. However, our definition allows an inconsistent state to be handled differently as long as the function $\mathcal{F}_{KB}$ is progressive and it does not turn an inconsistent situation into a consistent one.

Otherwise, for any $H \in \Phi^1_{KB}(I, J)$, either H is entailed by L and (I, J) (relative to Ω), or for some rule r, H is derivable via r for all consistent extensions of (I, J).

On the other hand, that $H \in \Phi^2_{KB}(I, J)$ requires two conditions. The first is that for H to be potentially true, its negation should not be entailed, while in the second, to demonstrate that H is potentially true, either it is potentially entailed (part (a)), or it is potentially derivable (part (b)); note that here each literal in $body(r)$ may be entailed by a different consistent extension.

The extended AFT is applicable to the operator Φ_{KB}, due to the following lemma.

Lemma 3. *Φ_{KB} is an approximating operator of $\mathcal{K}_{KB}$.*

Example 3. Let $KB = (\{\neg A(a)\}, \Pi)$ where $\Pi = \{A(a) \leftarrow not\ B(a);\ B(a) \leftarrow B(a);\ C(a) \leftarrow\}$. Let $\Omega = \Phi_P = \{A, B, C\}$ and $\Phi_C = \{a\}$. The well-founded fixpoint of Φ_{KB} is computed by the sequence generated by the stable revision operator $St_{\Phi_{KB}}$:

$$(\emptyset, HB_\Pi) \Rightarrow (\{C(a)\}, \{C(a)\}) \Rightarrow (\{C(a), A(a)\}, \{C(a)\}) \Rightarrow \ldots$$

With the least element $(\emptyset, HB_\Pi)$, according to Def. 2, we compute the new lower and upper estimates as $(lfp(\Phi_{KB}{}^1(\cdot, HB_\Pi)), lfp(\Phi_{KB}{}^2(\emptyset, \cdot))) = (\{C(a)\}, \{C(a)\})$. Similarly, we get the next pair $(\{C(a), A(a)\}, \{C(a)\})$, which is inconsistent. The eventual fixpoint is determined the function $\mathcal{F}_{KB}$. For example, if $\mathcal{F}_{KB}$ maps the last pair in the sequence to itself, then this pair is the well-founded fixpoint of Φ_{KB}.

Note that KB even has a model, $\{C(a), B(a)\}$. Thus, existence of a model does not guarantee that the well-founded fixpoint is a consistent set of literals.

In the next two examples, we illustrate combined reasoning with closed world as well as open world assumptions.

Example 4. Consider $KB = (L, \Pi)$, where

$$L = \{\forall x\ Certified(x) \supset Disabled(x)\}$$
$$\Pi = \{Assist(a) \leftarrow Disabled(a), not\ Employed(a))\}$$

Assume $\Omega = \{Assist, Employed\}$ and $\Phi_C = \{a\}$. Thus $HB_\Pi = \{Assist(a), Employed(a)\}$. The computation starts with the least element $(\emptyset, HB_\Pi)$, and it is easy to verify that the well-founded fixpoint of Φ_{KB} is $(\emptyset, \emptyset)$, which implies that $Assist(a)$ and $Employed(a)$ are false. $Assist(a)$ is not potentially true because $Disabled(a)$ is not entailed by any consistent extension of $(\emptyset, \emptyset)$. Note that there is a model of L satisfying the rule in Π that extends the well-founded fixpoint of Φ_{KB}; namely $Certified(a)$ and $Disabled(a)$ in addition are false.[4]

---

[4] The reasoning here is analogue to parallel circumscription [11], where the predicates *Employed* and *Assist* are minimized with *Certified* and *Disabled* varying.

Example 5. Consider the following FOL-program $KB = (L, \Pi)$, where

$$L = \{\forall x B(x) \supset A(x), \neg A(a) \vee C(a)\}$$
$$\Pi = \{B(a) \leftarrow B(a); \quad A(a) \leftarrow (\neg C(a) \wedge B(a)); \quad R(a) \leftarrow not\ C(a), not\ A(a)\}$$

Let $\Phi_P = \{A, B, R\}$, $\Omega = \{A, B\}$, $\Phi_C = \{a\}$, and $HB_\Pi = \{A(a), B(a), R(a)\}$. Starting with $(\emptyset, HB_\Pi)$, we get $(lfp(\Phi_{KB}{}^1(\cdot, HB_\Pi)), lfp(\Phi_{KB}{}^2(\emptyset, \cdot))) = (\emptyset, \{R(a)\})$. That $R(a)$ is potentially true because $(\emptyset, \emptyset) \models_L not\ A(a)$ and $(\emptyset, \emptyset) \not\models_L C(a)$, thus every body literal of the last rule is entailed by some consistent extension. After the next iteration, we get $(\{R(a)\}, \{R(a)\})$ as the fixpoint. Note that the FOL-formula, $C(a)$, because it is not in HB_Π, is not involved, positively or negatively, in the well-founded semantics of KB.

The well-founded and answer set semantics based on the operator Φ_{KB} are called *standard*, because they are generalizations of the WFS and answer set semantics for normal logic programs.[5]

Theorem 2. *Let $KB = (\emptyset, \Pi)$ be a normal logic program. Then, the Φ-WFS of KB coincides with the WFS of Π, and Φ-answer set semantics coincides with the standard stable model semantics of Π.*

We consider the formulation of Φ_{KB} novel, because the central mechanism, namely part (b) of Def. 6, captures the notion of *unfounded* set for normal logic programs, in a much simpler manner than the $\hat{\mathcal{D}}$-well-founded semantics of [13].

Approximating Operator Θ: Enhanced Semantics. As eluded earlier, in the definition of the operator Φ_{KB}, to derive potentially true atoms, we apply a rule if each body literal is entailed by some consistent extension. An alternative is to require that all body literals are entailed by some, but the same, consistent extension. This leads to a different operator.

Definition 7. *(Operator Θ_{KB}) Let $KB = (L, \Pi)$ be an FOL-program, $(I, J) \in (2^{HB_\Pi})^2$, and $\mathcal{F}_{KB}$ a progressive function. The operator Θ_{KB} is defined as: If (I, J) is inconsistent with L, then $\Theta_{KB}(I, J)$ is inconsistent and $\Theta_{KB}(I, J) = \mathcal{F}_{KB}(I, J)$; otherwise,*

- $\Theta_{KB}^1(I, J) = \Phi_{KB}^1(I, J)$.
- $\forall H \in HB_\Pi, H \in \Theta_{KB}^2(I, J)$ *iff* $(I, J) \not\models_L \neg H$ *and one of the following holds*
  (a) $\exists (I', J')$ with $I \subseteq I' \subseteq J' \subseteq J$ s.t. $(I'J') \models_L H$.
  (b) $\exists r \in \Pi$ with $head(r) = H$ s.t. there is a consistent extension (I', J') of (I, J) such that $(I', J') \models_L body(r)$

Lemma 4. *The operator Θ_{KB} is an approximating operator of $\mathcal{K}_{KB}$.*

Clearly, the operator Θ_{KB}^2 is at least as strict as Φ_{KB}^2. Less potentially true atoms that we have more false atoms there are. Thus it is not difficult to show

---

[5] This is also in line with the well-founded semantics (WFS) for logic programs with monotone and anti-monotone aggregates [1], and the WFS for dl-programs [9], both of which are generalizations of the WFS for normal logic programs.

Proposition 1. *Let $KB = (L, \Pi)$ be an FOL-program.*

- *For all pairs of interpretations (I, J), $\Phi_{KB}(I, J) \subseteq_p \Theta_{KB}(I, J)$.*
- *$lfp(St_{\Phi_{KB}}) \subseteq_p lfp(St_{\Theta_{KB}})$.*

Hence, according to the extended AFT, Θ_{KB} is more precise than Φ_{KB}. However, the higher precision comes with a cost. The Θ-well-founded semantics is not a generalization of the WFS for normal logic programs.

Example 6. Consider $KB = (\emptyset, \Pi)$, where Π is

$$P(a) \leftarrow Q(a), not\ Q(a). \quad Q(a) \leftarrow not\ Q(b). \quad Q(b) \leftarrow not\ Q(a).$$

and $\Omega = \{P, Q\}$. The well-founded fixpoint of Θ_{KB} is computed by the sequence

$$(\emptyset, HB_\Pi) \Rightarrow (\emptyset, \{Q(a), Q(b)\}) \Rightarrow (\emptyset, \{Q(a), Q(b)\})$$

Note that $\{\neg P(a)\}$ is not a 3-valued model in Kleene's 3-valued logic, as the first rule is not satisfied.

In contrast, the Φ-well-founded semantics has all atoms *undefined*, as the well-founded fixpoint of Φ_{KB} is $(\emptyset, \{Q(a), Q(b), P(a)\})$, which is generated as follows: Given $(\emptyset, HB_\Pi)$, to compute the new upper estimate, we start with $(\emptyset, \emptyset)$, and then get $(\emptyset, \{Q(a), Q(b)\})$, followed by a derivation of $P(a)$ as potentially true by the first rule, where each body literal is satisfied by a different consistent extension.

By Proposition 1, if the well-founded fixpoint of Φ_{KB} is inconsistent, so is the well-founded fixpoint of Θ_{KB}. Also, it is not difficult to show that Θ-answer sets are precisely Φ-answer sets, and vice versa. For instance, the FOL-program KB above has two answer sets, in both cases $P(a)$ is false. Then, the Θ-well-founded semantics is a better approximation to Φ-answer sets than the Φ-well-founded semantics.

Approximating Operator Ψ: Ultimate Semantics. Actually, the operator Θ_{KB} is half way towards what is called the *ultimate semantics* for logic programs [13]. We define a similar operator for FOL-programs.

Definition 8. (Operator Ψ_{KB}) *Let $KB = (L, \Pi)$ be an FOL-program, $(I, J) \in (2^{HB_\Pi})^2$, and $\mathcal{F}_{KB}$ a progressive function. The operator Ψ_{KB} is defined as: If (I, J) is inconsistent with L, then $\Psi_{KB}(I, J)$ is inconsistent and $\Psi_{KB}(I, J) = \mathcal{F}_{KB}(I, J)$; otherwise,*

- *$H \in \Psi^1_{KB}(I, J)$ iff one of the following holds*
  (a) $(I, J) \models_L H$.
  (b) For each consistent extension (I', J') of (I, J), $\exists r \in \Pi$ with $head(r) = H$ s.t. $(I', J') \models_L body(r)$.
- *$\Psi^2_{KB}(I, J) = \Theta^2_{KB}(I, J)$.*

The operator Ψ^1_{KB} differs from the previous two in that different rules may be used to derive H for different consistent extensions. For example, with $KB = (\emptyset, \Pi)$ where Π consists of

$$P(a) \leftarrow Q(a); \ P(a) \leftarrow not\ Q(a); \ Q(a) \leftarrow not\ R(a); \ R(a) \leftarrow not\ Q(a).$$

With all predicates in Ω, the well-founded fixpoint of Ψ_{KB} is $(\{P(a)\}, \{Q(a), R(a)\})$. $P(a)$ is derived by different rules for different consistent extensions of $(\emptyset, HB_\Pi)$.

Lemma 5. *The operator Ψ_{KB} is an approximating operator of $\mathcal{K}_{KB}$.*

It is clear that Ψ_{KB} is more relaxed than operator Θ_{KB} in deriving true atoms. Combined with Proposition 1, we can show the following

Proposition 2. *Let $KB = (L, \Pi)$ be an FOL-program.*

- *For all pairs of interpretations (I, J), $\Phi_{KB}(I, J) \subseteq_p \Theta_{KB}(I, J) \subseteq_p \Psi_{KB}(I, J)$.*
- *$lfp(St_{\Phi_{KB}}) \subseteq_p lfp(St_{\Theta_{KB}}) \subseteq_p lfp(St_{\Psi_{KB}})$.*

Since $\Psi_{KB}^2(I, J) = \Theta_{KB}^2(I, J)$, the same program in Example 6 shows that the Ψ-well-founded semantics is not a generalization of the WFS for normal logic programs.

To some researchers, the ultimate semantics sometimes behaves counter-intuitively. E.g., the single-rule normal program, $\{P(a) \leftarrow P(a)\}$, is not strongly equivalent to $\emptyset$. In addition, for normal logic programs the ultimate semantics has higher complexity than the standard semantics [6]. Nevertheless, the ultimate semantics is interesting because it provides the best possible approximation, in terms of information content.

4   Related Work and Discussion

Well-Supported Answer Set Semantics for FOL-Programs: In [17], the well-supported answer set semantics is defined, based on the notion of 2-valued *up to satisfaction*, which is similar to the definition of Φ_{KB}^1, but the latter is based on 3-valued consistent extensions. However, we are able to show the following.

Theorem 3. *Let $KB = (L, \Pi)$ be an FOL-program and $I \subseteq HB_\Pi$ a model of KB. The exact pair (I, I) is a stable fixpoint of Φ_{KB} iff I is a well-supported answer set of KB.*

The extended AFT provides an uniform treatment to the semantics of FOL-programs. We now know that there are at least two principled ways (based on operators Φ_{KB} and Θ_{KB} respectively) to approximate well-supported answer sets, which can be characterized by different approximating operators.

In the extended AFT, an inconsistent pair on a bilattice does not approximate any exact pair. Thus, since operator Θ_{KB} is more precise than operator Φ_{KB}, if we know that the Θ-well-founded semantics is inconsistent then we need not try to compute well-supported answer sets, since they do not exist. This observation may help design an implementation for the well-supported answer set semantics. None of the these insights are possible by the original AFT, because none of these operators can be defined only on consistent pairs of a bilattice.

Well-Founded Semantics: The most relevant work on the well-founded semantics for combing rules with DLs are [9,10]. The former embeds *dl-atoms* in rule bodies to serve as queries to the underlying ontology, and it does not allow the predicate in a rule head to be shared in the ontology. In both approaches, syntactic restrictions are posted so that the least fixpoint is always constructed over sets of consistent literals. It is also a unique feature in our approach that combined reasoning with closed world and open world is supported. We have seen in Section 3 that this is not a trivial issue for the construction of a least fixpoint.

Combining ASP with DLs or Classic Logic: The original AFT applies to dl-programs [7], which can be represented by HEX-programs [8] and aggregate programs [13], where an approximating operator can be defined so that the well-founded fixpoint defines the well-founded semantics [9] and the exact stable fixpoints define the well-supported answer set semantics [16] (also see [20]).

FO(ID) is formulated to integrate rules into classical logic FO in manner of putting rules on top of FO [19]. A program in FO(ID) has a clear knowledge representation "task" - the rule component is used to define concepts, whereas the FO component may assert additional properties of the defined concepts. All formulas in FO(ID) are interpreted under closed world assumption. Thus, FOL-programs and FO(ID) have fundamental differences in basic ideas. On semantics, FOL-formulas can be interpreted under open world and closed world flexibly. On modeling, the rule set in FO(ID) is built on ontologies, thus information can only flow from a first order theory to rules. But in FOL-programs, the first order theory and rules are tightly integrated, and information can flow from each other bilaterally.

Approximation Fixpoint Theory: In [6], the authors also show that the theory of consistent approximations can be applied to the entire bilattice $\mathcal{L}^2$, under the assumption that an approximating operator $\mathcal{A}$ is *symmetric*, which is defined as: for every $(x, y) \in \mathcal{L}^2$, $\mathcal{A}^1(x, y) = \mathcal{A}^2(y, x)$. This symmetry behavior guarantees that the restriction of $\mathcal{A}$ to $\mathcal{L}^c$ is an operator on $\mathcal{L}^c$, hence it does not allow a transition from a consistent state to an inconsistent one. The theory developed in this paper does not make this assumption. In fact, all three approximating operators defined in Section 3 of this paper are asymmetric. Finally, the theory of consistent approximations has been extended to HEX-programs [2] that allow disjunctive heads in rules.

The extended AFT developed in this paper supports two features, one for transitions from a consistent state to an inconsistent one, and the other for transitions from an inconsistent state to a possibly deeper inconsistent one. In this paper we utilize the first to define the semantics of FOL-programs. For potential applications of the second feature, one may consider inference systems based on non-classical logic (e.g., paraconsistent logic) or explore the idea of belief revision. In both cases we want to infer nontrivial conclusions in case of an inconsistent theory. This is an interesting future direction.

Acknowledgements. This work is partially supported by the National Natural Science Foundation of China (NSFC) grants 61373035 and 61373165, and by National High-tech R&D Program of China (863 Program) grant 2013AA013204. The second author is also partially supported by NSERC discovery grant CS413.

References

1. Alviano, M., Calimeri, F., Faber, W., Leone, N., Perri, S.: Unfounded sets and well-founded semantics of answer set programs with aggregates. J. Artif. Intell. Res. 42 (2011)
2. Antic, C., Eiter, T., Fink, M.: Hex semantics via approximation fixpoint theory. In: Cabalar, P., Son, T.C. (eds.) LPNMR 2013. LNCS, vol. 8148, pp. 102–115. Springer, Heidelberg (2013)

3. Bi, Y., You, J.H., Feng, Z.: A generalization of approximation fixpoint theory and its application to FOL-programs. Tech. rep., College of Computer Science and Technology, Tianjin University, Tianjin, China (June 2014),
   http://xinwang.tju.edu.cn/drupal/sites/default/files/
   basic_page/full-proof.pdf
4. de Bruijn, J., Eiter, T., Tompits, H.: Embedding approaches to combining rules and ontologies into autoepistemic logic. In: Proc. KR 2008, pp. 485–495 (2008)
5. de Bruijn, J., Pearce, D., Polleres, A., Valverde, A.: Quantified equilibrium logic and hybrid rules. In: Marchiori, M., Pan, J.Z., Marie, C.d.S. (eds.) RR 2007. LNCS, vol. 4524, pp. 58–72. Springer, Heidelberg (2007)
6. Denecker, M., Marek, V., Truszczynski, M.: Ultimate approximation and its application in nonmonotonic knowledge representation systems. Information and Computation 192(1), 84–121 (2004)
7. Eiter, T., Ianni, G., Lukasiewicz, T., Schindlauer, R., Tompits, H.: Combining answer set programming with description logics for the semantic web. Artifical Intelligence 172(12-13), 1495–1539 (2008)
8. Eiter, T., Ianni, G., Schindlauer, R., Tompits, H.: A uniform integration of higher-order reasoning and external evaluations in answer-set programming. In: Proc. IJCAI 2005, pp. 90–96 (2005)
9. Eiter, T., Lukasiewicz, T., Ianni, G., Schindlauer, R.: Well-founded semantics for description logic programs in the semantic web. ACM Transactions on Computational Logic 12(2) (2011)
10. Lukasiewicz, T.: A novel combination of answer set programming with description logics for the semantic web. IEEE TKDE 22(11), 1577–1592 (2010)
11. McCarthy, J.: Circumscription - a form of non-monotonic reasoning. Artifical Intelligence 13(27-39), 171–172 (1980)
12. Motik, B., Rosati, R.: Reconciling description logics and rules. Journal of the ACM 57(5), 1–62 (2010)
13. Pelov, N., Denecker, M., Bruynooghe, M.: Well-founded and stable semantics of logic programs with aggregates. Theory and Practice of Logic Programming 7, 301–353 (2007)
14. Rosati, R.: On the decidability and complexity of integrating ontologies and rules. Journal of Web Semantics 3(1), 61–73 (2005)
15. Rosati, R.: DL+log: Tight integration of description logics and disjunctive datalog. In: Proc. KR 2006, pp. 68–78 (2006)
16. Shen, Y.D.: Well-supported semantics for description logic programs. In: Proc. IJCAI 2011, pp. 1081–1086 (2011)
17. Shen, Y.-D., Wang, K.: Extending logic programs with description logic expressions for the semantic web. In: Aroyo, L., Welty, C., Alani, H., Taylor, J., Bernstein, A., Kagal, L., Noy, N., Blomqvist, E. (eds.) ISWC 2011, Part I. LNCS, vol. 7031, pp. 633–648. Springer, Heidelberg (2011)
18. Tarski, A.: A lattice-theoretical fixpoint theorem and its applications. Pacific Journal of Mathematics 5(2), 285–309 (1955)
19. Vennekens, J., Denecker, M., Bruynooghe, M.: FO(ID) as an extension of DL with rules. Ann. Math. Artif. Intell. 58(1-2), 85–115 (2010)
20. You, J.H., Shen, Y.D., Wang, K.: Well-supported semantics for logic programs with generalized rules. In: Erdem, E., Lee, J., Lierler, Y., Pearce, D. (eds.) Correct Reasoning. LNCS, vol. 7265, pp. 576–591. Springer, Heidelberg (2012)

Query Answering over Contextualized RDF/OWL Knowledge with Forall-Existential Bridge Rules: Attaining Decidability Using Acyclicity

Mathew Joseph[1,2], Gabriel Kuper[2], and Luciano Serafini[1]

[1] DKM, FBK-IRST, Trento, Italy
[2] DISI, University Of Trento, Trento, Italy
{mathew,serafini}@fbk.eu, kuper@disi.unitn.it

Abstract. The recent outburst of context-dependent knowledge on the Semantic Web (SW) has led to the realization of the importance of the quads in the SW community. Quads, which extend a standard RDF triple, by adding a new parameter of the 'context' of an RDF triple, thus informs a reasoner to distinguish between the knowledge in various contexts. Although this distinction separates the triples in an RDF graph into various contexts, and allows the reasoning to be decoupled across various contexts, bridge rules need to be provided for inter-operating the knowledge across these contexts. We call a set of quads together with the bridge rules, a quad-system. In this paper, we discuss the problem of query answering over quad-systems with expressive forall-existential bridge rules. It turns out the query answering over quad-systems is undecidable, in general. We derive a decidable class of quad-systems, namely *context-acyclic* quad-systems, for which query answering can be done using forward chaining. Tight bounds for data and combined complexity of query entailment has been established for the derived class.

Keywords: Contextualized RDF/OWL knowledge, Contextualized Query Answering, Quads, Forall-Existential Rules, Semantic Web, Knowledge Representation.

1 Introduction

One of the major recent changes in the SW community is the transformation from a *triple* to a *quad* as its primary knowledge carrier. As a consequence, more and more triple stores are becoming *quad* stores. Some of the popular quad-stores are 4store[1], Openlink Virtuoso[2], and some of the current popular triple stores like Sesame[3] internally keep track of the context by storing arrays of four names (c, s, p, o) (further denoted as $c : (s, p, o)$), where c is an identifier that stands for the context of the triple (s, p, o). Some of the recent initiatives in this direction have also extended existing formats like N-Triples to N-Quads. The latest Billion triples challenge datasets (BTC 2012) have all been released in the N-Quads format.

---

[1] http://4store.org
[2] http://virtuoso.openlinksw.com/rdf-quad-store/
[3] http://www.openrdf.org/

R. Kontchakov and M.-L. Mugnier (Eds.): RR 2014, LNCS 8741, pp. 60–75, 2014.
© Springer International Publishing Switzerland 2014

One of the main benefits of quads over triples are that they allow users to specify various attributes of meta-knowledge that further qualify knowledge [8], and also allow users to query for this meta knowledge [30]. Examples of these attributes, which are also called *context dimensions* [27], are provenance, creator, intended user, creation time, validity time, geo-location, and topic. Having defined various contexts in which triples are dispersed, one can declare in another meta-context mc, statements such as mc: $(c_1$, creator, John$)$, mc: $(c_1$, expiryTime, "jun-2013"$)$ that talk about the knowledge in context c_1, in this case its creator and expiry time. Another benefit of such a contextualized approach is that it opens possibilities of interesting ways for querying a contextualized knowledge base. For instance, if context c_1 contains knowledge about Football World Cup 2014 and context c_2 about Football Euro Cup 2012. Then the query "who beat Italy in both Euro Cup 2012 and World Cup 2014" can be formalized as the conjunctive query:

$$c_1: (x, \text{beat}, \text{Italy}) \wedge c_2: (x, \text{beat}, \text{Italy}), \text{ where } x \text{ is a variable.}$$

As the knowledge can be separated context wise and simultaneously be fed to separate reasoning engines, this approach increases both efficiency and scalability. Besides the above flexibility, *bridge rules* [4] can be provided for inter-interoperating the knowledge in different contexts. Such rules are primarily of the form:

$$c : \phi(\boldsymbol{x}) \rightarrow c' : \phi'(\boldsymbol{x})$$

where ϕ, ϕ' are both atomic concept (role) symbols, c, c' are contexts. The semantics of such a rule is that if, for any $\boldsymbol{a}$, $\phi(\boldsymbol{a})$ holds in context c, then $\phi'(\boldsymbol{a})$ should hold in context c', where $\boldsymbol{a}$ is a unary/binary vector dependending on whether ϕ, ϕ' are concept/role symbols. Although such bridge rules serve the purpose of specifying knowledge interoperability from a source context c to a target context c', in many practical situations there is the need of interoperating multiple source contexts with multiple target targets, for which the bridge rules of the form (1) is inadequate. Besides, one would also want the ability of creating new values in target contexts for the bridge rules.

In this work, we consider *forall-existential bridge rules* that allows conjunctions and existential quantifiers in them, and hence is more expressive than those, in DDL [4] and McCarthy et al. [29]. A set of quads together with such bridge rules is called a *quad-system*. The main contributions of this work can be summarized as:

1. We provide a basic semantics for contextual reasoning over quad-systems, and study contextualized conjunctive query answering over them. For query answering, we use the notion of a *distributed chase*, which is an extension of a standard *chase* [22,1] that is widely used in databases and KR for the same.
2. We show that conjunctive query answering over quad-systems, in general, is undecidable. We derive a class of quad-systems called *context acyclic* quad-systems, for which query answering is decidable and can be done by forward chaining. We give both data and combined complexity of conjunctive query entailment for the same.

The paper is structured as follows: In section 2, we formalize the idea of contextualized quad-systems, giving various definitions and notations for setting the background.

In section 3, we formalize the query answering on quad-systems, define notions such as distributed chase that is further used for query answering, and give the undecidability results of query entailment for unrestricted quad-systems. In section 4, we present context acyclic quad-systems and its properties. We give an account of relevant related works in section 5, and conclude in section 6. A version of this paper with detailed proofs is available at [23].

2 Contextualized Quad-Systems

In this section, we formalize the notion of a quad-system and its semantics. For any vector or sequence x, we denote by $\|x\|$ the number of symbols in x, and by $\{x\}$ the set of symbols in x. For any sets A and B, $A \rightarrow B$ denotes the set of all functions from set A to set B. Given the set of URIs $\mathbf{U}$, the set of blank nodes $\mathbf{B}$, and the set of literals $\mathbf{L}$, the set $\mathbf{C} = \mathbf{U} \uplus \mathbf{B} \uplus \mathbf{L}$ are called the set of (RDF) constants. Any $(s, p, o) \in \mathbf{C} \times \mathbf{C} \times \mathbf{C}$ is called a generalized RDF triple (from now on, just triple). A graph is defined as a set of triples. A *Quad* is a tuple of the form $c\colon (s, p, o)$, where (s, p, o) is a triple and c is a URI[4], called the *context identifier* that denotes the context of the RDF triple. A *quad-graph* is defined as a set of quads. For any quad-graph Q and any context identifier c, we denote by $graph_Q(c)$ the set $\{(s, p, o)|c\colon (s, p, o) \in Q\}$. We denote by $Q_{\mathcal{C}}$ the quad-graph whose set of context identifiers is $\mathcal{C}$. Let $\mathbf{V}$ be the set of variables, any element of the set $\mathbf{C}^{\mathbf{V}} = \mathbf{V} \cup \mathbf{C}$ is a *term*. Any $(s, p, o) \in \mathbf{C}^{\mathbf{V}} \times \mathbf{C}^{\mathbf{V}} \times \mathbf{C}^{\mathbf{V}}$ is called a *triple pattern*, and an expression of the form $c\colon (s, p, o)$, where (s, p, o) is a triple pattern, c a context identifier, is called a *quad pattern*. A triple pattern t, whose variables are elements of the vector x or elements of the vector y is written as $t(x, y)$. For any function $f\colon A \rightarrow B$, the *restriction* of f to a set A', is the mapping $f|_{A'}$ from $A' \cap A$ to B s.t. $f|_{A'}(a) = f(a)$, for each $a \in A \cap A'$. For any triple pattern $t = (s, p, o)$ and a function μ from $\mathbf{V}$ to a set A, $t[\mu]$ denotes $(\mu'(s), \mu'(p), \mu'(o))$, where μ' is an extension of μ to $\mathbf{C}$ s.t. $\mu'|_{\mathbf{C}}$ is the identity function. For any set of triple patterns G, $G[\mu]$ denotes $\bigcup_{t \in G} t[\mu]$. For any vector of constants $a = \langle a_1, \ldots, a_{\|a\|} \rangle$, and vector of variables x of the same length, x/a is the function μ s.t. $\mu(x_i) = a_i$, for $1 \leq i \leq \|a\|$. We use the notation $t(a, y)$ to denote $t(x, y)[x/a]$.

Bridge rules (BRs). Bridge rules (BR) enables knowledge propagation across contexts. Formally, a BR is an expression of the form:

$$\forall x \forall z \ [c_1\colon t_1(x, z) \wedge \ldots \wedge c_n\colon t_n(x, z) \rightarrow \exists y \ c_1'\colon t_1'(x, y) \wedge \ldots \wedge c_m'\colon t_m'(x, y)] \quad (1)$$

where $c_1, \ldots, c_n, c_1', \ldots, c_m'$ are context identifiers, x, y, z are vectors of variables s.t. $\{x\}, \{y\}$, and $\{z\}$ are pairwise disjoint. $t_1(x, z), \ldots, t_n(x, z)$ are triple patterns which do not contain blank-nodes, and whose set of variables are from x or z. $t_1'(x, y)$, $\ldots, t_m'(x, y)$ are triple patterns, whose set of variables are from x or y, and also does not contain blank-nodes. For any BR, r, of the form (1), $body(r)$ is the set of quad patterns $\{c_1\colon t_1(x, z), \ldots, c_n\colon t_n(x, z)\}$, and $head(r)$ is the set of quad patterns $\{c_1'\colon t_1'(x, y), \ldots c_m'\colon t_m'(x, y)\}$.

---

[4] Although, in general a context identifier can be a constant, for the ease of notation, we restrict them to be a URI.

Definition 1 (Quad-System). *A* quad-system *QS_C is defined as a pair $\langle Q_C, R \rangle$, where Q_C is a quad-graph, whose set of context identifiers is C, and R is a set of BRs.*

For any quad-graph Q_C (BR r), its symbols size $\|Q_C\|$ ($\|r\|$) is the number of symbols required to print Q_C (r). Hence, $\|Q_C\| \approx 4*|Q_C|$, where $|Q_C|$ denotes the cardinality of the set Q_C. Note that $|Q_C|$ equals the number of quads in Q_C. For a BR r, $\|r\| \approx 4*k$, where k is the number of quad-patterns in r. For a set of BRs R, its size $\|R\|$ is given as $\Sigma_{r \in R}\|r\|$. For any quad-system $QS_C = \langle Q_C, R \rangle$, its size $\|QS_C\| = \|Q_C\| + \|R\|$.

Semantics. In order to provide a semantics for enabling reasoning over a quad-system, we need to use a local semantics for each context to interpret the knowledge pertaining to it. Since the primary goal of this paper is a decision procedure for query answering over quad-systems based on forward chaining, we consider the following desiderata for the choice of the local semantics:

- there exists a set of inference rules and an operation lclosure() that computes the deductive closure of a graph w.r.t to the local semantics using the inference rules.
- given a finite graph as input, the lclosure() operation, terminates with a finite graph as output in polynomial time whose size is polynomial w.r.t. to the input set.

Some of the alternatives for the local semantics satisfying the above mentioned criterion are Simple, RDF, RDFS [19], OWL-Horst [20] etc. Assuming that a local semantics has been fixed, for any context c, we denote by $I^c = \langle \Delta^c, \cdot^c \rangle$ an interpretation structure for the local semantics, where Δ^c is the interpretation domain, $\cdot^c$ the corresponding interpretation function. Also $\models_{local}$ denotes the local satisfaction relation between a local interpretation structure and a graph. Given a quad graph Q_C, a *distributed interpretation structure* is an indexed set $\mathcal{I}^C = \{I^c\}_{c \in C}$, where I^c is a local interpretation structure, for each $c \in C$. We define the satisfaction relation $\models$ between a distributed interpretation structure $\mathcal{I}^C$ and a quad-system QS_C as:

Definition 2 (Model of a Quad-System). *A distributed interpretation structure $\mathcal{I}^C = \{I^c\}_{c \in C}$ satisfies a quad-system $QS_C = \langle Q_C, R \rangle$, in symbols $\mathcal{I}^C \models QS_C$, iff all the following conditions are satisfied:*

1. $I^c \models_{local} graph_{Q_C}(c)$, for each $c \in C$;
2. $a^{c_i} = a^{c_j}$, for any $a \in \mathbf{C}$, $c_i, c_j \in C$;
3. for each BR $r \in R$ of the form (1) and for each $\sigma \in \mathbf{V} \to \Delta^C$, where $\Delta^C = \bigcup_{c \in C} \Delta^c$, if

$$I^{c_1} \models_{local} t_1(\boldsymbol{x}, \boldsymbol{z})[\sigma], ..., I^{c_n} \models_{local} t_n(\boldsymbol{x}, \boldsymbol{z})[\sigma],$$

then there exists function $\sigma' \supseteq \sigma$, s.t.

$$I^{c'_1} \models_{local} t'_1(\boldsymbol{x}, \boldsymbol{y})[\sigma'], ..., I^{c'_m} \models_{local} t'_m(\boldsymbol{x}, \boldsymbol{y})[\sigma'].$$

Condition 1 in the above definition ensures that for any model $\mathcal{I}^C$ of a quad-graph, each $I^c \in \mathcal{I}^C$ is a local model of the set of triples in context c. Condition 2 ensures that any constant c is rigid, i.e. represents the same resource across a quad-graph, irrespective of the context in which it occurs. Condition 3 ensure that any model of a quad-system satisfies each BR in it. Any $\mathcal{I}^C$ s.t. $\mathcal{I}^C \models QS_C$ is said to be a model of QS_C. A quad-system QS_C is said to be *consistent* if there exists a model $\mathcal{I}^C$, s.t. $\mathcal{I}^C \models QS_C$, and otherwise said to be *inconsistent*. For any quad-system $QS_C = \langle Q_C, R \rangle$, it can be the case that $graph_{Q_C}(c)$ is locally consistent, for each $c \in C$, whereas QS_C is not consistent. This is because the set of BRs R adds more knowledge to the quad-system, and restricts the set of models that satisfy the quad-system.

Definition 3 (Quad-System Entailment). *(a) A quad-system QS_C entails a quad $c\colon (s, p, o)$, in symbols $QS_C \models c\colon (s, p, o)$, iff for any distributed interpretation structure $\mathcal{I}^C$, if $\mathcal{I}^C \models QS_C$ then $\mathcal{I}^C \models \langle\{c\colon (s, p, o)\}, \emptyset\rangle$. (b) A quad-system QS_C entails a quad-graph $Q'_{C'}$, in symbols $QS_C \models Q'_{C'}$ iff $QS_C \models c\colon (s, p, o)$ for any $c\colon (s, p, o) \in Q'_{C'}$. (c) A quad-system QS_C entails a BR r iff for any $\mathcal{I}^C$, if $\mathcal{I}^C \models QS_C$ then $\mathcal{I}^C \models \langle\emptyset, \{r\}\rangle$. (d) For a set of BRs R, $QS_C \models R$ iff $QS_C \models r$, for every $r \in R$. (e) Finally, a quad-system QS_C entails another quad-system $QS'_{C'} = \langle Q'_{C'}, R'\rangle$, in symbols $QS_C \models QS'_{C'}$ iff $QS_C \models Q'_{C'}$ and $QS_C \models R'$.*

We call the decision problems (DPs) corresponding to the entailment problems (EPs) in (a), (b), (c), (d), and (e) as *quad EP, quad-graph EP, BR EP, BRs EP, and quad-system EP*, respectively.

3   Query Answering on Quad-Systems

In the realm of quad-systems, the classical conjunctive queries or select-project-join queries are slightly extended to what we call *Contextualized Conjunctive Queries* (CCQs). A CCQ $CQ(x)$ is an expression of the form:

$$\exists y \; q_1(x, y) \wedge \ldots \wedge q_p(x, y) \tag{2}$$

where q_i, for $i = 1, \ldots, p$ are quad patterns over vectors of *free variables* x and *quantified variables* y. A CCQ is called a boolean CCQ if it does not have any free variables. For any CCQ $CQ(x)$ and a vector a of constants s.t. $\|x\| = \|a\|$, $CQ(a)$ is boolean. A vector a is an *answer* for a CCQ $CQ(x)$ w.r.t. structure $\mathcal{I}_C$, in symbols $\mathcal{I}_C \models CQ(a)$, iff there exists assignment $\mu\colon \{y\} \to \mathbf{B}$ s.t. $\mathcal{I}_C \models \bigcup_{i=1,\ldots,p} q_i(a, y)[\mu]$. A vector a is a *certain answer* for a CCQ $CQ(x)$ over a quad-system QS_C, iff $\mathcal{I}_C \models CQ(a)$, for every model $\mathcal{I}_C$ of QS_C. Given a quad-system QS_C, a CCQ $CQ(x)$, and a vector a, DP of determining whether $QS_C \models CQ(a)$ is called the *CCQ EP*. It can be noted that the other DPs over quad-systems that we have seen are reducible to CCQ EP. Hence, in this paper, we primarily focus on the CCQ EP.

dChase of a Quad-System. In order to do query answering over a quad-system, we employ what has been called in the literature, a *chase* [22,1], specifically, we adopt notion of the *skolem chase* given in Marnette [28] and Cuenca Grau et al [9]. In order to fit the framework of quad-systems, we extend the standard notion of chase to a *distributed chase*, abbreviated *dChase*. In the following, we show how the dChase of a quad-system can be constructed.

For any BR r of the form (1), the *skolemization* $sk(r)$ is the result of replacing each $y_i \in \{y\}$ with a globally unique Skolem function f_i^r, s.t. $f_i^r\colon \mathbf{C}^{\|x\|} \to \mathbf{B}_{sk}$, where $\mathbf{B}_{sk}$ is a fresh set of blank nodes called *skolem blank nodes*. Intuitively, for every distinct vector a of constants, with $\|a\| = \|x\|$, $f_i^r(a)$ is a fresh blank node, whose node id is a hash of a. Let $f^r = \langle f_1^r, \ldots, f_{\|y\|}^r \rangle$ be a vector of distinct Skolem functions; for any BR

r the form (1), with slight abuse (Datalog notation) we write its skolemization $sk(r)$ as follows:

$$c_1 : t_1(\boldsymbol{x}, \boldsymbol{z}), ..., c_n : t_n(\boldsymbol{x}, \boldsymbol{z}) \rightarrow c'_1 : t'_1(\boldsymbol{x}, \boldsymbol{f}^r), ..., c'_m : t'_m(\boldsymbol{x}, \boldsymbol{f}^r) \qquad (3)$$

Moreover, a skolemized BR r of the form (3) can be replaced by the following semantically equivalent set of formulas, whose symbol size is worst case quadratic w.r.t $\|r\|$:

$$\{c_1 : t_1(\boldsymbol{x}, \boldsymbol{z}), ..., c_n : t_n(\boldsymbol{x}, \boldsymbol{z}) \rightarrow c'_1 : t'_1(\boldsymbol{x}, \boldsymbol{f}^r), \qquad (4)$$
$$...,$$
$$c_1 : t_1(\boldsymbol{x}, \boldsymbol{z}), ..., c_n : t_n(\boldsymbol{x}, \boldsymbol{z}) \rightarrow c'_m : t'_m(\boldsymbol{x}, \boldsymbol{f}^r)\}$$

Note that each BR in the above set has exactly one quad pattern with optional function symbols in its head part. Also note that a BR with out function symbols can be replaced with a set of BRs with single quad-pattern heads. Hence, w.l.o.g, we assume that any BR in a skolemized set $sk(R)$ of BRs is of the form (4). For any quad-graph Q_C and a skolemized BR r of the form (4), *application* of r on Q_C, denoted by $r(Q_C)$, is given as:

$$r(Q_C) = \bigcup_{\mu \in \mathbf{V} \rightarrow \mathbf{C}} \{ c'_1 : t'_1(\boldsymbol{x}, \boldsymbol{f}^r)[\mu] \mid c_1 : t_1(\boldsymbol{x}, \boldsymbol{z})[\mu] \in Q_C, ..., c_n : t_n(\boldsymbol{x}, \boldsymbol{z})[\mu] \in Q_C \}$$

For any set of skolemized BRs R, application of R on Q_C is given by: $R(Q_C) = \bigcup_{r \in R} r(Q_C)$. For any quad-graph Q_C, we define:

$$\mathsf{lclosure}(Q_C) = \bigcup_{c \in C} \{ c : (s, p, o) \mid (s, p, o) \in \mathsf{lclosure}(graph_{Q_C}(c)) \}$$

For any quad-system $QS_C = \langle Q_C, R \rangle$, *generating BRs* R_F is the set of BRs in $sk(R)$ with function symbols, and the *non-generating BRs* is the set $R_I = sk(R) \setminus R_F$. Let $dChase_0(QS_C) = \mathsf{lclosure}(Q_C)$; for $i \in \mathbb{N}$, $dChase_{i+1}(QS_C) =$

$$\mathsf{lclosure}(dChase_i(QS_C) \cup R_I(dChase_i(QS_C))), \quad \text{if } R_I(dChase_i(QS_C)) \not\subseteq \\ dChase_i(QS_C);$$

$$\mathsf{lclosure}(dChase_i(QS_C) \cup R_F(dChase_i(QS_C))), \quad \text{otherwise};$$

The dChase of QS_C, denoted $dChase(QS_C)$, is given as:

$$dChase(QS_C) = \bigcup_{i \in \mathbb{N}} dChase_i(QS_C)$$

Intuitively, $dChase_i(QS_C)$ can be thought of as the state of $dChase(QS_C)$ at the end of iteration i. It can be noted that, if there exists i s.t. $dChase_i(QS_C) = dChase_{i+1}(QS_C)$, then $dChase(QS_C) = dChase_i(QS_C)$. An iteration i, s.t. $dChase_i(QS_C)$ is computed by the application of the set of (resp. non-)generating BRs R_F (resp. R_I), on $dChase_{i-1}(QS_C)$ is called a (resp. non-)*generating iteration*. The dChase $dChase(QS_C)$ of a consistent quad-system QS_C is a *universal model* [10] of the quad-system, i.e. it is a model of QS_C, and for any model $\mathcal{I}_C$ of QS_C, there is a homomorphism from $dChase(QS_C)$ to $\mathcal{I}_C$. Hence, for any boolean CCQ $CQ()$, $QS_C \models CQ()$

iff there exists a map $\mu \colon \mathbf{V}(CQ) \to \mathbf{C}$ s.t. $\{CQ()\}[\mu] \subseteq dChase(QS_C)$. We call the sequence $dChase_0(QS_C), dChase_1(QS_C), ...,$ the *dChase sequence* of QS_C. The following lemma shows that in a dChase sequence of a quad-system, the result of a single generating iteration and a subsequent number of non-generating iterations causes only an exponential blow up in size.

Lemma 1. *For a quad-system $QS_C = \langle Q_C, R \rangle$, the following holds: (i) if $i \in \mathbb{N}$ is a generating iteration, then $\|dChase_i(QS_C)\| = \mathcal{O}(\|dChase_{i-1}(QS_C)\|^{\|R\|})$, (ii) suppose $i \in \mathbb{N}$ is a generating iteration, and for any $j \geq 1, i+1, ..., i+j$ are non-generating iterations, then $\|dChase_{i+j}(QS_C)\| = \mathcal{O}(\|dChase_{i-1}(QS_C)\|^{\|R\|})$, (iii) for any iteration k, $dChase_k(QS_C)$ can be computed in time $\mathcal{O}(\|dChase_{k-1}(QS_C)\|^{\|R\|})$.*

Proof (sketch).

(i) R can be applied on $dChase_{i-1}(QS_C)$ by grounding R to the set of constants in $dChase_{i-1}(QS_C)$, the number of such groundings is of the order $\mathcal{O}(\|dChase_{i-1}(QS_C)\|^{\|R\|})$, $\|R(dChase_{i-1}(QS_C))\| = \mathcal{O}(\|R\| * \|dChase_{i-1}(QS_C)\|^{\|R\|})$. Since lclosure only increases the size polynomially, $\|dChase_i(QS_C)\| = \mathcal{O}(\|dChase_{i-1}(QS_C)\|^{\|R\|})$.

(ii) From (i) we know that $\|R(dChase_{i-1}(QS_C))\| = \mathcal{O}(\|dChase_{i-1}(QS_C)\|^{\|R\|})$. Since, no new constant is introduced in any subsequent non-generating iterations, and since any quad contains only four constants, the set of constants in any subsequent dChase iteration is $\mathcal{O}(4 * \|dChase_{i-1}(QS_C)\|^{\|R\|})$. Since only these many constants can appear in positions c, s, p, o of any quad generated in the subsequent iterations, the size of $dChase_{i+j}(QS_C)$ can only increase polynomially, which means that $\|dChase_{i+j}(QS_C)\| = \mathcal{O}(\|dChase_{i-1}(QS_C)\|^{\|R\|})$.

(iii) Since any dChase iteration k involves the following two operations: (a) lclosure(), and (b) computing $R(dChase_{k-1}(QS_C))$. (a) can be done in PTIME w.r.t to its input. (b) can be done in the following manner: ground R to the set of constants in $dChase_{i-1}(QS_C)$; then for each grounding g, if $body(g) \subseteq dChase_{i-1}(QS_C)$, then add $head(g)$ to $R(dChase_{k-1}(QS_C))$. Since, the number of such groundings is of the order $\mathcal{O}(\|dChase_{k-1}(QS_C)\|^{\|R\|})$, and checking if each grounding is contained in $dChase_{k-1}(QS_C)$, can be done in time polynomial in $\|dChase_{k-1}(QS_C)\|$, the time taken for (b) is $\mathcal{O}(\|dChase_{k-1}(QS_C)\|^{\|R\|})$. Hence, any iteration k can be done in time $\mathcal{O}(\|dChase_{k-1}(QS_C)\|^{\|R\|})$. □

Although, we now know how to compute the dChase of a quad-system, which can be used for deciding CCQ EP, it turns out that for the class of quad-systems whose BRs are of the form (1), which we call *unrestricted quad-systems*, the dChase can be infinite. This raises the question if there are other approaches that can be used, for instance similar problem arises in DLs with value creation, due to the presence of existential quantifiers, whereas the approaches like the one in Glim et al. [15] provides an algorithm for CQ entailment based on query rewriting.

Theorem 1. *The CCQ EP over unrestricted quad-systems is undecidable.*

Proof (sketch). We show that the well known undecidable problem of non-emptiness of intersection of context-free grammars (CFGs) is reducible to the CCQ entailment problem. Given two CFGs, $G_1 = \langle V_1, T, S_1, P_1 \rangle$ and $G_2 = \langle V_2, T, S_2, P_2 \rangle$, where V_1, V_2 are the set of variables, T s.t. $T \cap (V_1 \cup V_2) = \emptyset$ is the set of terminals. $S_1 \in V_1$ is the start symbol of G_1, and P_1 are the set of PRs of the form $v \rightarrow w$, where $v \in V$, w is a sequence of the form $w_1...w_n$, where $w_i \in V_1 \cup T$. Similarly s_2, P_2 is defined. Deciding whether the language generated by the grammars $L(G_1)$ and $L(G_2)$ have non-empty intersection is known to be undecidable [18].

Given two CFGs $G_1 = \langle V_1, T, S_1, P_1 \rangle$ and $G_2 = \langle V_2, T, S_2, P_2 \rangle$, we encode grammars G_1, G_2 into a quad-system $QS_c = \langle Q_c, R \rangle$, with only a single context identifier c. Each PR $r = v \rightarrow w \in P_1 \cup P_2$, with $w = w_1 w_2 w_3 .. w_n$, is encoded as a BR of the form: $c: (x_1, w_1, x_2), c: (x_2, w_2, x_3), ..., c: (x_n, w_n, x_{n+1}) \rightarrow c: (x_1, v, x_{n+1})$, where $x_1, .., x_{n+1}$ are variables. For each terminal symbol $t_i \in T$, R contains a BR of the form: $c: (x, \texttt{rdf:type}, C) \rightarrow \exists y\, c: (x, t_i, y), c: (y, \texttt{rdf:type}, C)$ and Q_c is the singleton: $\{c: (a, \texttt{rdf:type}, C)\}$. It can be observed that:

$$QS_c \models \exists y\, c: (a, S_1, y) \wedge c: (a, S_2, y) \Leftrightarrow L(G_1) \cap L(G_2) \neq \emptyset$$

We refer the reader to [23] for the complete proof. □

4   Context Acyclic Quad-Systems: A Decidable Class

In the previous section, we saw that query answering on unrestricted quad-systems is undecidable, in general. We in the following define a class of quad-systems for which query entailment is decidable. The class has the property that algorithms based on forward chaining, for deciding query entailment, can straightforwardly be implemented (by minor extensions) on existing quad stores. It should be noted that the technique we propose is reminiscent of the *Weak acyclicity* [12,11] technique used in the realm of Datalog+-.

Consider a BR r of the form: $c_1: t_1(x, z), c_2: t_2(x, z) \rightarrow \exists y\, c_3: t_3(x, y), c_4: t_4(x, y)$. Since such a rule triggers propagation of knowledge in a quad-system, specifically triples from the source contexts c_1, c_2 to the target contexts c_3, c_4 in a quad-system. As shown in Fig. 1, we can view a BR as a propagation rule across distinct compartments of knowledge, divided as contexts. For any BR of the form (1), each context in the set $\{c'_1, ..., c'_m\}$ is said to depend on the set of contexts $\{c_1, ..., c_n\}$. In a quad-system $QS_C = \langle Q_C, R \rangle$, for any $r \in R$, of the form (1), any context

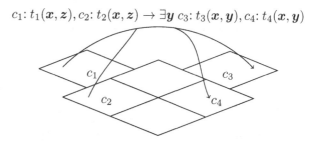

$$c_1: t_1(x, z), c_2: t_2(x, z) \rightarrow \exists y\, c_3: t_3(x, y), c_4: t_4(x, y)$$

Fig. 1. Bridge Rule: A mechanism for specifying propagation of knowledge among contexts.

whose identifier is in the set $\{c \mid c: (s, p, o) \in head(r), s$ or p or o is an existentially quantified variable$\}$, is called a *triple generating context* (TGC). One can analyze the set of BRs in a quad-system QS_C using a context dependency graph, which is a directed graph, whose nodes are context identifiers in $\mathcal{C}$, s.t. the nodes corresponding to TGCs are marked with a $*$, and whose edges are constructed as follows: for each BR of the form (1), there exists an edge from each c_i to c'_j, for $i = 1, ..., n, j = 1, ..., m$. A quad-system is said to be *context acyclic*, iff its context dependency graph does not contain cycles involving TGCs.

Example 1. Consider a quad-system, whose set of BRs R are:

$$c_1: (x_1, x_2, \mathsf{U}_1) \rightarrow \exists y_1 \, c_2: (x_1, x_2, y_1), c_3: (x_2, \mathtt{rdf:type, rdf:Property}) \quad (5)$$
$$c_2: (x_1, x_2, z_1) \rightarrow c_1: (x_1, x_2, \mathsf{U}_1) \quad (6)$$
$$c_3: (x_1, x_2, x_3) \rightarrow c_1: (x_1, x_2, x_3)$$

where U_1 be a URI, whose corresponding dependency graph is shown in Fig. 2. Note that the node corresponding to the triple generating context c_2 is marked with a '$*$' symbol. Since the cycle (c_1, c_2, c_1) in the quad-system contains c_2 which is a TGC, the quad-system is not context acyclic.

In a context acyclic quad-system $QS_{\mathcal{C}}$, since there exists no cyclic path through any TGC node in the context dependency graph, there exists a set of TGCs $\mathcal{C}' \subseteq \mathcal{C}$ s.t. for any $c \in \mathcal{C}'$, there exists no incoming path[5] from a TGC to c. We call such TGCs, *level-1 TGCs*. In other words, a TGC c is a level-1 TGC, if for any $c' \in \mathcal{C}$, there exists an incoming path from c' to c, implies c' is not a TGC. For $l \geq 1$, a level-l+1 TGC c is a TGC that has an incoming path from a level-l TGC, and for any incoming path from a level-l' TGC to c, is s.t. $l' \leq l$. Extending the notion of level also to the non-TGCs, we say that any non-TGC that does not have any incoming paths from a TGC is at level-0; we say that any non-TGC $c \in \mathcal{C}$ is at level-l, if there exists an incoming path from a level-l TGC to c, and for any incoming path from a level-l' TGC to c, is s.t. $l' \leq l$. Hence, the set of contexts in a context acyclic quad-system can be partitioned using the above notion of levels.

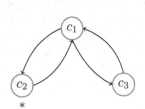

Fig. 2. Context Dependency graph

Definition 4. *For a quad-system $QS_{\mathcal{C}}$, a context $c \in \mathcal{C}$ is said to be saturated in an iteration i, iff for any quad of the form $c: (s, p, o)$, $c: (s, p, o) \in dChase(QS_{\mathcal{C}})$ implies $c: (s, p, o) \in dChase_i(QS_{\mathcal{C}})$.*

Intuitively, context c is saturated in the dChase iteration i, if no new quad of the form $c: (s, p, o)$ will be generated in any $dChase_k(QS_{\mathcal{C}})$, for any $k > i$. The following lemma gives the relation between the saturation of a context and the required number of dChase iterations, for a context acyclic quad-system.

Lemma 2. *For any context acyclic quad-system, the following holds: (i) any level-0 context is saturated before the first generating iteration, (ii) any level-1 TGC is saturated after the first generating iteration, (iii) any level-k context is saturated before the $k + 1$th generating iteration.*

---

[5] Assume that paths have at least one edge.

Proof. Let $QS_C = \langle Q_C, R \rangle$ be the quad-system, whose first generating iteration is i.

(i) for any level-0 context c, any BR $r \in R$, and any quad-pattern of the form $c: (s, p, o)$, if $c: (s, p, o) \in head(r)$, then for any c' s.t. $c': (s', p', o')$ occurs in $body(r)$ implies that c' is a level-0 context and r is a non-generating BR. Also, since c' is a level-0 context, the same applies to c'. Hence, it turns out that only non-generating BRs can bring triples to any level-0 context. Since at the end of iteration $i-1$, $dChase_{i-1}(QS_C)$ is closed w.r.t. the set of non-generating BRs (otherwise, by construction of dChase, i would not be a generating iteration). This implies that c is saturated before the first generating iteration i.

(ii) for any level-1 TGC c, any BR $r \in R$, and any quad-pattern $c: (s, p, o)$, if $c: (s, p, o) \in head(r)$, then for any c' s.t. $c': (s', p', o')$ occurs in $body(r)$ implies that c' is a level-0 context (Otherwise level of c would be greater than 1). This means that only contexts from which triples get propagated to c are level-0 contexts. From (i) we know that all the level-0 contexts are saturated before ith iteration, and since during the ith iteration R_F is applied followed by the lclosure() operation (R_I need not be applied, since $dChase_{i-1}(QS_C)$ is closed w.r.t. R_I), c is saturated after iteration i, the 1st generating iteration.

(iii) can be obtained from generalization of (i) and (ii), and from the fact that any level-k context can only have incoming paths from contexts whose levels are less than or equal to k. $\square$

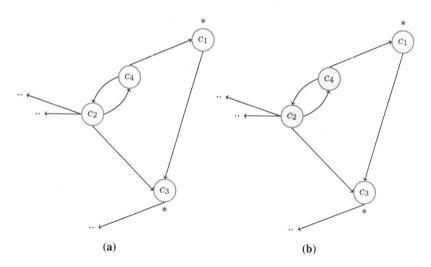

(a)   (b)

Fig. 3. Saturation of Contexts

Example 2. Consider the dependency graph in Fig. 3a, where .. indicates part of the graph that is not under the scope of our discussion. The TGCs nodes c_1 and c_3 are marked with a $*$. It can be seen that both c_2 and c_4 are level-0 contexts, since they do not have any incoming paths from TGCs. Since the only incoming paths to context c_1 are from c_2 and c_4, which are not TGCs, c_1 is a level-1 TGC. Context c_3 is a level-2 TGC, since it has an incoming path from the level-1 TGC c_1, and has no incoming path from a TGC whose level is greater than 1. Since the level-0 contexts only have incoming paths from level-0 contexts and only appear on the head part of non-generating BRs, before first generating iteration, all the level-0 TGCs becomes saturated, as the set of non-generating BRs R_I has been exhaustively applied. This situation is reflected in Fig. 3b, where the saturated nodes are shaded with gray. Note that after the first and second generating iterations c_1 and c_3 also become saturated, respectively.

The following lemma shows that for context acyclic quad-systems, there exists a finite bound on the size and computation time of its dChase.

Lemma 3. *For any context acyclic quad-system $QS_C = \langle Q_C, R \rangle$, the following holds: (i) the number of dChase iterations is finite, (ii) size of the dChase $\|dChase(QS_C)\|$ $= \mathcal{O}(2^{2^{\|QS_C\|}})$, (iii) computing $dChase(QS_C)$ is in 2EXPTIME, (iv) if R and the set of schema triples in Q_C is fixed, then $\|dChase(QS_C)\|$ is a polynomial in $\|QS_C\|$, and computing $dChase(QS_C)$ is in PTIME.*

Proof. (i) Since QS_C is context-acyclic, all the contexts can be partitioned according to their levels. Also, the number of levels k is s.t. $k \leq |\mathcal{C}|$. Hence, applying lemma 1, before the $k + 1$th generating iteration all the contexts becomes saturated, and $k + 1$th generating iteration do not produce any new quads, terminating the dChase computation process.

(ii) In the dChase computation process, since by lemma 1, any generating iteration and a sequence of non-generating iterations can only increase the dChase size exponentially in $\|R\|$, the size of the dChase before $k + 1$th generating iteration is $\mathcal{O}(\|dChase_0(QS_C)\|^{\|R\|^k})$, which can be written as $\mathcal{O}(\|QS_C\|^{\|R\|^k})$ (†). As seen in (i), there can only be $|\mathcal{C}|$ generating iterations, and a sequence of non-generating iterations. Hence, applying $k = |\mathcal{C}|$ to (†), and taking into account the fact that $|\mathcal{C}| \leq \|QS_C\|$, the size of the dChase $\|dChase(QS_C)\| = \mathcal{O}(2^{2^{\|QS_C\|}})$.

(iii) Since in any dChase iteration except the final one, atleast one new quad should be produced and the final dChase can have at most $\mathcal{O}(2^{2^{\|QS_C\|}})$ quads (by ii), the total number of iterations are bounded by $\mathcal{O}(2^{2^{\|QS_C\|}})$ (†). Since from lemma 1, we know that for any iteration i, computing $dChase_i(QS_C)$ is of the order $\mathcal{O}(\|dChase_{i-1}(QS_C)\|^{\|R\|})$. Since, $\|dChase_{i-1}(QS_C)\|$ can at most be $\mathcal{O}(2^{2^{\|QS_C\|}})$, computing $dChase_i(QS_C)$ is of the order $\mathcal{O}(2^{\|R\| * 2^{\|QS_C\|}})$. Also since $\|R\| \leq \|QS_C\|$, any iteration requires $\mathcal{O}(2^{2^{\|QS_C\|}})$ time (‡). From (†) and (‡), we can conclude that the time required for computing dChase is in 2EXPTIME.

(iv) In (ii) we saw that the size of the dChase before $k + 1$th generating iteration is given by $\mathcal{O}(\|QS_C\|^{\|R\|^k})$ (◇). Since by hypothesis $\|R\|$ is a constant and also the size of the dependency graph and the levels in it. Hence, the expression $\|R\|^k$ in (◇) amounts to a constant z. Hence, $\|dChase(QS_C)\| = \mathcal{O}(\|QS_C\|^z)$. Hence, the size of $dChase(QS_C)$ is a polynomial in $\|QS_C\|$.

Also, since in any dChase iteration except the final one, atleast one quad should be produced and the final dChase can have at most $\mathcal{O}(\|QS_C\|^z)$ quads, the total number of iterations are bounded by $\mathcal{O}(\|QS_C\|^z)$ (†). Also from lemma 1, we know that any dChase iteration i, computing $dChase_i(QS_C)$ involves two steps: (a) computing $R(dChase_{i-1}(QS_C))$, and (b) computing lclosure(), which can be done in

PTIME in the size of its input. Since computing $R(dChase_{i-1}(QS_C))$ is of the order $\mathcal{O}(\|dChase_{i-1}(QS_C)\|^{\|R\|})$, where $|R|$ is a constant and $\|dChase_{i-1}(QS_C)\|$ is a polynomial is $\|QS_C\|$, each iteration can be done in time polynomial in $\|QS_C\|$ (‡). From (†) and (‡), it can be concluded that dChase can be computed in PTIME.   □

Lemma 4. *For any context acyclic quad-system, the following holds: (i) data complexity of CCQ entailment is in* PTIME *(ii) combined complexity of CCQ entailment is in* 2EXPTIME.

Proof. For a context acyclic quad-system $QS_C = \langle Q_C, R \rangle$, since $dChase(QS_C)$ is finite, a boolean CCQ $CQ()$ can naively be evaluated by grounding the set of constants in the chase to the variables in the $CQ()$, and then checking if any of these groundings are contained in $dChase(QS_C)$. The number of such groundings can at most be $\|dChase(QS_C)\|^{\|CQ()\|}$ (†).

(i) Since for data complexity, the size of the BRs $\|R\|$, the set of schema triples, and $\|CQ()\|$ is fixed to constant. From lemma 3 (iv), we know that under the above mentioned settings the dChase can be computed in PTIME and is polynomial in the size of QS_C. Since $\|CQ()\|$ is fixed to a constant, and from (†), binding the set of constants in $dChase(QS_C)$ on $CQ()$ still gives a number of bindings that is worst case polynomial in the size of QS_C. Since membership of these bindings can checked in the polynomially sized dChase in PTIME, the time required for CCQ evaluation is in PTIME.

(ii) Since in this case $\|dChase(QS_C)\| = \mathcal{O}(2^{2^{\|QS_C\|}})$ (‡), from (†) and (‡), binding the set of constants in $\|dChase(QS_C)\|$ to variables in $CQ()$ amounts to $\mathcal{O}(2^{\|CQ()\| * 2^{\|QS_C\|}})$ bindings. Since the size of dChase is double exponential in $\|QS_C\|$, checking the membership of each of these bindings can be done in 2EXPTIME. Hence, the combined complexity is in 2EXPTIME. □

Theorem 2. *For any context acyclic quad-system, the following holds: (i) The data complexity of CCQ entailment is* PTIME-*complete, (ii) The combined complexity of CCQ entailment is* 2EXPTIME-*complete.*

For PTIME-hardness of data complexity, it can be shown that the well known problem of 3HornSat, the satisfiability of propositional Horn formulas with atmost 3 literals, and for 2EXPTIME-hardness for the combined complexity, it can be shown that the word problem of a double exponentially time bounded deterministic turing machine, which is a well known 2EXPTIME-hard problem, is reducible to the CCQ entailment problem (see [23] for detailed proof).

Reconsidering the quad-system in example 1, which is not context acyclic. Suppose that the contexts are enabled with RDFS inferencing, i.e lclosure() = rdfsclosure(). During dChase construction, since any application of rule (5) can only create a triple in c_2 in which the skolem blank node is in the object position, where as the application of rule (6), does not propogate constants in object postion to c_1. Although at a first look, the dChase might seem to terminate, but since the application of the following RDFS inference rule in c_2: $(s, p, o) \rightarrow (o, \text{rdf:type}, \text{rdfs:Resource})$, derives a quad of the form c_2: $(\_:b, \text{rdf:type}, \text{rdfs:Resource})$, where $\_:b$ is the skolem blank-node created by the application of rule (5). Now by application of rule (6) leads to c_1: $(\_:b, \mathtt{rdf:type}, U_1)$. Since rule (5) is applicable on c_1: $(\_:b, \mathtt{rdf:type}, U_1)$, which again brings a new skolem blank node to c_2, and hence the dChase construction doesn't terminate. Hence, as seen above the notion of context acyclicity can alarm us about such infinite cases.

5   Related Work

Contexts and Distributed Logics. The work on contexts began in the 80s when Mc-Carthy [21] proposed context as a solution to the generality problem in AI. After this various studies about logics of contexts mainly in the field of KR was done by Guha [17], *Distributed First Order Logics* by Ghidini et al. [13] and *Local Model Semantics* by Giunchiglia et al. [14]. Primarily in these works contexts are formalized as a first or-der/propositional theories, and bridge rules were provided to inter-operate the various contexts. Some of the initial works on contexts relevant to semantic web were the ones like *Distributed Description Logics* [4] by Borgida et al., *E-connections* [26] by Kutz et al., *Context-OWL* [5] by Bouqet et al., and the work of CKR [31,24] by Serafini et al. These were mainly logics based on DLs, which formalized contexts as OWL KBs, whose semantics is given using a distributed interpretation structure with additional se-mantic conditions that suits varying requirements. Compared to these works, the bridge rules we consider are much more expressive with conjunctions and existential variables that supports value/blank-node creation.

$\forall\exists$ *rules, TGDs, Datalog+- rules.* Query answering over rules with universal existen-tial quantifiers in the context of databases/KR, where these rules are called tuple gen-erating dependencies (TGDs)/Datalog+- rules, was done by Beeri and Vardi [3] even in the early 80s, where the authors show that the query entailment problem in gen-eral is undecidable. However, recently many classes of such rules have been identified for which query answering is decidable. These includes (a) fragments s.t. the result-ing models have bounded tree widths, called bounded treewidth sets (BTS), such as Weakly guarded rules [7], Frontier guarded rules [2], (b) fragments called finite unifi-cation sets (FUS), such as 'sticky' rules [6,16], and (c) fragments called finite extension sets (FES), where sufficient conditions are enforced to ensure finiteness of the chase and its termination. The approach used for query answering in FUS is to rewrite the input query w.r.t. to the TGDs to another query that can be evaluated directly on the set of instances, s.t. the answers for the former query and latter query coincides. The approach is called the *query rewriting approach.* FES classes uses certain termination guarantying tests that check whether certain sufficient conditions are satisfied by the structure of TGDs. A large number of classes in FES are based on tests that detects 'acyclicity conditions' by analyzing the information flow between the TGD rules. *Weak acyclicity* [12,11], was one of the first such notions, and was extended to *joint acyclic-ity* [25], *super weak acyclicity* [28], and *model faithful acyclicity* [9]. The most similar approach to ours is the weak acyclicity technique, where the structure of the rules is ana-lyzed using a dependency graph that models the propagation of constants across various predicates positions, and restricting the dependency graph to be acyclic. Although this technique can be used in our scenario by translating a quad-system to a set of TGDs;

Table 1. Complexity info for various quad-system fragments

Quad-System Fragment	dChase size w.r.t input quad-system	Data Complexity of CCQ entailment	Combined Complexity of CCQ entailment
Unrestricted Quad-Systems	Infinite	Undecidable	Undecidable
Context acylic Quad-Systems	Double exponential	PTIME-complete	2EXPTIME-complete

if the obtained translation is weakly acyclic, then one could use existing algorithms for chase computation for the TGDs to compute the chase, the query entailment check can be done by querying the obtained chase. However, our approach has the advantage of straightforward implementability on existing quad-stores.

6   Summary and Conclusion

In this paper, we study the problem of query answering over contextualized RDF knowledge. We show that the problem in general is undecidable, and present a decidable class called context acyclic quad-systems. Table 1 summarizes the main results obtained. We can show that the notion of context acyclicity, introduced in section 4 can be used to extend the currently established tools for contextual reasoning to give support for expressive BRs with conjuction and existentials with decidability guarantees. We view the results obtained in this paper as a general foundation for contextual reasoning and query answering over contextualized RDF knowledge formats such as Quads, and can straightforwardly be used to extend existing Quad stores to encorporate for-all existential BRs of the form (1).

References

1. Abiteboul, S., Hull, R., Vianu, V.: Foundations of Databases. Addison-Wesley (1995)
2. Baget, J., Mugnier, M., Rudolph, S., Thomazo, M.: Walking the Complexity Lines for Generalized Guarded Existential Rules. In: IJCAI, pp. 712–717 (2011)
3. Beeri, C., Vardi, M.Y.: The Implication Problem for Data Dependencies. In: Even, S., Kariv, O. (eds.) ICALP 1981. LNCS, vol. 115, pp. 73–85. Springer, Heidelberg (1981)
4. Borgida, A., Serafini, L.: Distributed Description Logics: Assimilating Information from Peer Sources. J. Data Semantics 1, 153–184 (2003)
5. Bouquet, P., Giunchiglia, F., van Harmelen, F., Serafini, L., Stuckenschmidt, H.: C-OWL: Contextualizing Ontologies. In: Fensel, D., Sycara, K., Mylopoulos, J. (eds.) ISWC 2003. LNCS, vol. 2870, pp. 164–179. Springer, Heidelberg (2003)
6. Calì, A., Gottlob, G., Pieris, A.: Query Answering under Non-guarded Rules in Datalog+/-. In: Hitzler, P., Lukasiewicz, T. (eds.) RR 2010. LNCS, vol. 6333, pp. 1–17. Springer, Heidelberg (2010)
7. Cali, A., Gottlob, G., Lukasiewicz, T., Marnette, B., Pieris, A.: Datalog+/-: A Family of Logical Knowledge Representation and Query Languages for New Applications. In: 25th Annual IEEE Symposium on Logic in Computer Science (LICS), pp. 228–242 (July 2010)

8. Carroll, J., Bizer, C., Hayes, P., Stickler, P.: Named graphs, provenance and trust. In: Proc. of the 14th Int.l. Conf. on WWW, pp. 613–622. ACM, New York (2005)

9. Cuenca Grau, B., Horrocks, I., Krötzsch, M., Kupke, C., Magka, D., Motik, B., Wang, Z.: Acyclicity Conditions and their Application to Query Answering in Description Logics. In: Proceedings of the 13th International Conference on Principles of Knowledge Representation and Reasoning (KR 2012), pp. 243–253. AAAI Press (2012)

10. Deutsch, A., Nash, A., Remmel, J.: The chase revisited. In: Proceedings of the Twenty-Seventh ACM SIGMOD-SIGACT-SIGART Symposium on Principles of Database Systems, PODS 2008, pp. 149–158 (2008)

11. Deutsch, A., Tannen, V.: Reformulation of XML Queries and Constraints. In: Calvanese, D., Lenzerini, M., Motwani, R. (eds.) ICDT 2003. LNCS, vol. 2572, pp. 225–238. Springer, Heidelberg (2002)

12. Fagin, R., Kolaitis, P.G., Miller, R.J., Popa, L.: Data Exchange: Semantics and Query Answering. Theoretical Computer Science 28(1), 89–124 (2005)

13. Ghidini, C., Serafini, L.: Distributed first order logics. In: Frontiers of Combining Systems 2, Studies in Logic and Computation, pp. 121–140. Research Studies Press (1998)

14. Giunchiglia, F., Ghidini, C.: Local models semantics, or contextual reasoning = locality + compatibility. Artificial Intelligence 127 (2001)

15. Glimm, B., Lutz, C., Horrocks, I., Sattler, U.: Answering conjunctive queries in the $\mathcal{SHIQ}$ description logic. In: Proceedings of the IJCAI 2007, pp. 299–404. AAAI Press (2007)

16. Gottlob, G., Manna, M., Pieris, A.: Polynomial Combined Rewritings for Existential Rules. In: KR 2014: International Conference on Principles of Knowledge Representation and Reasoning (2014)

17. Guha, R.: Contexts: a Formalization and some Applications. Ph.D. thesis, Stanford (1992)

18. Harrison, M.A.: Introduction to Formal Language Theory, 1st edn. Addison-Wesley Longman Publishing Company, Inc., Boston (1978)

19. Hayes, P. (ed.): RDF Semantics. W3C Recommendation (February 2004)

20. ter Horst, H.J.: Completeness, decidability and complexity of entailment for RDF Schema and a semantic extension involving the OWL vocabulary. Web Semantics: Science, Services and Agents on the WWW 3(2-3), 79–115 (2005)

21. McCarthy, J.: Generality in AI. Comm. of the ACM 30(12), 1029–1035 (1987)

22. Johnson, D.S., Klug, A.C.: Testing containment of conjunctive queries under functional and inclusion dependencies. Computer and System Sciences 28, 167–189 (1984)

23. Joseph, M., Kuper, G., Serafini, L.: Query Answering over Contextualized RDF/OWL Knowledge with Forall-Existential Bridge Rules: Attaining Decidability using Acyclicity (full version). Tech. rep., CoRR Technical Report arXiv:1406.0893, Arxiv e-Print archive (2014), http://arxiv.org/abs/1406.0893

24. Joseph, M., Serafini, L.: Simple reasoning for contextualized RDF knowledge. In: Proc. of Workshop on Modular Ontologies (WOMO 2011) (2011)

25. Krötzsch, M., Rudolph, S.: Extending decidable existential rules by joining acyclicity and guardedness. In: Walsh, T. (ed.) Proceedings of the 22nd International Joint Conference on Artificial Intelligence (IJCAI 2011), pp. 963–968. AAAI Press/IJCAI (2011)

26. Kutz, O., Lutz, C., Wolter, F., Zakharyaschev, M.: E-Connections of Abstract Description Systems. Artificial Intelligence 156(1), 1–73 (2004)

27. Lenat, D.: The Dimensions of Context Space. Tech. rep., CYCorp (1998) published online, http://www.cyc.com/doc/context-space.pdf

28. Marnette, B.: Generalized schema-mappings: from termination to tractability. In: Proceedings of the Twenty-Eighth ACM SIGMOD-SIGACT-SIGART Symposium on Principles of Database Systems, PODS 2009, pp. 13–22. ACM, New York (2009)

29. McCarthy, J., Buvac, S., Costello, T., Fikes, R., Genesereth, M., Giunchiglia, F.: Formalizing Context (Expanded Notes) (1995)
30. Schueler, B., Sizov, S., Staab, S., Tran, D.T.: Querying for meta knowledge. In: WWW 2008: Proceeding of the 17th International Conference on World Wide Web, pp. 625–634. ACM, New York (2008)
31. Serafini, L., Homola, M.: Contextualized knowledge repositories for the semantic web. Web Semantics: Science, Services and Agents on the World Wide Web (2012)

Computing Datalog Rewritings for Disjunctive Datalog Programs and Description Logic Ontologies

Mark Kaminski, Yavor Nenov, and Bernardo Cuenca Grau

Department of Computer Science
University of Oxford, UK

Abstract. We study the closely related problems of rewriting disjunctive datalog programs and non-Horn DL ontologies into plain datalog programs that entail the same facts for every dataset. We first propose the class of *markable* disjunctive datalog programs, which is efficiently recognisable and admits polynomial rewritings into datalog. Markability naturally extends to $\mathcal{SHI}$ ontologies, and markable ontologies admit (possibly exponential) datalog rewritings. We then turn our attention to resolution-based rewriting techniques. We devise an enhanced rewriting procedure for disjunctive datalog, and propose a second class of $\mathcal{SHI}$ ontologies that admits exponential datalog rewritings via resolution. Finally, we focus on conjunctive query answering over disjunctive datalog programs. We identify classes of queries and programs that admit datalog rewritings and study the complexity of query answering in this setting. We evaluate the feasibility of our techniques over a large corpus of ontologies, with encouraging results.

1 Introduction

Answering conjunctive queries is a key reasoning problem for many applications of ontologies. Query answering can sometimes be implemented via rewriting into datalog, where a rewriting of a query q w.r.t. an ontology $\mathcal{O}$ is a datalog program $\mathcal{P}$ that preserves the answers to q for any dataset. Rewriting queries into datalog not only ensures tractability in data complexity—an important requirement in data-intensive applications—but also enables the reuse of scalable rule-based reasoners such as OWLIM [4], Oracle's Data Store [21], and RDFox [16].

Datalog rewriting techniques have been investigated in depth for Horn Description Logics (i.e., DLs whose ontologies can be normalised as first-order Horn clauses), and optimised algorithms have been implemented in systems such as Requiem [18], Clipper [6], and Rapid [20]. Techniques for non-Horn DLs, however, have been studied to a lesser extent, and only for atomic queries.

If we restrict ourselves to atomic queries, rewritability for non-Horn DL ontologies is strongly related to the rewritability of disjunctive datalog programs into datalog: every $\mathcal{SHIQ}$ ontology can be transformed into a (positive) disjunctive datalog program that entails the same facts for every dataset (and hence

R. Kontchakov and M.-L. Mugnier (Eds.): RR 2014, LNCS 8741, pp. 76–91, 2014.
© Springer International Publishing Switzerland 2014

preserves answers to all atomic queries) [8].[1] It is well-known that disjunctive datalog programs cannot be generally rewritten into plain datalog. In particular, datalog rewritings may not exist even for disjunctive programs that correspond to ontologies expressed in the basic DL $\mathcal{ELU}$ [11,5], and sufficient conditions that ensure rewritability were identified in [9]. Deciding datalog rewritability of atomic queries w.r.t. $\mathcal{SHI}$ ontologies was proved NExpTime-complete in [3].

In our previous work [10], we proved a characterisation of datalog rewritability for disjunctive programs based on linearity: a restriction that requires each rule to contain at most one IDB atom in the body. It was shown that every linear disjunctive program can be polynomially rewritten into plain datalog; conversely, every datalog program can be polynomially translated into an equivalent linear disjunctive datalog program. We then proposed *weakly linear disjunctive datalog*, which extends both datalog and linear disjunctive datalog, and which admits polynomial datalog rewritings. In a weakly linear program, the linearity requirement is relaxed: instead of applying to all IDB predicates, it applies only to those that "depend" on a disjunctive rule.

A different approach to rewriting disjunctive programs into datalog by means of a resolution-based procedure was proposed in [5]. The procedure works by saturating the input disjunctive program $\mathcal{P}$ such that in each resolution step at least one of the premises is a non-Horn rule; if this process terminates, the procedure outputs the subset of datalog rules in the saturation, which is guaranteed to be a rewriting of $\mathcal{P}$. The procedure was shown to terminate for so-called *simple* disjunctive programs; furthermore, it was shown that ontologies expressed in certain logics of the DL-Lite$_{bool}$ family [1] can be transformed into disjunctive programs that satisfy the simplicity condition.

If we wish to go beyond atomic queries and consider general conjunctive queries, it is no longer possible to obtain query-independent datalog rewritings. Lutz and Wolter [12] showed that for *any* non-Horn ontology (or disjunctive program) $\mathcal{O}$ there exists a conjunctive query q such that answering the (fixed) q w.r.t. (fixed) $\mathcal{O}$ and an input dataset is co-NP-hard; thus, under standard complexity-theoretic assumptions no datalog rewriting for such q and $\mathcal{O}$ exists. To the best of our knowledge, no rewriting techniques for arbitrary CQs w.r.t. non-Horn ontologies and programs have been developed.

In this paper, we propose significant enhancements over existing techniques for rewriting atomic queries [10,5], which we then extend to the setting of arbitrary conjunctive queries. Furthermore, we evaluate the practical feasibility of our techniques over a large corpus of non-Horn ontologies. Specifically, our contributions are as follows.

In Section 3, we propose the class of *markable* disjunctive datalog programs, in which the weak linearity condition from [10] is further relaxed. We show that our extended class of programs is efficiently recognisable and that each markable program admits a polynomial datalog rewriting. These results can be readily applied to ontology reasoning. We first consider the "intersection"

---

[1] Disjunctive datalog typically allows for negation-as-failure, which we don't consider since we focus on monotonic reasoning.

between OWL 2 and disjunctive datalog (which we call RL$^{\sqcup}$), and show that fact entailment over RL$^{\sqcup}$ ontologies corresponding to a markable program is tractable in combined complexity (and hence no harder than in OWL 2 RL [15]). We then lift the markability condition to ontologies, and show that markable $\mathcal{SHI}$-ontologies admit a (possibly exponential) datalog rewriting.

In Section 4, we refine the resolution-based rewriting procedure from [5] by further requiring that only atoms involving disjunctive predicates can participate in resolution inferences. This refinement can significantly reduce the number of inferences drawn during saturation, without affecting correctness. We then focus on ontologies, and propose an extension of the logics in the DL-Lite$_{bool}$ family that admits (possibly exponential) datalog rewritings.

In Section 5, we shift our attention to conjunctive queries and propose classes of queries and disjunctive datalog programs that admit datalog rewritings. Furthermore, we discuss the implications of these results to ontology reasoning.

We have implemented and evaluated our techniques on a large ontology repository. Our results show that many realistic non-Horn ontologies can be rewritten into datalog. Furthermore, we have tested the scalability of query answering over the programs obtained using our techniques, with promising results.

The proofs of our technical results can be found in an extended version of the paper available online: https://krr-nas.cs.ox.ac.uk/2014/RR/report.pdf

2   Preliminaries

We consider standard notions of terms, atoms, literals, formulae, sentences, and entailment. A *fact* is a ground atom and a *dataset* is a finite set of facts. We assume that equality $\approx$ is an ordinary predicate and that each set of formulae contains the axiomatisation of $\approx$ as a congruence relation for its signature. Clauses, substitutions, most general unifiers (MGUs), clause subsumption, tautologies, binary resolution, and factoring are as usual [2]. Clause C θ-*subsumes* D if C subsumes D and C has no more literals than D. Clause C is *redundant* in a set of clauses if C is tautological or if C is θ-subsumed by another clause in the set. A *condensation* of a clause C is a minimal subset that is subsumed by C.

A *rule* r is a function-free sentence $\forall \boldsymbol{x} \forall \boldsymbol{z}.[\varphi(\boldsymbol{x}, \boldsymbol{z}) \to \psi(\boldsymbol{x})]$ where tuples of variables $\boldsymbol{x}$ and $\boldsymbol{z}$ are disjoint, $\varphi(\boldsymbol{x}, \boldsymbol{z})$ is a conjunction of distinct equality-free atoms, and $\psi(\boldsymbol{x})$ is a disjunction of distinct atoms. Formula φ is the *body* of r, and ψ is the *head*. Quantifiers in rules are omitted. We assume that rules are safe. A rule is *datalog* if $\psi(\boldsymbol{x})$ has at most one atom, and it is *disjunctive* otherwise. A *program* $\mathcal{P}$ is a finite set of rules; it is *datalog* if it consists only of datalog rules, and *disjunctive* otherwise. We assume that rules in $\mathcal{P}$ do not share variables. For convenience, we treat $\top$ and $\bot$ in a non-standard way as a unary and a nullary predicate, respectively. Given a program $\mathcal{P}$, $\mathcal{P}_{\top}$ is the program with a rule $P(x_1, \ldots, x_n) \to \top(x_i)$ for each predicate P in $\mathcal{P}$ and each $1 \le i \le n$, and a rule $\to \top(a)$ for each constant a in $\mathcal{P}$. We assume that $\mathcal{P}_{\top} \subseteq \mathcal{P}$ and $\top$ does not occur in head position in $\mathcal{P} \setminus \mathcal{P}_{\top}$. We define $\mathcal{P}_{\bot}$ as consisting of a rule with $\bot$ as body and empty head. We assume $\mathcal{P}_{\bot} \subseteq \mathcal{P}$ and no rule in $\mathcal{P} \setminus \mathcal{P}_{\bot}$ has an

Table 1. Normalised axioms. A, B are atomic or $\top$, C atomic or $\bot$, and R, S, T atomic.

1.	$\sqcap_{i=1}^{n} A_i \sqsubseteq \bigsqcup_{j=1}^{m} C_j$	$\bigwedge_{i=1}^{n} A_i(x) \to \bigvee_{j=1}^{m} C_j(x)$
2.	$\exists R.A \sqsubseteq B$	$R(x,y) \wedge A(y) \to B(x)$
3.	$A \sqsubseteq \mathsf{Self}(R)$	$A(x) \to R(x,x)$
4.	$\mathsf{Self}(R) \sqsubseteq A$	$R(x,x) \to A(x)$
5.	$R \sqsubseteq S$	$R(x,y) \to S(x,y)$
6.	$R \sqsubseteq S^-$	$R(x,y) \to S(y,x)$
7.	$R \circ S \sqsubseteq T$	$R(x,z) \wedge S(z,y) \to T(x,y)$
8.	$A \sqsubseteq\, \geq m\, R.B$	$A(x) \to \exists^{\geq m} y.(R(x,y) \wedge B(y))$
9.	$A \sqsubseteq\, \leq m\, R.B$	$A(z) \wedge \bigwedge_{i=0}^{m} R(z,x_i) \wedge B(x_i) \to \bigvee_{0 \leq i < j \leq m} x_i \approx x_j$

empty head or $\bot$ in the body. Thus, $\mathcal{P} \cup \mathcal{D} \models \top(a)$ for every a in $\mathcal{P} \cup \mathcal{D}$, and $\mathcal{P} \cup \mathcal{D}$ is unsatisfiable iff $\mathcal{P} \cup \mathcal{D} \models \bot$. Head predicates in $\mathcal{P} \setminus \mathcal{P}_\top$ are *intensional* (or *IDB*) in $\mathcal{P}$. All other predicates (including $\top$) are *extensional* (*EDB*). An atom is intensional (extensional) if so is its predicate. A rule is *linear* if it has at most one IDB body atom. A program $\mathcal{P}$ is linear if all its rules are.

We assume familiarity with DLs. W.l.o.g. we consider normalised axioms as in Table 1. An ontology $\mathcal{O}$ is a finite set of axioms. An ontology $\mathcal{O}$ is $\mathcal{SHIQ}$ if each axiom of type 7 satisfies $R = S = T$;[2] it is $\mathcal{SHI}$ if it is $\mathcal{SHIQ}$, it does not contain axioms of type 9, and each axiom of type 8 satisfies $m = 1$; it is $\mathcal{ALCHI}$ if it is $\mathcal{SHI}$ and it has no axiom of type 7; it is RL$^\sqcup$ if it does not contain axioms of type 8, and it is RL if it is RL$^\sqcup$ and $m = 1$ for each axiom of type 1 and 9. Programs obtained from RL$^\sqcup$ ontologies have rules with bounded number of variables: fact entailment is PTIME-complete for RL and co-NP-complete for RL$^\sqcup$ (in combined complexity).[3]

A *conjunctive query* (*CQ*) q is a datalog rule of the form $\varphi(\boldsymbol{x}, \boldsymbol{y}) \to A_q(\boldsymbol{x})$, with A_q a distinguished query predicate uniquely associated with q. A CQ is Boolean if A_q is propositional, and it is *atomic* if $\varphi(\boldsymbol{x}, \boldsymbol{y})$ consists of a single atom. A (disjunctive) program $\mathcal{P}$ is a *rewriting* of q w.r.t. a set of sentences $\mathcal{F}$ if for each dataset $\mathcal{D}$ over the signature of $\mathcal{F}$ and each tuple of constants $\boldsymbol{a}$ we have $\mathcal{F} \cup \mathcal{D} \cup \{q\} \models A_q(\boldsymbol{a})$ iff $\mathcal{P} \cup \mathcal{D} \models A_q(\boldsymbol{a})$. Program $\mathcal{P}$ is a rewriting of $\mathcal{F}$ if for each dataset $\mathcal{D}$ and each fact α over the signature of $\mathcal{F}$ we have $\mathcal{F} \cup \mathcal{D} \models \alpha$ iff $\mathcal{P} \cup \mathcal{D} \models \alpha$. Clearly, $\mathcal{P}$ is a rewriting of $\mathcal{F}$ if and only if $\mathcal{P}$ is a rewriting of every atomic query over the signature of $\mathcal{F}$. Hudstadt et al. [8] developed an algorithm for transforming a $\mathcal{SHIQ}$ ontology into a disjunctive program that preserves entailment of facts over non-transitive relations. This technique was extended in [5] to preserve all facts. Thus, every $\mathcal{SHIQ}$ ontology $\mathcal{O}$ admits a disjunctive datalog rewriting $\mathsf{DD}(\mathcal{O})$, which can be of exponential size.

---

[2] $\mathcal{SHIQ}$ enforces additional restrictions to ensure decidability, which we omit here.
[3] RL$^\sqcup$ and RL allow for nominals, which we omit. All our results immediately extend.

$$\mathcal{P}_0 = \{ C(x) \rightarrow B(x) \vee G(x) \quad (1)$$
$$G(y) \wedge E(x,y) \rightarrow B(x) \quad (2)$$
$$B(y) \wedge E(x,y) \rightarrow G(x) \quad (3)$$
$$E(y,x) \rightarrow E(x,y) \} \quad (4)$$

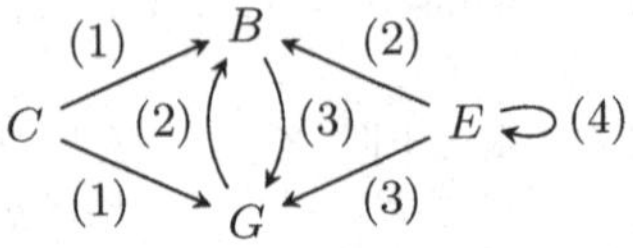

Fig. 1. A weakly linear disjunctive datalog program

3 Datalog Rewritings Based on Linearity

In [10], we proposed the class of *weakly linear programs* (WL), which extends both datalog and linear disjunctive datalog. In a WL program predicates are partitioned into disjunctive (i.e., those whose extension may depend on a disjunctive rule) and datalog (those that depend solely on datalog rules). A program is WL if all rules have at most one occurrence of a disjunctive predicate in the body.

Definition 3.1. *The* dependency graph $G_{\mathcal{P}} = (V, E, \mu)$ *of a program* $\mathcal{P}$ *is the smallest edge-labeled digraph such that:*

1. *V contains every predicate occurring in $\mathcal{P}$;*
2. *$r \in \mu(P, Q)$ whenever $P, Q \in V$, $r \in \mathcal{P} \setminus \mathcal{P}_\top$, P occurs in the body of r, and Q occurs in the head of r; and*
3. *$(P, Q) \in E$ whenever $\mu(P, Q)$ is nonempty.*

A predicate Q depends on a rule $r \in \mathcal{P}$ if $G_{\mathcal{P}}$ has a path that ends in Q and involves an r-labeled edge. Predicate Q is datalog *if it only depends on datalog rules; otherwise, Q is* disjunctive. *Program $\mathcal{P}$ is* weakly linear *(WL for short) if each rule body in $\mathcal{P}$ has at most one occurrence of a disjunctive predicate.*

Consider the disjunctive program $\mathcal{P}_0$ and its dependency graph depicted in Fig. 1. Predicate C is EDB, predicates B and G depend on Rule (1) and hence are disjunctive, whereas E depends only on Rule (4) and hence it is datalog. Each rule has at most one disjunctive body atom and the program is WL.

WL programs admit a polynomial rewriting [10]. Roughly speaking, they are translated into datalog by "moving" all disjunctive body atoms to the head and all disjunctive head atoms to the body while replacing their predicates with fresh ones of higher arity; the new predicates are "initialised" using additional rules.

Markable Programs. We next propose the class of *markable* disjunctive datalog programs, which extends WL programs. A key feature of a markable program is that one can identify a subset of disjunctive predicates, called *marked predicates*, such that the program can be translated into datalog by "moving" only those disjunctive atoms in a rule whose predicates are marked.

Definition 3.2. *Let $\mathcal{P}$ be a disjunctive program. A marking of $\mathcal{P}$ is a set M of disjunctive predicates in $\mathcal{P}$ such that:*

1. *Every rule in $\mathcal{P}$ has at most one body atom $Q(t)$ with $Q \in M$.*
2. *Every rule in $\mathcal{P}$ has at most one head atom $Q(t)$ with $Q \notin M$.*
3. *If $Q \in M$ and P is reachable from Q in $G_{\mathcal{P}}$, then $P \in M$.*

A predicate Q is marked *by M if $Q \in M$. An atom is* marked *if so is its predicate. A disjunctive program is* markable *if it has a marking.*

Markability generalises weak linearity in the following sense.

Proposition 3.3. *A disjunctive program $\mathcal{P}$ is WL if and only if the set of all disjunctive predicates in $\mathcal{P}$ constitutes a marking of $\mathcal{P}$.*

Let $\mathcal{P}_1$ extend $\mathcal{P}_0$ with the following rules:

$$V(x) \to C(x) \vee U(x) \qquad (5) \qquad\qquad C(x) \wedge U(x) \to \bot \qquad (6)$$

The dependency graph is given next. Note that $C, U, B,$ and G are disjunctive as they depend on Rule (5). Thus, (6) has two disjunctive body atoms and $\mathcal{P}_1$ is not WL. The program, however, has markings $\{C, B, G\}$ and $\{U, B, G\}$.

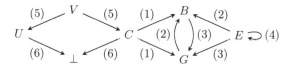

   Checking markability of a disjunctive program $\mathcal{P}$ is amenable to efficient implementation via reduction to 2-SAT. To this end, we first associate with every predicate Q in $\mathcal{P}$ a distinct propositional variable X_Q. Then, for each rule $\varphi \wedge P_1(s_1) \wedge \cdots \wedge P_n(s_n) \to Q_1(t_1) \vee \cdots \vee Q_m(t_m) \in \mathcal{P}$, where φ is the conjunction of all datalog atoms in the rule, we associate the following binary clauses:

1. $\neg X_{P_i} \vee \neg X_{P_j}$ for all $1 \le i < j \le n$;
2. $\neg X_{P_i} \vee X_{Q_j}$ for all $1 \le i \le n$ and $1 \le j \le m$;
3. $X_{Q_i} \vee X_{Q_j}$ for all $1 \le i < j \le m$.

Clauses of the form (1) indicate that at most one body atom in the rule may be marked. By (2), if a body atom is marked, then so must be all head atoms. Finally, (3) ensures that at most one head atom may be unmarked. The resulting set $\mathcal{N}$ of propositional clauses is quadratic in the size of $\mathcal{P}$. Moreover, $\mathcal{N}$ is satisfiable if and only if $\mathcal{P}$ has a marking, and every model I of $\mathcal{N}$ yields a marking $M_I = \{ Q \mid Q \text{ occurs in } \mathcal{P} \text{ and } X_Q \in I \}$. Since 2-SAT is solvable in linear time, we obtain the following.

Proposition 3.4. *Markability can be checked in time quadratic in the size of the input program.*

Datalog Rewritability of Markable Programs. We now show that markable programs are rewritable into datalog by means of a quadratic translation Ξ_M, which extends the translation for weakly linear programs given in [10].

Consider $\mathcal{P}_1$ and the marking $M = \{B, G, U\}$. We introduce fresh binary predicates $\overline{B}^Y$, $\overline{G}^Y$, and $\overline{U}^Y$ for every disjunctive predicate Y. Intuitively, if a fact $\overline{B}^G(c, d)$ holds in $\Xi_M(\mathcal{P}_1) \cup \mathcal{D}$ then proving $B(c)$ suffices for proving $G(d)$ in $\mathcal{P}_1 \cup \mathcal{D}$ (or, in other words, we have $\mathcal{P}_1 \cup \mathcal{D} \models B(c) \to G(d)$). Analogously, for the unmarked disjunctive predicate C we introduce fresh binary predicates C^Y for each disjunctive predicate Y; these predicates have a different intuitive interpretation: if a fact $C^U(c, d)$ holds in $\Xi_M(\mathcal{P}_1) \cup \mathcal{D}$ then $\mathcal{P}_1 \cup \mathcal{D}$ entails $C(c) \vee U(d)$. To "initialise" the extension of the fresh predicates we need the following rules for every $X \in M$ and every disjunctive predicate Y.

$$\top(x) \to \overline{X}^X(x, x) \quad (7) \qquad \top(y) \wedge C(x) \to C^Y(x, y) \quad (9)$$

$$\overline{X}^Y(x, y) \wedge X(x) \to Y(y) \quad (8) \qquad C^C(x, x) \to C(x) \quad (10)$$

These rules encode the intended meaning of the auxiliary predicates. For example, Rule (8) states that if $X(c)$ holds for some constant c and this is sufficient to prove $Y(d)$ for some d, then $Y(d)$ holds. The key step is to "flip" the direction of all rules in $\mathcal{P}_1$ involving the marked predicates B, G and U by moving all marked atoms from the head to the body and vice versa while at the same time replacing their predicates with the relevant auxiliary predicates. Thus, Rule (2) leads to the following rules in $\Xi_M(\mathcal{P}_1)$ for each disjunctive predicate Y:

$$\overline{B}^Y(x, z) \wedge E(x, y) \to \overline{G}^Y(y, z)$$

These rules are natural consequences of Rule (2) under the intended meaning of the auxiliary predicates: if we can prove a goal $Y(z)$ by proving first $B(x)$, and $E(x, y)$ holds, then by Rule (2) we deduce that proving $G(y)$ suffices to prove $Y(z)$. In contrast to (2), Rule (1) contains no disjunctive body atoms. We "flip" this rule as follows, for each disjunctive predicate Y:

$$C(x) \wedge \overline{B}^Y(x, y) \wedge \overline{G}^Y(x, y) \to Y(y)$$

Similarly to the previous case, this rule follows from Rule (1): if $C(x)$ holds and we can establish that $Y(y)$ can be proved from $B(x)$ and also from $G(x)$, then $Y(y)$ must hold. In contrast to marked atoms, unmarked atoms are not moved. So, Rules (5) and (6) yield the following rules for each disjunctive predicate Y:

$$V(x) \wedge \overline{U}^Y(x, y) \to C^Y(x, y) \qquad\qquad C^Y(x, y) \to \overline{U}^Y(x, y)$$

And indeed, these rules are consequences of Rule (5) and (6), respectively, under the intended meaning of the auxiliary predicates: $V(x)$ and $U(x) \to Y(y)$ imply $C(x) \vee Y(y)$ by Rule (5), while $C(x) \vee Y(y)$ and $U(x)$ imply $Y(y)$ by Rule (6).

Definition 3.5. *Let $\mathcal{P}$ be a disjunctive program, Σ the set of disjunctive predicates in $\mathcal{P} \setminus \mathcal{P}_\top$, and $M \subseteq \Sigma$ a marking of $\mathcal{P}$. For each $(P, Q) \in \Sigma^2$, let P^Q*

and $\overline{P}^Q$ be fresh predicates of arity $\mathsf{arity}(P) + \mathsf{arity}(Q)$. Then, $\varXi_M(\mathcal{P})$ is the datalog program with the rules given next, where φ is the conjunction of all datalog atoms in a rule, $\varphi_\top$ is the least conjunction of $\top$-atoms that makes a rule safe, all predicates P_i, Q_j are in $\varSigma$, and $\boldsymbol{y}, \boldsymbol{z}$ are disjoint vectors of fresh variables:

1. every rule in $\mathcal{P}$ that contains no disjunctive predicates;
2. a rule $\varphi_\top \wedge \varphi \wedge \bigwedge_{j=1}^{m} Q_j^R(\boldsymbol{t}_j, \boldsymbol{y}) \wedge \bigwedge_{i=1}^{n} \overline{P}_i^R(\boldsymbol{s}_i, \boldsymbol{y}) \to \overline{Q}^R(\boldsymbol{t}, \boldsymbol{y})$ for every rule $r = \varphi \wedge Q(\boldsymbol{t}) \wedge \bigwedge_{j=1}^{m} Q_j(\boldsymbol{t}_j) \to \bigvee_{i=1}^{n} P_i(\boldsymbol{s}_i) \in \mathcal{P} \setminus \mathcal{P}_\top$ and every $R \in \varSigma$, where $Q(\boldsymbol{t})$ is the unique marked body atom of r;
3. a rule $\varphi_\top \wedge \varphi \wedge \bigwedge_{j=1}^{m} Q_j^R(\boldsymbol{t}_j, \boldsymbol{y}) \wedge \bigwedge_{i=1}^{n} \overline{P}_i^R(\boldsymbol{s}_i, \boldsymbol{y}) \to R(\boldsymbol{y})$ for every rule $r = \varphi \wedge \bigwedge_{j=1}^{m} Q_j(\boldsymbol{t}_j) \to \bigvee_{i=1}^{n} P_i(\boldsymbol{s}_i) \in \mathcal{P} \setminus \mathcal{P}_\top$ and each $R \in \varSigma$, where r has no marked body atoms and no unmarked head atoms;
4. a rule $\varphi_\top \wedge \varphi \wedge \bigwedge_{j=1}^{m} Q_j^R(\boldsymbol{t}_j, \boldsymbol{y}) \wedge \bigwedge_{i=1}^{n} \overline{P}_i^R(\boldsymbol{s}_i, \boldsymbol{y}) \to P^R(\boldsymbol{s}, \boldsymbol{y})$ for every rule $r = \varphi \wedge \bigwedge_{j=1}^{m} Q_j(\boldsymbol{t}_j) \to P(\boldsymbol{s}) \vee \bigvee_{i=1}^{n} P_i(\boldsymbol{s}_i) \in \mathcal{P} \setminus \mathcal{P}_\top$ and each $R \in \varSigma$, where r has no marked body atoms, and $P(\boldsymbol{s})$ is the unique unmarked head atom;
5. a rule $\varphi_\top \to \overline{R}^R(\boldsymbol{y}, \boldsymbol{y})$ for every $R \in M$;
6. a rule $Q(\boldsymbol{z}) \wedge \overline{Q}^R(\boldsymbol{z}, \boldsymbol{y}) \to R(\boldsymbol{y})$ for every pair $(Q, R) \in M \times \varSigma$;
7. a rule $\varphi_\top \wedge Q(\boldsymbol{z}) \to Q^R(\boldsymbol{z}, \boldsymbol{y})$ for every pair $(Q, R) \in (\varSigma \setminus M) \times \varSigma$;
8. a rule $R^R(\boldsymbol{y}, \boldsymbol{y}) \to R(\boldsymbol{y})$ for every $R \in \varSigma \setminus M$.

The transformation is quadratic and the arity of predicates is at most doubled. For $\mathcal{P}_1$ and the marking $M = \{B, G, U\}$, we obtain the datalog program $\varXi_M(\mathcal{P}_1)$ consisting of the following rules, where $X \in M$ and Y is disjunctive:

$$C(x) \wedge \overline{B}^Y(x, y) \wedge \overline{G}^Y(x, y) \to Y(y) \quad (1')$$
$$\overline{B}^Y(x, z) \wedge E(x, y) \to \overline{G}^Y(y, z) \quad (2')$$
$$\overline{G}^Y(x, z) \wedge E(x, y) \to \overline{B}^Y(y, z) \quad (3')$$
$$V(x) \wedge \overline{U}^Y(x, y) \to C^Y(x, y) \quad (5')$$
$$C^Y(x, y) \to \overline{U}^Y(x, y) \quad (6')$$

$$E(y, x) \to E(x, y) \quad (4)$$
$$\top(x) \to \overline{X}^X(x, x) \quad (7)$$
$$X(x) \wedge \overline{X}^Y(x, y) \to Y(y) \quad (8)$$
$$\top(y) \wedge C(x) \to C^Y(x, y) \quad (9)$$
$$C^C(x, x) \to C(x) \quad (10)$$

In total, this yields 41 rules. Additionally, $\varXi_M(\mathcal{P}_1)$ contains the rules in $\varXi_M(\mathcal{P}_1)_\perp$ and an axiomatisation of $\approx$ (which can be omitted since $\approx$ does not occur in the above rules). Correctness of $\varXi_M$ is established by the following theorem.

Theorem 3.6. Let $\mathcal{P}$ be a disjunctive program and let M be a marking of $\mathcal{P}$. Then $\varXi_M(\mathcal{P})$ is a polynomial datalog rewriting of $\mathcal{P}$.

$\varXi_M(\mathcal{P})$ preserves answers to all atomic queries over $\mathcal{P}$. If we only want to query a specific predicate Q, we can compute a smaller program, which is linear in the size of $\mathcal{P}$ and preserves the extension of Q. If Q is datalog, each proof in $\mathcal{P}$ of a fact about Q involves only datalog rules, and if Q is disjunctive each such proof involves only fresh predicates X^Q and $\overline{X}^Q$. Thus, in $\varXi_M$ we can dispense with all rules involving auxiliary predicates X^R or $\overline{X}^R$ for $R \neq Q$ (if Q is datalog the rewriting has no auxiliary predicates).

Theorem 3.7. *Let P be a program, M a marking of P, S a set of predicates, and P' obtained from $\Xi_M(P)$ by removing all rules with a predicate X^R or $\overline{X}^R$ for $R \notin S \cup \{\bot\}$. Then P' is a rewriting of P w.r.t. all atomic queries over S.*

Rewriting Ontologies. Our results are directly applicable to $RL^{\sqcup}$. In [10], we showed tractability of fact entailment for the class of $RL^{\sqcup}$ ontologies corresponding to WL programs. The following theorem extends this result to the more general class of markable programs.

Theorem 3.8. *Checking $\mathcal{O} \cup \mathcal{D} \models \alpha$, for $\mathcal{O}$ an $RL^{\sqcup}$ ontology that corresponds to a markable program, is PTIME-complete w.r.t. data and combined complexity.*

We next lift the markability condition from disjunctive programs to $\mathcal{SHI}$ ontologies. Observe that the notions of dependency graph and markability naturally extend to sets of first-order clauses (written as rules where function symbols are allowed). We define a predicate to be *disjunctive in $\mathcal{O}$* if it is disjunctive in the set $\mathcal{F}_\mathcal{O}$ of clauses obtained by skolemisation; we call $\mathcal{O}$ *markable* if so is $\mathcal{F}_\mathcal{O}$; and we call a set of predicates a *marking of $\mathcal{O}$* if it is a marking of $\mathcal{F}_\mathcal{O}$.

Example 3.9. Consider the ontology $\mathcal{O}_1$ and its corresponding clauses $\mathcal{F}_{\mathcal{O}_1}$:

$$\mathcal{O}_1 = \{\mathsf{Person} \sqsubseteq \mathsf{Man} \sqcup \mathsf{Woman}, \mathsf{Person} \sqsubseteq \exists \mathsf{parent}.\mathsf{Person},$$
$$\exists \mathsf{married}.\mathsf{Person} \sqsubseteq \mathsf{Person}, \mathsf{Woman} \sqsubseteq \mathsf{Person}, \mathsf{Man} \sqsubseteq \mathsf{Person}\}$$
$$\mathcal{F}_{\mathcal{O}_1} = \{\mathsf{Person}(x) \rightarrow \mathsf{Man}(x) \vee \mathsf{Woman}(x), \mathsf{Person}(x) \rightarrow \mathsf{parent}(x, f(x)),$$
$$\mathsf{Person}(x) \rightarrow \mathsf{Person}(f(x)), \mathsf{Person}(y) \wedge \mathsf{married}(x, y) \rightarrow \mathsf{Person}(x),$$
$$\mathsf{Woman}(x) \rightarrow \mathsf{Person}(x), \mathsf{Man}(x) \rightarrow \mathsf{Person}(x)\}$$

Ontology $\mathcal{O}_1$ is markable since the set $\{\mathsf{Person}, \mathsf{Man}, \mathsf{Woman}\}$ is a marking of $\mathcal{F}_{\mathcal{O}_1}$.

As already mentioned, every normalised $\mathcal{SHI}$ ontology can be rewritten into disjunctive datalog by means of a resolution-based calculus [8,5]. The following lemma establishes that binary resolution and factoring preserve markability.

Lemma 3.10. *Let M be a marking of a set of clauses $\mathcal{F}$, and let $\mathcal{F}'$ be obtained from $\mathcal{F}$ using binary resolution and factoring. Then M is a marking of $\mathcal{F}'$.*

Thus, markable $\mathcal{SHI}$ ontologies admit a (possibly exponential) rewriting.

Theorem 3.11. *Let $\mathcal{O}$ be a $\mathcal{SHI}$ ontology and let M be a marking of $\mathcal{O}$. Then M is a marking of $\mathsf{DD}(\mathcal{O})$ and $\Xi_M(\mathsf{DD}(\mathcal{O}))$ is a datalog rewriting of $\mathcal{O}$ (where $\mathsf{DD}(\mathcal{O})$ is defined as in [5]).*

Corollary 3.12. *Checking $\mathcal{O} \cup \mathcal{D} \models \alpha$, for $\mathcal{O}$ a markable $\mathcal{SHI}$ ontology is PTIME-complete w.r.t. data and in EXPTIME w.r.t. combined complexity.*

Procedure 1. Compile-Horn

Input: $\mathcal{S}$: set of clauses
Output: $\mathcal{S}_H$: set of Horn clauses

 1: $\mathcal{S}_H := \{C \in \mathcal{S} \mid C$ is a Horn clause and not a tautology$\}$
 2: $\mathcal{S}_{\overline{H}} := \{C \in \mathcal{S} \mid C$ is a non-Horn clause and not a tautology$\}$
 3: **repeat**
 4:    $\mathcal{F} :=$ factors of each $C_1 \in \mathcal{S}_{\overline{H}}$ non-redundant in $\mathcal{S}_H \cup \mathcal{S}_{\overline{H}}$
 5:    $\mathcal{R} :=$ resolvents of each $C_1 \in \mathcal{S}_{\overline{H}}$ and $C_2 \in \mathcal{S}_{\overline{H}} \cup \mathcal{S}_H$ not redundant in $\mathcal{S}_H \cup \mathcal{S}_{\overline{H}}$
 6:    **for each** $C \in \mathcal{F} \cup \mathcal{R}$ **do**
 7:        $C' :=$ the condensation of C
 8:        Delete from $\mathcal{S}_H$ and $\mathcal{S}_{\overline{H}}$ all clauses θ-subsumed by C'
 9:        **if** C' is Horn **then** $\mathcal{S}_H := \mathcal{S}_H \cup \{C'\}$
10:        **else** $\mathcal{S}_{\overline{H}} := \mathcal{S}_{\overline{H}} \cup \{C'\}$
11: **until** $\mathcal{F} \cup \mathcal{R} = \emptyset$
12: **return** $\mathcal{S}_H$

---

4 Resolution-Based Rewritings

Resolution provides an alternative technique for rewriting disjunctive programs into datalog [5]. Procedure 1 saturates the input program $\mathcal{P}$ under binary resolution and positive factoring, with the restriction that two Horn clauses are never resolved together. The procedure is compatible with redundancy elimination techniques such as tautology elimination, subsumption and condensation. If it terminates, the procedure returns the subset of Horn clauses (equivalently, datalog rules) in the saturation, which is guaranteed to be a rewriting of $\mathcal{P}$.

We show that the separation between disjunctive and datalog predicates (Definition 3.1) can be exploited to refine this procedure. The idea is to further refine resolution by ensuring that the resolved atoms involve a disjunctive predicate.

Definition 4.1. Compile-Horn-Restricted *is obtained from Procedure 1 by adding to the definition of* $\mathcal{R}$ *in step 5 the additional restriction that the predicate in the atoms being resolved must be disjunctive in* $\mathcal{S}$.

Correctness of Compile-Horn-Restricted relies on the observation that resolutions on datalog predicates can always be delegated to the datalog reasoner and hence do not have to be performed as part of the rewriting process.

Theorem 4.2. *If* Compile-Horn-Restricted *terminates on a disjunctive program* $\mathcal{P}$ *with a program* $\mathcal{P}'$, *then* $\mathcal{P}'$ *is a datalog rewriting of* $\mathcal{P}$.

The class of disjunctive programs over which Compile-Horn-Restricted terminates is incomparable with the class of markable programs. Moreover, the rewritings produced by both approaches are quite different. Markable programs lead to polynomial rewritings, in which the arity of predicates is increased; rewritings computed via resolution can be much larger, but since all the datalog rules in the rewriting are logically entailed by the original program, the arity of predicates stays the same. In Section 6 we will discuss practical implications.

Rewriting Ontologies. The procedure Compile-Horn was shown to terminate for a class of programs called *simple* [5]; furthermore, DL-Lite$_{bool}^{\mathcal{H},+}$ ontologies are transformed into disjunctive programs that satisfy the simplicity condition using the algorithm by Hustadt, Motik and Sattler [8]. We now extend this result by devising a sufficient condition for datalog rewritability of $\mathcal{SHI}$ ontologies via Compile-Horn-Restricted. Since transitivity axioms can be eliminated from $\mathcal{SHI}$ ontologies by a polynomial transformation while preserving fact entailment (see [8,5]), it suffices to formulate our condition for $\mathcal{ALCHI}$.[4] First, we adapt the notion of simple rules in [5] as follows.

Definition 4.3. *An axiom of the form* $\exists R.A \sqsubseteq B$ *is* simple w.r.t. *a set of predicates* S *(or* S-simple*) if* $A \notin S$. *An ontology* $\mathcal{O}$ *is* S-simple *if so is every axiom of the form* $\exists R.A \sqsubseteq B$ *in* $\mathcal{O}$.

Note that ontology $\mathcal{O}_1$ from Example 3.9 is not simple w.r.t. its disjunctive predicates due to axiom $\exists married.Person \sqsubseteq Person$. If, however, we replace this axiom with $Man \sqcap Woman \rightarrow \bot$, we obtain a simple ontology, which in turn is no longer markable. The following theorem then generalises the result in [5] to a sufficient condition for datalog rewritability of $\mathcal{ALCHI}$ ontologies.

Theorem 4.4. *Let* $\mathcal{O}$ *be an* $\mathcal{ALCHI}$ *ontology that is simple w.r.t. its disjunctive predicates. Then* Compile-Horn-Restricted *terminates on* $\mathsf{DD}(\mathcal{O})$ *with a datalog rewriting of* $\mathcal{O}$.

5 Conjunctive Queries

By the results in [12], disjunctive programs cannot be rewritten to datalog in a query-independent way while preserving answers to CQs. Nonetheless, rewriting techniques for atomic queries can still be used to answer specific queries, which can be appended to the program as additional rules.

Rewriting CQs Using Markability. This observation immediately suggests how our markability condition in Section 3 can be applied to rewriting CQs.

Proposition 5.1. *Let* $\mathcal{P}$ *be a disjunctive program, let* M *be a marking of* $\mathcal{P}$, *and let* q *be a CQ with at most one atom marked by* M. *Then,* $\Xi_M(\mathcal{P} \cup \{q\})$ *is a rewriting of* q *w.r.t.* $\mathcal{P}$.

Indeed, M constitutes a marking of $\mathcal{P} \cup \{q\}$ if and only if q contains at most one body atom marked by M. From this, we obtain the following result, which applies equally to disjunctive programs and $RL^{\sqcup}$ ontologies.

---

[4] Note that neither Compile-Horn nor Compile-Horn-Restricted are well-suited for dealing with (axiomatised) equality. Both will diverge on every disjunctive program with equality due to the congruence axioms $P(x) \wedge x \approx y \rightarrow P(y)$ with P disjunctive.

Proposition 5.2. *Let $\mathcal{F}$ be a disjunctive program (or an $RL^\sqcup$ ontology), let $\mathbf{M}$ be the set of all (minimal) markings of $\mathcal{F}$, and let q be a (Boolean) CQ. If there is some $M \in \mathbf{M}$ that marks at most one atom of q, then answering the (fixed) q w.r.t. (fixed) $\mathcal{F}$ and an arbitrary dataset is a tractable problem.*

Example 5.3. Consider the following $RL^\sqcup$ ontology $\mathcal{O}$ and query q:[5]

$$\mathcal{O} = \{A \sqsubseteq B \sqcup C\}$$
$$q = R(x,y) \wedge R(y,z_1) \wedge R(y,z_2) \wedge B(z_1) \wedge C(z_2) \rightarrow A_q(x)$$

The empty ontology is a rewriting of $\mathcal{O}$, which can be determined using markability or resolution. Indeed, for every dataset $\mathcal{D}$ and fact α we have $\mathcal{O} \cup \mathcal{D} \models \alpha$ iff $\mathcal{D} \models \alpha$. The empty ontology, however, is not a rewriting of q, as witnessed by the following dataset $\mathcal{D}$, for which $\mathcal{O} \cup \mathcal{D} \cup \{q\} \models A_q(a)$ but $\mathcal{D} \cup \{q\} \not\models A_q(a)$:

$$\{R(a,b_1), R(a,b_2), R(b_1,c_1), R(b_1,c_2), R(b_2,c_2), R(b_2,c_3), B(c_1), A(c_2), C(c_3)\}$$

Clearly, $M = \{B\}$ is a marking of $\mathcal{O}$, and q contains one marked atom. Then $\mathcal{P} = \Xi_M(\mathcal{O} \cup \{q\})$ has the following rules, with $X \in \{B, A_q\}$ and $Y \in \{B, C, A_q\}$:

$$A(x) \wedge \overline{B}^Y(x,y) \rightarrow C^Y(x,y) \tag{11}$$

$$\overline{A}_q^Y(x,u) \wedge R(x,y) \wedge R(y,z_1) \wedge R(y,z_2) \wedge C^Y(z_2,u) \rightarrow \overline{B}^Y(z_1,u) \tag{12}$$

$$\top(x) \rightarrow \overline{X}^X(x,x) \tag{13}$$

$$X(x) \wedge \overline{X}^Y(x,y) \rightarrow Y(y) \tag{14}$$

$$\top(y) \wedge C(x) \rightarrow C^Y(x,y) \tag{15}$$

$$C^C(x,x) \rightarrow C(x) \tag{16}$$

Figure 2 shows a derivation of $A_q(a)$ from $\mathcal{P} \cup \mathcal{D}$.

Although this approach is immediately applicable to disjunctive programs and hence to $RL^\sqcup$ ontologies, it only transfers to $\mathcal{SHI}(\mathcal{Q})$ ontologies if q corresponds to a normalised $\mathcal{SHI}(\mathcal{Q})$ axiom. The reduction in [8,5] from $\mathcal{SHI}(Q)$ to disjunctive datalog is only complete for inputs equivalent to $\mathcal{SHIQ}$ ontologies.

Rewriting CQs via Resolution. The resolution-based approach naturally extends to a class of CQs satisfying certain conditions closely related to simplicity.

Definition 5.4. *Let S be a set of unary and binary predicates. A CQ q is S-simple if for some variable x in q all of the following conditions are satisfied:*

1. *if q is not Boolean, then $A_q(x)$ is the head atom of q;*
2. *Every S-atom (i.e., atom whose predicate is in S) in q is of the form $B(x)$, $R(x,x)$, $S(x,y)$, or $T(y,x)$; and*

---

[5] This example is based on a personal communication with Carsten Lutz.

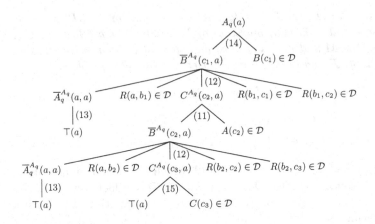

Fig. 2. Derivation of $A_q(a)$ from $\Xi_M(\mathcal{O} \cup \{q\}) \cup \mathcal{D}$ in Example 5.3

3. *every variable $y \neq x$ occurs in at most one S-atom in q.*

Example 5.5. Consider the following $RL^\sqcup$ ontology $\mathcal{O}$ and queries q_1, q_2:

$$\mathcal{O} = \{\mathsf{Person} \sqsubseteq \mathsf{Man} \sqcup \mathsf{Woman}, \exists\mathsf{married}.\mathsf{Person} \sqsubseteq \mathsf{Person}\}$$
$$q_1 = \mathsf{Man}(x) \wedge \mathsf{married}(x, y) \to A_{q_1}(x)$$
$$q_2 = \mathsf{Man}(x) \wedge \mathsf{married}(x, y) \wedge \mathsf{Woman}(y) \to A_{q_2}(x)$$

Ontology $\mathcal{O}$ is simple w.r.t. the set $S = \{\mathsf{Man}, \mathsf{Woman}\}$ of the disjunctive predicates in $\mathcal{O}$. Query q_1 is S-simple while q_2 is not. It is straightforward to verify that Compile-Horn-Restricted terminates on $\mathcal{O} \cup \{q_1\}$ but not on $\mathcal{O} \cup \{q_2\}$.

Theorem 5.6. *Let $\mathcal{O}$ be an $RL^\sqcup$ ontology that is simple w.r.t. the set S of the disjunctive predicates in $\mathcal{O}$. Then Procedure* Compile-Horn-Restricted *terminates on $\mathcal{O} \cup \{q\}$ with a datalog rewriting of q w.r.t. $\mathcal{O}$ for every S-simple CQ q.*

Consequently, answering any (fixed) CQ q over any (fixed) ontology $\mathcal{O}$ satisfying the conditions of Theorem 5.6 is a tractable problem.

6   Evaluation

Rewritability Experiments. We have evaluated whether realistic ontologies can be rewritten into datalog using our approaches. We analysed 118 ontologies that use disjunctive constructs from BioPortal, the Protégé library, and the corpus in [7]. To transform ontologies into disjunctive datalog we used KAON2 [14], which succeeded to compute disjunctive programs for 103 ontologies.[6] Out of the

---

[6] We doctored the ontologies to remove constructs outside $\mathcal{SHIQ}$. The modified ontologies can be found on
https://krr-nas.cs.ox.ac.uk/2014/RR/ontologies.tar.bz2

Table 2. Average times for answering UOBM's 15 standard queries

	Linearity			Resolution		HermiT		Pellet	
	dlog	disj	err	all	err	all	err	all	err
U01	<1s	12s	1	<1s	1	47s	0	147s	5
U04	<1s	87s	1	1s	1	57s	1	—	—
U07	<1s	168s	2	2s	1	122s	1	—	—
U10	<1s	53s	5	3s	1	196s	1	—	—

103 disjunctive programs, 32 were WL, and 35 were markable. Furthermore, 26 programs could be rewritten using Compile-Horn, and 27 could be rewritten using Compile-Horn-Restricted.[7] In both cases, the average time for computing a rewriting was below 1s (where the average is taken over the successful runs). Despite the potentially exponential blowup, the increase in program size was modest in practice: 49% for Compile-Horn and 34% for Compile-Horn-Restricted on average w.r.t. the number of rules.

Many of the programs obtained by KAON2 contained equality, and hence could not be rewritten by means of resolution (see Section 4). Hence, we additionally considered simplified versions of the 103 programs where we removed all rules containing equality. Out of these, 33 turned out to be WL, and 36 were markable; as expected the effect of equality on linearity-based approaches is rather minor. In contrast, resolution-based approaches were significantly more effective than before: Compile-Horn succeeded in 39 cases, and Compile-Horn-Restricted in 41 cases. Again, computing a successful rewriting took less that 1s on average in both cases. The increase in program size was even smaller than before: 16% for Compile-Horn and 6% for Compile-Horn-Restricted on average.

Both Compile-Horn and Compile-Horn-Restricted succeeded on some ontologies that were not simple w.r.t. disjunctive predicates. At the same time, being worst-case exponential, both algorithms failed to rewrite (within 1h) one simple ontology and two that were simple w.r.t. disjunctive predicates.

The intersection between the programs rewritable using markability and resolution turned out to be quite large: in the general case, there were 16 programs that could be rewritten by only one approach, and in the equality-free case only 5. Still, taken together, the two approaches succeeded to rewrite 39 programs (38%) in the general case and 41 programs (40%) in the equality-free case. Moreover, on average, 73% of the predicates were datalog, and so could be queried using a datalog engine even if the disjunctive program was not rewritable. Finally, we found that 20 out of the 103 ontologies were $RL^\sqcup$, out of which 17 were markable. Of the remaining three, two could be rewritten via resolution.

CQ Answering. We have also tested scalability of CQ answering over the UOBM benchmark [13]. We considered the $RL^\sqcup$ subset of UOBM without

---

[7] We ran the rewritability experiments on a laptop with a 2.5GHz Intel Core i5 processor and 8GB RAM, and set a timeout of 1h per ontology.

equality,[8] and generated datasets for 1 to 10 universities (denoted as U01-U10). Furthermore, we considered the 15 standard queries in the benchmark. While not markable, our test ontology can be converted to a markable (in fact, WL) program by the algorithm in [10]. Moreover, it is rewritable using Compile-Horn-Restricted (but not using Compile-Horn). We used RDFox as a datalog engine, and measured performance against HermiT [17] and Pellet [19]. We used a server with two Intel Xeon E5-2643 processors and 128GB RAM. Systems were compared on individual queries with a timeout of 10min per query. We ran RDFox on 16 threads. Table 2 shows average query answering times, and number of queries on which a system failed.[9] The time spent on computing the rewritings[10] is not included into the query answering times since query rewriting can be done in a data-independent way.

Pellet could only answer queries on U01. It timed out on 5 queries, and was much slower on the remaining queries than the other systems.

Using the linearity-based approach we could answer queries for all datasets. From the 15 test queries, 7 were disjunctive (i.e., contained at least one disjunctive atom), and 8 were datalog. One disjunctive query could not be rewritten. Datalog queries were answered instantaneously (<1s) for all datasets. Disjunctive queries were much harder, and performance on those was comparable to HermiT. Memory-outs were encountered for U07 (1 query) and U10 (4 queries). For all rewritable queries, computing the rewriting (including the conversion to a WL program) took less that 1s.

The resolution-based approach was clearly superior to the others. Only one query could not be rewritten, and all the remaining queries could be answered almost instantaneously even for the largest dataset (query rewriting itself took 26s on average). In contrast to the linearity-based approach, rewritings obtained by resolution introduce no predicates of higher arity, and thus lead to smaller materialisations. Also, there was no significant difference in query answering times for datalog and disjunctive queries. Once again, the increase in program size was modest (4.6 times on average). Notably, computing the rewritings took 26s on average—considerably longer than with the linearity-based approach.

7   Conclusion

We have proposed enhanced techniques for rewriting disjunctive datalog programs and DL ontologies into plain datalog programs. Our techniques enable the use of scalable datalog engines for data reasoning, and our experiments suggest practical feasibility of our approach. In the near future, we are planning to extend our results for CQ answering to capture larger classes of queries.

Acknowledgements. This work was supported by the Royal Society, the EPSRC projects Score!, ExODA, and MaSI[3], and the FP7 project OPTIQUE.

---

[8] Equality makes the resolution-based approach non-applicable.
[9] Average times do not reflect queries on which a system failed.
[10] Query rewriting was performed with a 1h timeout per query.

References

1. Artale, A., Calvanese, D., Kontchakov, R., Zakharyaschev, M.: The DL-Lite family and relations. J. Artif. Intell. Res. 36, 1–69 (2009)
2. Bachmair, L., Ganzinger, H.: Resolution theorem proving. In: Handbook of Automated Reasoning, pp. 19–99. Elsevier and MIT Press (2001)
3. Bienvenu, M., ten Cate, B., Lutz, C., Wolter, F.: Ontology-based data access: A study through disjunctive datalog, CSP, and MMSNP. In: PODS, pp. 213–224 (2013) arXiv:1301.6479
4. Bishop, B., Kiryakov, A., Ognyanoff, D., Peikov, I., Tashev, Z., Velkov, R.: OWLIM: A family of scalable semantic repositories. Semantic Web 2(1), 33–42 (2011)
5. Cuenca Grau, B., Motik, B., Stoilos, G., Horrocks, I.: Computing datalog rewritings beyond Horn ontologies. In: IJCAI, pp. 832–838 (2013) arXiv:1304.1402
6. Eiter, T., Ortiz, M., Šimkus, M., Tran, T.K., Xiao, G.: Query rewriting for Horn-SHIQ plus rules. In: AAAI, pp. 726–733 (2012)
7. Gardiner, T., Tsarkov, D., Horrocks, I.: Framework for an automated comparison of description logic reasoners. In: Cruz, I., Decker, S., Allemang, D., Preist, C., Schwabe, D., Mika, P., Uschold, M., Aroyo, L.M. (eds.) ISWC 2006. LNCS, vol. 4273, pp. 654–667. Springer, Heidelberg (2006)
8. Hustadt, U., Motik, B., Sattler, U.: Reasoning in description logics by a reduction to disjunctive datalog. J. Autom. Reasoning 39(3), 351–384 (2007)
9. Kaminski, M., Cuenca Grau, B.: Sufficient conditions for first-order and datalog rewritability in $\mathcal{ELU}$. In: DL, pp. 271–293 (2013)
10. Kaminski, M., Nenov, Y., Cuenca Grau, B.: Datalog rewritability of disjunctive datalog programs and its applications to ontology reasoning. In: AAAI (2014) arXiv:1404.3141
11. Krisnadhi, A., Lutz, C.: Data complexity in the $\mathcal{EL}$ family of description logics. In: Dershowitz, N., Voronkov, A. (eds.) LPAR 2007. LNCS (LNAI), vol. 4790, pp. 333–347. Springer, Heidelberg (2007)
12. Lutz, C., Wolter, F.: Non-uniform data complexity of query answering in description logics. In: KR, pp. 297–307 (2012)
13. Ma, L., Yang, Y., Qiu, Z., Xie, G.T., Pan, Y., Liu, S.: Towards a complete OWL ontology benchmark. In: Sure, Y., Domingue, J. (eds.) ESWC 2006. LNCS, vol. 4011, pp. 125–139. Springer, Heidelberg (2006)
14. Motik, B.: Reasoning in Description Logics using Resolution and Deductive Databases. Ph.D. thesis, Univesität Karlsruhe (TH), Karlsruhe, Germany (2006)
15. Motik, B., Cuenca Grau, B., Horrocks, I., Wu, Z., Fokoue, A., Lutz, C.: OWL 2 web ontology language profiles. W3C Recommendation (2009)
16. Motik, B., Nenov, Y., Piro, R., Horrocks, I., Olteanu, D.: Parallel materialisation of datalog programs in centralised, main-memory RDF systems. In: AAAI (2014)
17. Motik, B., Shearer, R., Horrocks, I.: Hypertableau reasoning for description logics. J. Artif. Intell. Res. 36, 165–228 (2009)
18. Pérez-Urbina, H., Motik, B., Horrocks, I.: Tractable query answering and rewriting under description logic constraints. J. Appl. Log. 8(2), 186–209 (2010)
19. Sirin, E., Parsia, B., Cuenca Grau, B., Kalyanpur, A., Katz, Y.: Pellet: A practical OWL-DL reasoner. J. Web Sem. 5(2), 51–53 (2007)
20. Trivela, D., Stoilos, G., Chortaras, A., Stamou, G.B.: Optimising resolution-based rewriting algorithms for DL ontologies. In: DL, pp. 464–476 (2013)
21. Wu, Z., Eadon, G., Das, S., Chong, E.I., Kolovski, V., Annamalai, M., Srinivasan, J.: Implementing an inference engine for RDFS/OWL constructs and user-defined rules in Oracle. In: ICDE, pp. 1239–1248 (2008)

Querying Temporal Databases via OWL 2 QL

Szymon Klarman and Thomas Meyer

Centre for Artificial Intelligence Research,
CSIR Meraka and University of KwaZulu-Natal, South Africa
{sklarman,tmeyer}@csir.co.za

Abstract. SQL:2011, the most recently adopted version of the SQL query language, has unprecedentedly standardized the representation of temporal data in relational databases. Following the successful paradigm of ontology-based data access, we develop a practical approach to querying the SQL:2011-based temporal data model via the semantic layer of OWL 2 QL. The interval-based *temporal query language* (TQL), which we propose for this task, is based on naturally characterizable combinations of temporal logic with conjunctive queries. As the central contribution, we present rules for sound and complete rewriting of TQL queries into two-sorted first-order logic, and consequently, into corresponding SQL queries, which can be evaluated in any existing relational database management system compliant with the SQL:2011 temporal data model. Importantly, the proposed rewriting is based on the direct reuse of the standard rewriting techniques for conjunctive queries under OWL 2 QL. This renders our approach modular and easily implementable. As a notable corollary, we show that the data complexity of TQL query answering remains in AC^0, i.e., as in the usual, non-temporal case.

1 Introduction

The ability to manage the temporal aspects of information is critical for a variety of applications. One natural and prevailing scenario is that of representing and querying the *validity time* of data, i.e., the time during which data is deemed true about the application domain. The significance of this task is particularly visible in the area of semantic technologies, where the systematically growing number of proposed solutions, building on different levels of the Semantic Web architecture and differing in the flavour and depth of temporal reasoning they support, aim at addressing essentially the same problem [15,14,5,22,7,2]. A very similar proliferation of proposals was witnessed in the 1990s in the field of temporal databases. Intensive attempts to extend the traditional relational data model and SQL with temporal features inspired then a large body of candidate specifications, including such extensions as TSQL2, SQL3 or SQL/Temporal [24], which eventually failed to be adopted by the database community due to the persistent lack of consensus as to the preferred approach. Only very recently, that discussion has been picked up again and a compromise temporal extension has eventually found its way into SQL:2011 [20] — the newest standardization

R. Kontchakov and M.-L. Mugnier (Eds.): RR 2014, LNCS 8741, pp. 92–107, 2014.
© Springer International Publishing Switzerland 2014

of the SQL query language. This unprecedented circumstance offers an interesting opportunity to address the problem of reasoning with temporal semantic data from yet another angle, namely, by relating it via known links between relational databases and semantic technologies to its analogue in the database world, thus using the SQL:2011 standard as a leverage for the solution. In this paper, we contribute precisely to this research agenda by proposing a novel, temporal extension to the framework of ontology-based data access.

Ontology-based data access (OBDA) is a popular paradigm of managing information, which combines the data storage and querying capabilities offered by relational database management systems (RDBMSs) with the semantically enhanced view on the data provided by ontologies. The ontology language OWL 2 QL, based on the DL-*Lite* family of Description Logics, is a profile OWL 2 designed specifically to support optimally balanced OBDA. In a nutshell, conjunctive queries, posed over data under an OWL 2 QL ontology, can be rewritten into first-order logic using the ontology's axioms, then translated to SQL and answered within an RDBMS, in such a way as if the ontology was mediating in the process [8]. As large portions of data available through the Web are in fact still hosted in relational datastores, OBDA provides a crucial channel for accessing this data from the level of the Semantic Web applications.

In this work, we establish an analogical OBDA interface between the semantic layer of OWL 2 QL and temporal data model endorsed by SQL:2011. The interval-based *temporal query language* (TQL), which we propose for this task, is based on naturally characterizable combinations of temporal logic with conjunctive queries, identified in [16]. TQL is tailored specifically to offer maximum expressivity while preserving the possibility of reuse central to OBDA first-order rewriting techniques and tools, developed specifically for the use with OWL 2 QL. While this technical compliance warrants the minimal implementation overhead for our approach, its well-defined logic foundations allow us to identify basic formal properties of TQL. In particular, we are able to demonstrate that under the finite time domain assumption the data complexity of query entailment remains in AC^0, i.e., as in the case of standard (non-temporal) conjunctive queries, even though the combined complexity increases to PSPACE-complete. As the main contribution, we develop a rewriting of TQL queries in the presence of OWL 2 QL ontologies into two-sorted first-order logic, and consequently to SQL, which opens the way to efficient query answering by means of existing, commercially supported RDBMSs, such as IBM DB2 10.1, Oracle Database 11g Workspace Manager, or Teradata — all of which have by now adopted certain variants of the SQL:2011 standard[1].

The paper is organized as follows. In the next section we lay down the preliminaries. In Section 3, we introduce TQL and define the query entailment problem. In Section 4, we present the TQL query rewriting rules and, in Section 5, we study the formal properties of query answering queries via this rewriting. The related work is discussed in Section 6 and the paper is concluded in Section 7. Some proofs are included in the online technical report [19].

---

[1] See http://www.cs.arizona.edu/~rts/sql3.html for an overview.

2   Basic Notions

We start by recapping the logic foundations of OBDA. Then we motivate and formally introduce the temporal extension of OBDA studied in this paper.

2.1   Ontology-Based Data Access

OWL 2 QL is a profile of OWL 2 based on the DL-*Lite* family of Description Logics (DLs) [8]. A DL vocabulary $\Sigma = (\mathsf{N_I}, \mathsf{N_C}, \mathsf{N_R})$ consists of countably infinite sets of individual names ($\mathsf{N_I}$), concept names ($\mathsf{N_C}$) and role names ($\mathsf{N_R}$). An ABox $\mathcal{A}$ is a finite set of assertions of type $A(a)$ and $r(a, b)$, for $a, b \in \mathsf{N_I}$, $A \in \mathsf{N_C}$ and $r \in \mathsf{N_R}$, which we also generically denote with $\alpha(a)$. A TBox $\mathcal{T}$ is a finite set of concept inclusions $B \sqsubseteq C$ and role inclusions $r \sqsubseteq s$, where B, C and r, s are possibly complex concepts and roles, respectively, built using logical constructors allowed in OWL 2 QL, such as $\exists r.\top, A \sqcap B, r^-$, whose particulars are not of importance for this work.[2] The semantics is given in terms of DL interpretations $\mathcal{I} = (\Delta^\mathcal{I}, \cdot^\mathcal{I})$, defined as usual [4]. An interpretation $\mathcal{I}$ is a model of $\mathcal{T}$ and $\mathcal{A}$, denoted as $\mathcal{I} \models \mathcal{T}, \mathcal{A}$, iff it satisfies every axiom in $\mathcal{T}$ and $\mathcal{A}$.

In OBDA, the instance data, represented as an ABox, is accessed via an ontology, given as a TBox, using a designated query language such as, most commonly considered in that context, the language of conjunctive queries [13]. Let $\mathsf{N_V}$ be a countably infinite set of variables. A *conjunctive query* (CQ) over a DL vocabulary Σ is a first-order formula:

$$q(\boldsymbol{y}) = \exists \boldsymbol{x}.(\bigwedge_{1 \leq j \leq n} \alpha_j(\boldsymbol{d_j}))$$

where $\boldsymbol{y}$ and $\boldsymbol{x}$ are sequences of, respectively, free and existentially bounded variables occurring in $q(\boldsymbol{y})$ and every atom $\alpha_j(\boldsymbol{d_j})$ is of the form $A(d)$ or $r(d_1, d_2)$, where $A \in \mathsf{N_C}$, $r \in \mathsf{N_R}$, and $d, d_1, d_2 \in \mathsf{N_I} \cup \mathsf{N_V}$. Whenever it is not confusing, we sometimes also abbreviate $q(\boldsymbol{y})$ to q. By $\mathsf{term}(q)$ we denote the set of all terms occurring in q and by $\mathsf{obj}(q)$ the set of all free variables. We call q grounded whenever $\mathsf{obj}(q) = \emptyset$. A grounded CQ q is satisfied in $\mathcal{I}$ iff there exists a mapping $\mu : \mathsf{term}(q) \mapsto \Delta^\mathcal{I}$, with $\mu(d) = d^\mathcal{I}$ for every $d \in \mathsf{N_I}$, such that $\mu(d) \in A^\mathcal{I}$ and $(\mu(d_1), \mu(d_2)) \in r^\mathcal{I}$ for every $A(d)$ and $r(d_1, d_2)$ in q. Further, we say that q is *entailed* by $\mathcal{T}, \mathcal{A}$, denoted as $\mathcal{T}, \mathcal{A} \models q$ iff q is satisfied in every model of $\mathcal{T}, \mathcal{A}$. Whenever $\emptyset, \mathcal{A} \models q$ we also write $\mathcal{A} \models q$. An *answer* to q is a mapping $\sigma : \mathsf{obj}(q) \mapsto \mathsf{N_I}$. By $\sigma(q)$ we denote the result of uniformly substituting every occurrence of x in q with $\sigma(x)$, for every $x \in \mathsf{obj}(q)$. An answer σ is called *certain* over $\mathcal{T}$, $\mathcal{A}$ iff $\sigma(q)$ is entailed by $\mathcal{T}, \mathcal{A}$.

A prominent property of CQs is their *first-order* (FO) *rewritability* in OWL 2 QL, formally defined as follows.

---

[2] See http://www.w3.org/TR/owl2-profiles/#OWL_2_QL for full details.

Definition 1 (FO Rewritability [8]). *For every CQ q and a TBox $\mathcal{T}$, there exists a FO formula $q^{\mathcal{T}}$, called the* FO *rewriting of q in $\mathcal{T}$, such that for every ABox $\mathcal{A}$ and answer σ to q, it holds that:*

$$\mathcal{T}, \mathcal{A} \models \sigma(q) \quad \text{iff} \quad db(\mathcal{A}) \Vdash \sigma(q^{\mathcal{T}}),$$

where $\Vdash$ is the FO satisfaction relation and $db(\mathcal{A})$ denotes $\mathcal{A}$ considered as a database/FO interpretation, i.e., a structure $(\mathsf{N}_\mathsf{I}, \cdot^{\mathcal{D}})$, where N_I is the data domain and $\cdot^{\mathcal{D}}$ is an interpretation function defined as $\alpha^{\mathcal{D}} = \{\boldsymbol{a} \mid \alpha(\boldsymbol{a}) \in \mathcal{A}\}$, for every $\alpha \in \mathsf{N}_\mathsf{C} \cup \mathsf{N}_\mathsf{R}$.

By the standard techniques the FO rewriting of q in $\mathcal{T}$ is a union of conjunctive queries, i.e., a formula $q^{\mathcal{T}}(\boldsymbol{y}) = \bigvee\limits_{1 \leq i \leq m} q_i(\boldsymbol{y})$, where every $q_i(\boldsymbol{y})$ is a CQ. FO rewritability is particularly significant from the practical perspective. It implies that answering CQs in OWL 2 QL can be effectively performed in existing RDBMSs via a translation to SQL, as the ontological component in the task can be always compiled out in the query rewritting, without loss of soundness or completeness. As a theoretical corollary, it follows also that the data complexity of query answering in this setup is AC^0, as in first-order logic (FOL).

In this work, we focus exclusively on OWL 2 QL TBoxes, even though some of the presented results should clearly transfer to other fragments of DLs warranting the FO rewritability property for CQs. Without always stating it explicitly, we assume that every TBox or ontology mentioned in the remainder of this paper is expressed in OWL 2 QL.

2.2 Ontology-Based Access to Temporal Data

In this paper, we study ontology-based access to temporal data, in the sense of an extension to the OBDA paradigm whose prototypical application could be illustrated with the following scenario.

Consider a temporal database (TDB) presented in Table 1, with columns `from` and `to` marking the limits of the validity periods of the respective records. Such databases can be naturally mapped to (virtual) temporal ABoxes, such as given in Table 2. Our goal is to define a dedicated language for querying temporal ABoxes, which would combine support for two essential functionalities: representation of temporal constraints over the validity periods of data and semantically enhanced access to that data via ontologies. For instance, given the ontology $\mathcal{T} = \{Emp \sqsubseteq Person, department \sqsubseteq worksAt, location \sqsubseteq basedIn\}$, the language should be able to support queries such as:

(**Q**) *Find all persons X and times Y, such that X worked in a department based in Barcelona during Y and in a department based in Madrid some time earlier.*

The expected set of answers should then include $e1$ as X with the associated period $[1999, 2000]$ as Y. The practical rationale behind ontology-based access to temporal data defined in this way is to eventually enable such queries to be translated to SQL and answered within existing RDBMSs.

Table 1. SQL:2011 temporal database

Emp				
id	name	department	from	to
e1	john	d1	1998	2000
e1	john	d3	2000	2003
e2	mark	d2	1999	2002

Dep				
id	type	location	from	to
d1	financial	madrid	1998	1999
d1	financial	barcelona	1999	2003
d2	hr	barcelona	2000	2003
d3	hq	london	2000	2003

Table 2. Temporal ABoxes corresponding to the temporal relations in Table 1

$[1998, 2000] : Emp(e1)$
$[1998, 2000] : name(e1, john)$
$[1998, 2000] : department(e1, d1)$
...

$[1998, 1999] : Dep(d1)$
$[1998, 1999] : type(d1, financial)$
$[1998, 1999] : location(d1, madrid)$
...

Formally, the temporal data model under consideration is grounded in the point- and (derived) interval-based time domains.

Definition 2 (Time Domain, Time Intervals). *A* time domain *is a tuple* $\mathfrak{T} = (T, <)$, *where* T *is a nonempty set of elements called* time points *and* $<$ *is an irreflexive, linear ordering on* T. *A* time interval $\tau = [\tau^-, \tau^+]$ *over* $\mathfrak{T}$ *is a set of time points* $\{t \in T \mid \tau^- \leq t \leq \tau^+\}$, *where* $\tau^-, \tau^+ \in T$, *such that* $\tau^- \leq \tau^+$. *The points* τ^- *and* τ^+ *are called the* beginning *and the* end *of* τ.[3] *The set of all time intervals over* $\mathfrak{T}$ *is denoted by* I.

In the SQL:2011 standard, every temporal relation extends a non-temporal one with two additional attributes storing the beginning and the end time of the validity period of a given tuple [20] — exactly as in the example from Table 1. Such a model supports a representation of so-called concrete TDBs, i.e., finite syntactic encodings of temporal data. The actual meaning of these encodings is captured by possibly infinite abstract TDBs [12]. In the OBDA setting these two notions translate naturally into the corresponding types of ABoxes.

Definition 3 (Concrete and Abstract Temporal ABoxes). *A temporal assertion is an expression* $\tau : \alpha(\boldsymbol{a})$, *where* $\alpha(\boldsymbol{a})$ *is an ABox assertion and* $\tau \in I$, *stating that* $\alpha(\boldsymbol{a})$ *is valid in every time point in* τ. *A* concrete temporal ABox *(CTA)* $\mathfrak{A}$ *is a finite set of temporal assertions. For a concrete temporal ABox* $\mathfrak{A}$ *there exists a corresponding* abstract temporal ABox *(ATA), obtained by means of a mapping* $\| \cdot \|$, *such that* $\|\mathfrak{A}\| = (\mathfrak{A}_t)_{t \in T}$, *where* $\mathfrak{A}_t = \{\alpha(\boldsymbol{a}) \mid \tau : \alpha(\boldsymbol{a}) \in \mathfrak{A} \text{ and } t \in \tau\}$.

---

[3] Note that SQL:2011 adopts a closed-open semantics for the validity periods, i.e., for any $\tau \in I$ it holds that $\tau^- \in \tau$ and $\tau^+ \notin \tau$. This a technically insignificant difference which we omit here for clarity of presentation.

The link between concrete temporal ABoxes and SQL:2011 TDBs is defined in a strict analogy to the non-temporal OBDA.

Definition 4 (TDB). *Let $\mathfrak{A}$ be a CTA over the time domain $\mathfrak{T} = (T, <)$. By $tdb(\mathfrak{A})$ we denote $\mathfrak{A}$ considered as a* temporal database *(TDB) over the signature $\Gamma = \{R_\alpha \mid \alpha \in N_C \cup N_R\}$, i.e., the structure $(N_I, T, <, \cdot^\mathcal{D})$, where N_I is the data domain, $(T, <)$ is the time domain, and $\cdot^\mathcal{D}$ is the interpretation defined as $R_\alpha^\mathcal{D} = \{(a, \tau^-, \tau^+) \mid \tau : \alpha(a) \in \mathfrak{A}\}$, for every $\alpha \in N_C \cup N_R$.*

3 Temporal Query Language

The *Temporal Query Language* (TQL), presented in this section, is defined using a generic construction method for temporal query languages in DLs, explored also in [16,3,7,18]. TQL is a combination of a temporal language with CQs, obtained by substituting CQs for the atoms in temporal formulas. This design allows for very flexible interleaving of data queries with temporal constraints, while benefiting from the expressive power of both components. As the temporal language we use first-order monadic logic of orders, which is at least as expressive as most common linear temporal logics [23]. Two specific characteristics of TQL, which distinguish it from other similar proposals, are:

- the use of an interval-based variant of the temporal language, rather than a point-based, which enables direct querying of concrete TDBs, without requiring intermediate translations, such as studied in [25];
- the use of the epistemic semantics for embedding CQs in the temporal language, as suggested in [16], which renders the language more expressive and computationally well-behaved, as explained further.

By I_V we denote a countably infinite set of variables ranging over I.

Definition 5 (TQL). *Temporal query language (TQL) is the smallest set of formulas induced by the grammar:*

$$\psi ::= [q](u) \mid u^* < v^* \mid \neg\psi \mid \psi_1 \wedge \psi_2 \mid \exists y.\psi$$

where q is a CQ, $u, v \in I \cup I_V$, $y \in I_V$, and $* \in \{-, +\}$. An i-substitution is a mapping $\pi : I \cup I_V \mapsto I$ such that $\pi(\tau) = \tau$, for every $\tau \in I$. By $\mathrm{obj}(\psi)$ we denote the set of free individual variables and by $\mathrm{int}(\psi)$ the set of free interval variables in ψ. A TQL formula ψ is grounded iff $\mathrm{obj}(\psi) = \mathrm{int}(\psi) = \emptyset$. The satisfaction relation for TQL formulas, w.r.t. a TBox $\mathcal{T}$, a CTA $\mathfrak{A}$, and an i-substitution π, is defined inductively as follows:

$$(\dagger) \quad
\begin{array}{lll}
\mathcal{T}, \mathfrak{A}, \pi \models [q](u) & \text{iff} & \mathcal{T}, \mathfrak{A}_t \models q, \text{ for every } t \in \pi(u), \\
\mathcal{T}, \mathfrak{A}, \pi \models u^* < v^* & \text{iff} & \pi(u)^* < \pi(v)^*, \\
\mathcal{T}, \mathfrak{A}, \pi \models \neg\psi & \text{iff} & \mathcal{T}, \mathfrak{A}, \pi \not\models \psi, \\
\mathcal{T}, \mathfrak{A}, \pi \models \psi_1 \wedge \psi_2 & \text{iff} & \mathcal{T}, \mathfrak{A}, \pi \models \psi_1 \text{ and } \mathcal{T}, \mathfrak{A}, \pi \models \psi_2, \\
\mathcal{T}, \mathfrak{A}, \pi \models \exists y.\psi & \text{iff} & \text{there exists } \tau \in I, \text{ such that} \\
& & \mathcal{T}, \mathfrak{A}, \pi[y \mapsto \tau] \models \psi,
\end{array}$$

where $\pi[y \mapsto \tau]$ denotes the i-substitution exactly as π except for that we fix $\pi(y) = \tau$. We say that $\mathcal{T}, \mathfrak{A}$ entail a grounded TQL formula ψ, denoted as $\mathcal{T}, \mathfrak{A} \models \psi$, iff there exists an i-substitution π, such that $\mathcal{T}, \mathfrak{A}, \pi \models \psi$.

TQL formulas with free variables can naturally serve as queries over concrete temporal ABoxes. We refer to such formulas as *concrete TQL queries* (CTQs). As an example of a CTQ, consider a rephrasing of the query (**Q**) from Section 2.2:

$$\psi(x, y) := [\exists z.(Person(x) \wedge worksAt(x, z) \wedge basedIn(z, barcelona))](y) \wedge$$
$$\exists v.(v^+ < y^- \wedge [\exists z.(worksAt(x, z) \wedge basedIn(z, madrid))](v))$$

As one of its answers, $\psi(x, y)$ should return $\{x \mapsto e1, y \mapsto [1999, 2000]\}$. The certain answer semantics for such queries is defined as expected.

Definition 6 (CTQ Answering). *Let $\mathcal{T}$ be a TBox, $\mathfrak{A}$ a CTA and ψ a CTQ with free variables $\mathsf{obj}(\psi)$ and $\mathsf{int}(\psi)$. An answer to ψ is a mapping σ such that $\sigma : \mathsf{obj}(\psi) \mapsto \mathsf{N}_\mathsf{I}$ and $\sigma : \mathsf{int}(\psi) \mapsto I$. By $\sigma(\psi)$ we denote the result of uniformly substituting every occurrence of x in ψ with $\sigma(x)$, for every $x \in \mathsf{obj}(\psi) \cup \mathsf{int}(\psi)$. An answer σ is called* certain *over $\mathcal{T}$, $\mathfrak{A}$ iff $\mathcal{T}, \mathfrak{A} \models \sigma(\psi)$.*

The key to the design of TQL is condition (†), in Definition 5, which ensures the epistemic interpretation of the CQs embedded in TQL queries. The formula $[q](\tau)$, for a grounded CQ q and $\tau \in I$, reads as "*q is entailed in all time points in τ*". Analogically, $\neg[q](\tau)$ is interpreted as negation-as-failure: "*it is not true that q is entailed in all time points in τ*". This approach of combining FO-based query languages has been originally proposed by Calvanese et al. [9], giving rise to a family of lightweight query languages, which permit well-behaved, modular answering algorithms and support the use of negation (interpreted as negation-as-failure), which otherwise easily leads to undecidability in the context of querying DL ontologies.

As a consequence, condition (†) can be effectively replaced with its equivalent, which involves the standard FO rewritability techniques in the sense of Definition 1:

$$\mathcal{T}, \mathfrak{A}, \pi \models [q](u) \quad \text{iff} \quad db(\mathfrak{A}_t) \Vdash q^{\mathcal{T}}, \text{ for every } t \in \pi(u),$$

where $q^{\mathcal{T}}$ is an FO rewriting of q in $\mathcal{T}$. What it eventually means, is that all occurrences of CQs in a CTQ can be replaced with their FO rewritings, so that the ontology $\mathcal{T}$ can be dropped while the formula can be evaluated exclusively over the temporal ABox seen as a sequence of FO interpretations. However, such point-based rewriting strategy, also explored in [7], is highly impractical when applied over concrete TDBs, as it necessitates either unfolding a concrete TDB into an abstract one, or a further translation from point-based to an interval-based language. Instead, here we pursue a direct interval-based approach, which requires a more sophisticated rewriting technique, capable of handling the well-known problems of computing temporal joins and coalescing, as explained in detail in the following section.

4   Query Rewriting

We start by defining a two-sorted first-order language, tailored specifically for talking about temporal relations of TDBs.

Definition 7 (2FO). *The language of two-sorted first-order logic (2FO) over the signature $\Gamma = \{R_1, R_2, \ldots\}$ is the smallest set of formulas induced by the grammar:*

$$\varphi ::= R(d_1, \ldots, d_n, t_1, t_2) \mid \neg\varphi \mid \varphi_1 \wedge \varphi_2 \mid t_1 < t_2 \mid \exists x.\varphi \mid \exists y.\varphi$$

where $R \in \Gamma$, $d_1, \ldots, d_n \in \mathsf{N_I} \cup \mathsf{N_V}$, $t_1, t_2 \in T \cup \mathsf{T_V}$, $x \in \mathsf{N_V}$ and $y \in \mathsf{T_V}$. A d-substitution δ is a mapping $\delta : \mathsf{N_I} \cup \mathsf{N_V} \mapsto \mathsf{N_I}$, such that $\delta(a) = a$ for every $a \in \mathsf{N_I}$. A t-substitution ν is a mapping $\nu : T \cup \mathsf{T_V} \mapsto T$, such that $\nu(t) = t$ for every $t \in T$. A 2FO formula φ is called grounded whenever it does not have any free variables. The satisfaction relation for 2FO formulas, w.r.t. a TDB $tdb(\mathfrak{A}) = (\mathsf{N_I}, T, <, \cdot^{\mathcal{D}})$, a d-substitution δ and a t-substitution ν, is defined inductively as follows:

$$
\begin{aligned}
tdb(\mathfrak{A}), \delta, \nu \Vdash R(d_1, \ldots, d_n, t_1, t_2) \quad &\text{iff} \quad (\delta(d_1), \ldots, \delta(d_n), \nu(t_1), \nu(t_2)) \in R^{\mathcal{D}}, \\
tdb(\mathfrak{A}), \delta, \nu \Vdash \neg\varphi \quad &\text{iff} \quad tdb(\mathfrak{A}), \delta, \nu \not\Vdash \varphi, \\
tdb(\mathfrak{A}), \delta, \nu \Vdash \varphi_1 \wedge \varphi_2 \quad &\text{iff} \quad tdb(\mathfrak{A}), \delta, \nu \Vdash \varphi_1 \text{ and } tdb(\mathfrak{A}), \delta, \nu \Vdash \varphi_2, \\
tdb(\mathfrak{A}), \delta, \nu \Vdash t_1 < t_2 \quad &\text{iff} \quad \nu(t_1) < \nu(t_2), \\
tdb(\mathfrak{A}), \delta, \nu \Vdash \exists x.\varphi \quad &\text{iff} \quad \text{for } x \in \mathsf{N_V}, \text{ there exists } a \in \mathsf{N_I} \text{ such} \\
& \qquad \text{that } tdb(\mathfrak{A}), \delta[x \mapsto a], \nu \Vdash \varphi, \\
tdb(\mathfrak{A}), \delta, \nu \Vdash \exists y.\varphi \quad &\text{iff} \quad \text{for } y \in \mathsf{T_V}, \text{ there exists } t \in T \text{ such that} \\
& \qquad tdb(\mathfrak{A}), \delta, \nu[y \mapsto t] \Vdash \varphi,
\end{aligned}
$$

where $\delta[x \mapsto a]$ ($\nu[y \mapsto t]$) denotes the substitution exactly as δ (ν) except for that we fix $\delta(x) = a$ ($\nu(y) = t$). We say that a 2FO formula φ is satisfied in $tdb(\mathfrak{A})$, denoted as $tdb(\mathfrak{A}) \Vdash \varphi$, iff there exist d-/t-substitutions δ, ν, such that $tdb(\mathfrak{A}), \delta, \nu \Vdash \varphi$.

Next, we define the rules for rewriting CTQs into 2FO formulas.

Definition 8 (2FO Rewriting of CTQs). *The 2FO rewriting of a CTQ ψ is a formula $\lceil\psi\rceil^{2FO}$ obtained from ψ by applying the transformation $\lceil\cdot\rceil^{2FO}$, defined inductively as follows:*

$$
\begin{aligned}
\lceil [q(\boldsymbol{d})](u) \rceil^{2FO} &= \exists t_1, t_2.(R_{q\mathcal{T}}^{coal}(\boldsymbol{d}, t_1, t_2) \wedge t_1 \leq u^- \wedge u^+ \leq t_2), \\
\lceil u^* < v^* \rceil^{2FO} &= u^* < v^*, \\
\lceil \neg\psi \rceil^{2FO} &= \neg\lceil\psi\rceil^{2FO}, \\
\lceil \psi_1 \wedge \psi_2 \rceil^{2FO} &= \lceil\psi_1\rceil^{2FO} \wedge \lceil\psi_2\rceil^{2FO}, \\
\lceil \exists y.\psi \rceil^{2FO} &= \exists y^-, y^+.(y^- \leq y^+ \wedge \lceil\psi\rceil^{2FO}),
\end{aligned}
$$

where the involved syntactic abbreviations are as follows:

$$
\begin{aligned}
R_{q\mathcal{T}}^{coal}(\boldsymbol{d}, u, v) \triangleq \exists t_1, t_2.R_{q\mathcal{T}}(\boldsymbol{d}, u, t_1) \wedge R_{q\mathcal{T}}(\boldsymbol{d}, t_2, v) \wedge \\
\neg\exists t_3, t_4.(R_{q\mathcal{T}}(\boldsymbol{d}, t_3, t_4) \wedge t_3 < u \wedge u \leq t_4) \wedge \\
\neg\exists t_3, t_4.(R_{q\mathcal{T}}(\boldsymbol{d}, t_3, t_4) \wedge t_3 \leq v \wedge v < t_4) \wedge \qquad\qquad (1) \\
\neg\exists t_3, t_4.(R_{q\mathcal{T}}(\boldsymbol{d}, t_3, t_4) \wedge u < t_3 \wedge t_4 \leq v \wedge \\
\neg\exists t_5, t_6.(R_{q\mathcal{T}}(\boldsymbol{d}, t_5, t_6) \wedge t_5 < t_3 \wedge t_3 \leq t_6))
\end{aligned}
$$

where for $q^{\mathcal{T}} = \bigvee\limits_{1 \leq i \leq m} q_i$:

$$R_{q^{\mathcal{T}}}(\boldsymbol{d}, u, v) \triangleq \bigvee\limits_{1 \leq i \leq m} R_{q_i}(\boldsymbol{d}, u, v) \tag{2}$$

and for every $q_i(\boldsymbol{d}) = \exists \boldsymbol{x}.(\bigwedge\limits_{1 \leq j \leq n} \alpha_j(\boldsymbol{d_j}))$:

$$R_{q_i}(\boldsymbol{d}, u, v) \triangleq \exists t_1, \ldots, t_{2n}.\exists \boldsymbol{x}. \bigwedge\limits_{1 \leq j \leq n} (R_{\alpha_j}(\boldsymbol{d_j}, t_j, t_{n+j})) \wedge$$
$$u = \max(t_1, \ldots, t_n) \wedge v = \min(t_{n+1}, \ldots, t_{2n}) \wedge u \leq v \tag{3}$$

where:

$$u = \max(t_1, \ldots, t_n) \triangleq \bigwedge\limits_{1 \leq i \leq n} ((\bigwedge\limits_{1 \leq j \leq n} t_j \leq t_i) \rightarrow u = t_i)$$
$$v = \min(t_1, \ldots, t_n) \triangleq \bigwedge\limits_{1 \leq i \leq n} ((\bigwedge\limits_{1 \leq j \leq n} t_i \leq t_j) \rightarrow v = t_i) \tag{4}$$

The pivotal part of CTQ rewriting is the translation of the embedded CQs. In this respect, the proposed approach builds directly on the standard FO rewritings of CQs, which are obtainable via existing techniques [8]. Given an FO rewriting $q^{\mathcal{T}}$ of a CQ q, all atoms in $q^{\mathcal{T}}$ are temporalized (3) and further incorporated in special formula templates (1)-(4), in order to meet two key challenges inherent to querying concrete TDBs:

- computing *temporal joins*, i.e., identifying maximal time intervals over which conjunctions of atoms are satisfied,
- applying *coalescing*, i.e., merging overlapping and adjacent intervals for the (intermediate) query results.

To illustrate these issues and the roles played by the formula templates, consider the example presented in Figure 1, which addresses the following setup:

CTA: $\mathfrak{A} = \{[1, 7] : B(a), [1, 3] : C(a), [3, 10] : C(a), [6, 12] : D(a)\}$,
TBox: $\mathcal{T} = \{D \sqsubseteq B\}$,
CTQ: $[q(a)](u)$, where $q(x) = B(x) \wedge C(x)$ and $u = [1, 10]$.

Under the assumed TBox, the FO rewriting of q can be formulated as $q^{\mathcal{T}}(x) = q_1(x) \vee q_2(x)$, where $q_1(x) = B(x) \wedge C(x)$ and $q_2(x) = D(x) \wedge C(x)$. By Definition 8, the satisfaction of condition $tdb(\mathfrak{A}) \Vdash [q(a)](u)$ is equivalent to verifying whether $(a, t_1, t_2) \in (R_{q^{\mathcal{T}}}^{coal})^{\mathcal{D}}$, for some $t_1, t_2 \in T$, such that $t_1 \leq u^- \leq u^+ \leq t_2$. Computing $(R_{q^{\mathcal{T}}}^{coal})^{\mathcal{D}}$ can be conceptually divided into three consecutive phases.

Firstly, we compute temporal joins over the sets of ABox assertions entailing each CQ q_i, i.e., we identify all intervals τ, such that there exist a set of assertions $S = \{\tau_j : \alpha_j(\boldsymbol{d_j}) \in \mathfrak{A}\}$, where the atoms $\alpha_j(\boldsymbol{d_j})$ in S provide the exact matches for the conjuncts of q_i, while $\bigcap_j \tau_j = \tau$. Relation R_{q_i}, defined via formula template (3), augmented with (4), which fixes an FO formalization of the functions max and min, selects exactly such intervals for each answer to each q_i. In our example, R_{q_1} contains tuples $(a, 1, 3)$ and $(a, 3, 7)$, while R_{q_2}

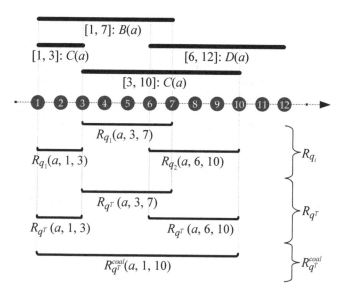

Fig. 1. Computing R_{q_i}, R_{q^T} and $R_{q^T}^{coal}$, for $q(x) = B(x) \wedge C(x)$ and $\mathcal{T} = \{D \sqsubseteq B\}$

tuple $(a, 6, 10)$. In the second phase, captured by the template (2), all such tuples become automatically instances of the common relation R_{q^T}. Finally, R_{q^T} is coalesced, resulting in the relation $R_{q^T}^{coal}$. This phase is executed by means of the template (1), which is based on a straightforward FO formalization of the coalescing mechanism, known in the context of TDBs [6]. Technically, for each tuple a comprising an answer to q, the relation $R_{q^T}^{coal}$ selects the minimal and maximal time points, such that for every point t between the two there exists a tuple $(a, t_1, t_2) \in (R_{q^T})^{\mathcal{D}}$ with $t_1 \leq t \leq t_2$. In the presented example, $R_{q^T}^{coal}$ contains exactly one tuple $(a, 1, 10)$, which is the result of coalescing tuples $(a, 1, 3), (a, 3, 7)$ and $(a, 6, 10)$ in R_{q^T}. Consequently, the CTQ $[q(a)](u)$ evaluates to true over $\mathcal{T}$ and $\mathfrak{A}$.

Since 2FO queries, in the shape defined above, can be naturally considered as formulas of the relational calculus, the final step of restating them into the SQL syntax can be carried out following broadly adopted translation strategies [17]. In practice, to ensure appropriate quantification over the time domain in RDBMSs, one would usually need to include it as an explicitly represented (and stored in the TDB) monadic relation R_T, such that $R_T^{\mathcal{D}} = T$, which should be further introduced in the translation of the temporal quantifiers from the Definition 1, as follows:

$$\lceil \exists y.\psi \rceil^{2FO} \quad = \quad \exists y^-, y^+.(R_T(y^-) \wedge R_T(y^+) \wedge y^- \leq y^+ \wedge \lceil \psi \rceil^{2FO}).$$

Accordingly, every free temporal variable $x \in \mathsf{T}_V$ occurring in a 2FO query should be additionally guarded by the restriction $R_T(x)$ in the resulting translation, in order to be properly handled by the substitution mechanisms implemented in typical RDBMSs.

Agreeably, even though the presented rewriting is mathematically correct, as we demonstrate in the following part, it can be expected to be suboptimal in terms of the query answering efficiency. For instance, the naive coalescing mechanism, captured by the formula template (1), is known to be highly inefficient and can be substantially improved using more sophisticated approaches [26]. The usefulness of this and other potential optimizations, on which we also shortly remark at the end of the next section, can be ultimately assessed only on the grounds of empirical evaluation, which is outside the scope of the current work.

5   Query Answering via 2FO Rewriting

The next theorem ensures correctness of CTQ answering via the 2FO rewriting.

Theorem 1 (Correctness of 2FO Rewriting). *For every TBox $\mathcal{T}$, CTA $\mathfrak{A}$, CTQ ψ, and an answer σ to ψ, it holds that:*

$$\mathcal{T}, \mathfrak{A} \models \sigma(\psi) \quad \text{iff} \quad tdb(\mathfrak{A}) \Vdash \lceil \sigma(\psi) \rceil^{2FO}.$$

The proof, presented in the appendix, follows by structural induction over the syntactic cases addressed in the rewriting rules. In the technically most demanding case of $\lceil [q(\boldsymbol{d})](u) \rceil^{2FO}$ we essentially formalize the discussion from the previous section. The remaining ones are largely straightforward.

Importantly, the definitions of CTQs and their 2FO rewritings, guarantee that query answering remains insensitive to the particular ways the CTAs might encode temporal data, as long as the data is semantically equivalent. This result is due to the fact that the semantics of CTQs is defined in terms of the underlying ATAs, hence whenever $\|\mathfrak{A}\| = \|\mathfrak{A}'\|$, for any two CTAs $\mathfrak{A}$ and $\mathfrak{A}'$, the answers over them must always coincide.

Theorem 2 (CTQs are $\| \cdot \|$-Generic). *For every two CTAs $\mathfrak{A}$ and $\mathfrak{A}'$ over a time domain $\mathfrak{T} = (T, <)$ such that $\|\mathfrak{A}\| = \|\mathfrak{A}'\|$, a TBox $\mathcal{T}$, a CTQ ψ and an answer σ to ψ, it holds that:*

1. *$\mathcal{T}, \mathfrak{A} \models \sigma(\psi)$ iff $\mathcal{T}, \mathfrak{A}' \models \sigma(\psi)$,*
2. *$tdb(\mathfrak{A}) \Vdash \lceil \sigma(\psi) \rceil^{2FO}$ iff $tdb(\mathfrak{A}') \Vdash \lceil \sigma(\psi) \rceil^{2FO}$.*

Further, we address the complexity and algorithmic aspects of the CTQ entailment. We start by observing that, similarly as in the case of CQ rewriting, 2FO rewritings of CTQs can be in principle exponential in the size of the original queries. In fact, as the next proposition shows, this exponential blow-up is exclusively due to the embedded CQs, as the extra temporal layer itself adds only linearly to the overall size of the resulting 2FO formulas. This observation follows directly by scrutinizing the rewriting rules from Definition 8.

Proposition 1 (Size of Rewriting). *Let $\mathcal{T}$ be a TBox, and ψ a CTQ with $q_1, \ldots, q_n$ being all the CQs occurring in ψ. Then the size of $\lceil \psi \rceil^{2FO}$ is linear in the joint size of ψ and the FO rewritings $q_1^{\mathcal{T}}, \ldots, q_n^{\mathcal{T}}$ of the CQs.*

Algorithm 1. Incremental computation of relations $R_{q\mathcal{T}}^{coal}$

Input: CTA $\mathfrak{A}$, CTQ ψ
Output: A database $\mathcal{R} = (\mathsf{N}_\mathsf{I}, T, <, \cdot^{\mathcal{D}})$ over the signature $\{R_{q\mathcal{T}}^{coal} \mid q \in \mathsf{cq}(\psi)\}$
 1: **for all** $q \in \mathsf{cq}(\psi)$ **do**
 2:     **for all** $q_i(\boldsymbol{a}) \in \mathsf{cq}(q^\mathcal{T})$ **do**
 3:         **for all** $t_1, t_2 \in T$ such that $tdb(\mathfrak{A}) \Vdash R_{q_i}(\boldsymbol{a}, t_1, t_2)$ **do**
 4:             $(R_{q\mathcal{T}}^{coal})^\mathcal{D} := (R_{q\mathcal{T}}^{coal})^\mathcal{D} \cup \{(\boldsymbol{a}, t_1, t_2)\}$
 5:             **while** applicable **do** {*coalescing*}
 6:                 **for all** $(\boldsymbol{a}, t_1, t_2), (\boldsymbol{a}, t_3, t_4) \in (R_{q\mathcal{T}}^{coal})^\mathcal{D}$ **do**
 7:                     **if** $[t_1, t_2] \cap [t_3, t_4] \neq \emptyset$ **then**
 8:                         $(R_{q\mathcal{T}}^{coal})^\mathcal{D} := (R_{q\mathcal{T}}^{coal})^\mathcal{D} \setminus \{(\boldsymbol{a}, t_1, t_2), (\boldsymbol{a}, t_3, t_4)\}$
 9:                         $(R_{q\mathcal{T}}^{coal})^\mathcal{D} := (R_{q\mathcal{T}}^{coal})^\mathcal{D} \cup \{(\boldsymbol{a}, \min(t_1, t_3), \max(t_2, t_4))\}$
10:                     **end if**
11:                 **end for**
12:             **end while**
13:         **end for**
14:     **end for**
15: **end for**

Clearly, the fragment of two-sorted first-order logic used for the CTQ rewriting is as expressive as FOL. At the same time, it is obviously not more expressive than that, as one can apply a commonly known, linear reduction, involving the introduction of two designated predicates for representing the respective domains, which are used to guard the scopes of the two sorts of quantifiers (*cf.* the last paragraph of the previous section). Based on these observations, we obtain two results concerning the complexity of CTQ query entailment.

Corollary 1 (Data Complexity). *The data complexity of CTQ entailment in CTAs in the presence of TBoxes, over finite time domains, is in* AC^0.

The restriction to finite time domains in this and the following case is a natural strategy of ensuring that the first-order structures, over which queries are evaluated, are indeed finite and, consequently, that model-checking can be effectively performed over them. Note, that in terms of data complexity, deciding CTQ entailment via 2FO rewriting remains precisely as hard as deciding CQ entailment via FO rewriting. The combined complexity, however, increases from NP- to PSPACE-complete.

Theorem 3 (Combined Complexity). *The combined complexity of CTQ entailment over CTAs in the presence of TBoxes, over finite time domains, is* PSPACE-*complete.*

The result transfers from the entailment problem for boolean FO queries over relational databases, which is known to be PSPACE-complete [10]. However, due to the exponential blow-up in the size of the 2FO rewritings, explained in Proposition 1, answering CTQs in PSPACE requires a somewhat more sophisticated strategy than just a straightforward evaluation of the rewritten queries over a

TDB. In the approach described in the following proof, we decide the entailment of a grounded CTQ ψ, by first incrementally computing the interpretations of relations $R_{q^T}^{coal}$, for each CQ q embedded in ψ, and then evaluate over them a restricted 2FO rewriting of ψ of at most polynomial size.

Proof. Let $\mathsf{cq}(\psi)$ denote all CQs occurring in ψ and grounded as in ψ, and let $\mathsf{cq}(q^T)$, for every $q \in \mathsf{cq}(\psi)$, be the set of all CQs comprising the FO rewriting q^T. With every such $q_i(\boldsymbol{a}) \in \mathsf{cq}(q^T)$, we associate the (query) formula $R_{q_i}(\boldsymbol{a}, y_1, y_2)$ defined via the templates (3) and (4). Note that every such formula is linear in the size of ψ. Algorithm 1 constructs a TDB $\mathcal{R} = (\mathsf{N}_\mathsf{I}, T, <, \cdot^{\mathcal{D}})$ over the signature $\{R_{q^T}^{coal} \mid q \in \mathsf{cq}(\psi)\}$, by collecting and coalescing all matches to the formulas $R_{q_i}(\boldsymbol{a}, y_1, y_2)$ over $tdb(\mathfrak{A})$. Observe that by applying coalescing on $R_{q^T}^{coal}$ every time another tuple is added, we guarantee that the size of each $(R_{q^T}^{coal})^{\mathcal{D}}$, and therefore of $\mathcal{R}$, is linear in the size of T, and in fact cannot be larger then $\mathfrak{A}$. Therefore the algorithm runs in PSPACE. Once $\mathcal{R}$ is computed, we decide $\mathcal{R} \Vdash \lceil \psi \rceil_{\mathcal{R}}^{2FO}$, where $\lceil \cdot \rceil_{\mathcal{R}}^{2FO}$ is a rewriting specified exactly as $\lceil \cdot \rceil^{2FO}$ in Definition 8, but without employing any of the formula templates (1)-(4), and instead, retaining all $R_{q^T}^{coal}$ as actual predicates in the resulting 2FO formulas. Clearly, $\lceil \psi \rceil_{\mathcal{R}}^{2FO}$ is linear in the size of ψ. Since deciding $\mathcal{R} \Vdash \lceil \psi \rceil_{\mathcal{R}}^{2FO}$ is a variant of the (PSPACE-hard [10]) entailment problem for boolean FO queries over relational databases (observe, that $\lceil \cdot \rceil_{\mathcal{R}}^{2FO}$ permits n-ary predicates, for $n \geq 1$), it follows that deciding the CTQ entailment over CTAs in the presence of TBoxes, over finite time domains, is PSPACE-complete. Note that since the interpretations of $R_{q^T}^{coal}$ in both $\mathcal{R}$ and $tdb(\mathfrak{A})$ must coincide, the procedure preserves soundness and completeness of query answering. $\qquad\Box$

The query answering algorithm implied by the above decision procedure paves the way towards efficient practical implementations, based on the well-known database technique of *materialized view maintenance* [21], particularly in the context of large or often changing TDBs. By maintaining the intermediate views containing the answers to the embedded CQs, the approach should facilitate more localized and fine-grained updating of the answers in response to database updates and query refinements (be it on the CQ or temporal level), as well as forms of incremental, anytime query answering.

6   Related Work

This work is naturally related to the research on TDBs, conducted largely in the 1990s [11,12]. Although the advancements in that area provide some theoretical grounding for our proposal, they are obviously agnostic about the ontology-based approach on the problem, which we concentrate on here.

The research on temporal extensions to OBDA has been taken up only very recently by Borgwardt et al. [3,7] and Artale et al. [2]. In [3], the authors study the same prototypical scenario as addressed here, but focus on its more foundational aspects. They consider a more expressive DL $\mathcal{ALC}$ as the background ontology language and adopt a less restrictive definition of temporal queries.

This offers a richer setting, yet without apparent application prospects in the context of existing TDBs. The work in [7] is closer aligned with ours. It also considers DL-*Lite* ontologies and studies a rewriting approach based on the use of standard CQ rewriting techniques. However, the authors define their query language without involving the epistemic interpretation of embedded CQs and consider only its positive fragment. It is not difficult to show that under OWL 2 QL, such design leads to a query language strictly less expressive from the one proposed here. Observe, that due to the condition (†), in Definition 5, the class of negation-free CTQs preserves the semantics regardless of the use of the epistemic interpretation of CQs, and so it coincides with the fragment considered in [7]. Furthermore, the authors focus exclusively on abstract TDBs and do not address the problem of direct querying of concrete representations, which is the main goal of our work. In [2], the authors study an orthogonal approach to temporal OBDA. Instead of adding temporal features on the query level, they propose temporal extensions to OWL 2 QL, in the spirit of temporal DLs [1].

A number of other frameworks have been developed for managing the validity time of Semantic Web data represented natively in RDF(S)/OWL languages [15,14,5,22]. These proposals can be seen as parallel to ours, in that they address variants of the same problem, but focus on the use of dedicated Semantic Web technologies, such as SPARQL and RDF triplestores, instead of involving a semantic view on the data managed within traditional RDBMSs. This rises interesting questions about the formal correspondences between the employed reasoning methods, regardless of their practical implementations. Unfortunately, most of these approaches are strongly technology-driven and often fall short of indicating their links to the logic foundations of temporal querying.

7 Conclusions

In this work, we have proposed a principled, yet practical approach to lifting the popular paradigm of OBDA to temporal case. The presented language TQL allows for querying temporal data stored in SQL:2011-compliant databases via a semantic layer of OWL 2 QL ontologies. We believe that this proposal strikes a good balance between the theoretical strength of its formal foundations and the feasibility of practical applications, warranted by the possibility of answering TQL queries in commercially supported RDBMSs via a translation to SQL.

Considering the growing interest in temporal extensions of OBDA, it is critical to continue the formal study of temporal features supported by the SQL standards, temporal extensions to logic-based query languages and ontology languages intended for use in practical OBDA scenarios, and finally, the relationships holding between all of them. In this respect, as part of future research, it is necessary to establish stronger links between the approach pursued here and those proposed recently in [3,7,2], with the prospect of gaining a clearer view on the landscape of technical possibilities regarding the temporal OBDA. In particular, the scenario of querying temporal data, as considered here and in [3,7], under temporalized ontologies, such as introduced by Artale et al. [2], deserves further attention and in-depth study.

References

1. Artale, A., Kontchakov, R., Lutz, C., Wolter, F., Zakharyaschev, M.: Temporal-ising tractable description logics. In: Proceedings of the Fourteenth International Symposium on Temporal Representation and Reasoning (2007)
2. Artale, A., Kontchakov, R., Wolter, F., Zakharyaschev, M.: Temporal description logic for ontology-based data access. In: Proceedings of the International Joint Conference on Artificial Intelligence (IJCAI 2013) (2013)
3. Baader, F., Borgwardt, S., Lippmann, M.: Temporalizing ontology-based data access. In: Bonacina, M.P. (ed.) CADE 2013. LNCS, vol. 7898, pp. 330–344. Springer, Heidelberg (2013)
4. Baader, F., Calvanese, D., Mcguinness, D.L., Nardi, D., Patel-Schneider, P.F.: The description logic handbook: theory, implementation, and applications. Cambridge University Press (2003)
5. Batsakis, S., Stravoskoufos, K., Petrakis, E.G.M.: Temporal Reasoning for Supporting Temporal Queries in OWL 2.0. In: König, A., Dengel, A., Hinkelmann, K., Kise, K., Howlett, R.J., Jain, L.C. (eds.) KES 2011, Part I. LNCS, vol. 6881, pp. 558–567. Springer, Heidelberg (2011)
6. Böhlen, M.H., Snodgrass, R.T., Soo, M.D.: Coalescing in temporal databases. IEEE Computer 19, 35–42 (1996)
7. Borgwardt, S., Lippmann, M., Thost, V.: Temporal query answering in the description logic DL-Lite. In: Fontaine, P., Ringeissen, C., Schmidt, R.A. (eds.) FroCoS 2013. LNCS, vol. 8152, pp. 165–180. Springer, Heidelberg (2013)
8. Calvanese, D., De Giacomo, G., Lembo, D., Lenzerini, M., Rosati, R.: Tractable reasoning and efficient query answering in description logics: The DL-Lite family. J. of Automated Reasoning 39(3), 385–429 (2007)
9. Calvanese, D., Giacomo, G.D., Lembo, D., Lenzerini, M., Rosati, R.: Eql-lite: Effective first-order query processing in description logics. In: Proc. of IJCAI 2007 (2007)
10. Chandra, A.K., Merlin, P.M.: Optimal implementation of conjunctive queries in relational data bases. In: Proc. of the ACM Symposium on Theory of Computing (STOC 1977) (1977)
11. Chomicki, J.: Temporal query languages: A survey. In: Gabbay, D.M., Ohlbach, H.J. (eds.) ICTL 1994. LNCS, vol. 827, pp. 506–534. Springer, Heidelberg (1994)
12. Chomicki, J., Toman, D.: Temporal Databases. In: Handbook of Temporal Reasoning in Artificial Intelligence (Foundations of Artificial Intelligence), pp. 429–468. Elsevier Science Inc. (2005)
13. Glimm, B., Horrocks, I., Lutz, C., Sattler, U.: Conjunctive query answering for the description logic SHIQ. Journal of Artificial Intelligence Research (2008)
14. Grandi, F.: T-SPARQL: a TSQL2-like temporal query language for RDF. In: Proc. of the International Workshop on Querying Graph Structured Data (2010)
15. Gutierrez, C., Hurtado, C.A., Vaisman, A.A.: Introducing time into RDF. IEEE Transactions onn Knowledge and Data Engineering 19(2), 207–218 (2007)
16. Gutiérrez-Basulto, V., Klarman, S.: Towards a unifying approach to representing and querying temporal data in description logics. In: Krötzsch, M., Straccia, U. (eds.) RR 2012. LNCS, vol. 7497, pp. 90–105. Springer, Heidelberg (2012)
17. Kawash, J.: Complex quantification in Structured Query Language (SQL): A tutorial using relational calculus. Journal of Computers in Mathematics and Science Teaching 23(2), 169–190 (2004)

18. Klarman, S., Meyer, T.: Prediction and explanation over DL-Lite data streams. In: McMillan, K., Middeldorp, A., Voronkov, A. (eds.) LPAR-19 2013. LNCS, vol. 8312, pp. 536–551. Springer, Heidelberg (2013)

19. Klarman, S., Meyer, T.: Querying temporal databases via OWL 2 QL. Tech. rep., CAIR, UKZN/CSIR Meraka (2014),
http://klarman.synthasite.com/resources/KlaMeyRR14.pdf

20. Kulkarni, K., Michels, J.E.: Temporal features in SQL:2011. SIGMOD Rec. 41(3) (2012)

21. Mohania, M., Konomi, S., Kambayashi, Y.: Incremental maintenance of materialized views. In: Tjoa, A.M. (ed.) DEXA 1997. LNCS, vol. 1308, pp. 551–560. Springer, Heidelberg (1997)

22. Motik, B.: Representing and querying validity time in RDF and OWL: A logic-based approach. Journal of Web Semantics: Science, Services and Agents on the World Wide Web 57(5), 1–62 (2012)

23. Reynolds, M.: The complexity of decision problems for linear temporal logics. Journal of Studies in Logic 3(1) (2010)

24. Snodgrass, R.T., Böhlen, M.H., Jensen, C.S., Steiner, A.: Transitioning temporal support in TSQL2 to SQL3. In: Etzion, O., Jajodia, S., Sripada, S. (eds.) Dagstuhl Seminar 1997. LNCS, vol. 1399, pp. 150–194. Springer, Heidelberg (1998)

25. Toman, D.: Point vs. interval-based query languages for temporal databases. In: Proc. of the Symposium on Principles of Database Systems (PODS 1996) (1996)

26. Zhou, X., Wang, F., Zaniolo, C.: Efficient temporal coalescing query support in relational database systems. In: Bressan, S., Küng, J., Wagner, R. (eds.) DEXA 2006. LNCS, vol. 4080, pp. 676–686. Springer, Heidelberg (2006)

Towards Mapping Analysis
in Ontology-Based Data Access

Domenico Lembo[1], Jose Mora[1], Riccardo Rosati[1],
Domenico Fabio Savo[1], and Evgenij Thorstensen[2]

[1] Sapienza Università di Roma
`lastname@dis.uniroma1.it`
[2] University of Oslo
`evgenit@ifi.uio.no`

Abstract. In this paper we study mapping analysis in ontology-based data access (OBDA), providing an initial set of foundational results for this problem. We start by defining general, language-independent notions of mapping inconsistency, mapping subsumption, and mapping redundancy in OBDA. Then, we focus on specific mapping languages for OBDA and illustrate techniques for verifying the above properties of mappings.

1 Introduction

Ontology-based data access (OBDA) is a data integration paradigm that relies on a three-level architecture, constituted by the ontology, the data sources, and the mapping between the two [18]. The ontology is the specification of a conceptual view of the domain, and it is the system interface towards the user, whereas the mapping relates the elements of the ontology with the data at the sources.

In the past years, studies on OBDA have mainly concentrated on query answering, and various algorithms for it have been devised, as well as tools implementing them [6,20,7,16,4,27,22]. Intensional reasoning has been instead so far limited to the ontology level only. This means that currenlty available services of this kind in OBDA are exactly as for stand-alone ontologies (e.g., concept/role subsumption, classification, logical implication, etc.). As a consequence, in the specification of an OBDA system, a designer can rely only on classical off-the-shelf ontology reasoners (e.g., [26,25,24,12]), but she cannot find tools supporting the modeling of the other crucial component of the OBDA architecture, i.e., the mapping.

Both industrial and research OBDA projects (see, e.g., [13,1]) have experienced that mapping specification is a very complex activity, which requires a profound understanding of both the ontology and the data sources. Indeed, data sources are in general autonomous and pre-existing the OBDA application, and thus the way in which they are structured typically does not reflect the ontology, which is instead an independent representation of the domain of interest, rather than of the underlying data sources. To reconcile this "cognitive distance"

R. Kontchakov and M.-L. Mugnier (Eds.): RR 2014, LNCS 8741, pp. 108–123, 2014.

between the sources and the ontology, the mapping usually assumes a complex form, and it is expressed in terms of assertions that relate queries over the ontology to queries over the data sources.

This form of mappings has been widely studied in data integration and data exchange [17,9,2]. In these contexts, the research on mappings has been mainly focused on mapping composition or inversion, whereas very few efforts have been made towards the analysis of the specification, to verify, e.g., whether it is redundant or inconsistent per se (i.e., independently from the source data). In fact, in data integration and exchange, the integrated (a.k.a. global or target) schema is in general not as expressive as an ontology, and thus analysis checks are to some extent easier or less crucial than in OBDA.

In this paper we study mapping analysis in OBDA, with the aim of providing the designer with services that are useful to devise a well-founded OBDA specification. We introduce our novel definitions that formalize properties of interest for the mapping. In particular, we define when a mapping $\mathcal{M}$ is inconsistent w.r.t. an ontology $\mathcal{O}$ and a source schema $\mathcal{S}$, which intuitively means that retrieving data through all the assertions in $\mathcal{M}$ always leads to an inconsistent OBDA specification composed by $\mathcal{O}$, $\mathcal{M}$ and $\mathcal{S}$, whatever non-empty source instance is assigned to $\mathcal{S}$. Also, we define when a mapping $\mathcal{M}$ subsumes a mapping $\mathcal{M}'$ under $\mathcal{O}$ and $\mathcal{S}$, which intuitively means that the systems composed by $\mathcal{O}$, $\mathcal{S}$, and either $\mathcal{M}$ or $\mathcal{M} \cup \mathcal{M}'$ are equivalent (and thus $\mathcal{M}'$ is redundant in the specification). We point out that verifying such properties is indeed of crucial importance in real-life OBDA projects, when hundreds of mapping assertions are usually needed, and it is very likely that mappings are redundant or even inconsistent in the sense we have described above.

After defining the mapping analysis tasks we are interested in, we discuss techniques for verifying consistency, redundancy, and subsumption for (some generalizations of) the so-called GAV mapping [17], when some specific languages for querying the sources and the ontology are used in mapping assertions.

We organize our paper as follows. In Section 2 we give some preliminary definitions on OBDA and mapping languages. In Section 3 we provide our notions of consistency, subsumption, and redundancy for mappings. In Section 4 we study decidability of verification tasks for some mapping languages under the GAV paradigm. In Section 5 we add some preliminary discussions on mappings that go beyond GAV, and in Section 6 we conclude the paper.

2   Definitions

OBDA Specifications. An OBDA specification is a triple $\mathcal{J} = \langle \mathcal{O}, \mathcal{S}, \mathcal{M} \rangle$ where $\mathcal{O}$ is an ontology, $\mathcal{S}$ is a source schema, and $\mathcal{M}$ is a mapping between the two. $\mathcal{O}$ typically (although not necessarily) represents intensional knowledge and is specified in a language $\mathcal{L}_{\mathcal{O}}$, whereas $\mathcal{S}$ is specified in a language $\mathcal{L}_{\mathcal{S}}$. We denote with $\Sigma_{\mathcal{O}}$ and $\Sigma_{\mathcal{S}}$ the signature of $\mathcal{O}$ and $\mathcal{S}$, respectively, and we assume that both $\mathcal{L}_{\mathcal{O}}$ and $\mathcal{L}_{\mathcal{S}}$ are (fragments of) first-order logic (FOL). For instance, in a typical OBDA setting, $\mathcal{O}$ is a Description Logic TBox and $\mathcal{S}$ is a relational schema, possibly with integrity constraints [20]. Finally, the mapping $\mathcal{M}$ is a set of assertions of the form

$$\phi(\boldsymbol{x}) \rightsquigarrow \psi(\boldsymbol{x}) \tag{1}$$

where $\phi(\boldsymbol{x})$ is a query over Σ_S and $\psi(\boldsymbol{x})$ is a query over $\Sigma_{\mathcal{O}}$, both with free variables $\boldsymbol{x}$, which are called the *frontier variables*. The number of variables in $\boldsymbol{x}$ is the *arity* of the mapping assertion. Given a mapping assertion m of the form (1), we also use $FR(m)$ do denote the frontier variables $\boldsymbol{x}$, $head(m)$ to denote the query $\psi(\boldsymbol{x})$, and $body(m)$ to denote the query $\phi(\boldsymbol{x})$, and we assume that both such queries are specified in (some fragment of) FOL.

Example 1. We give here an example of OBDA specification that we will use as ongoing example throughout the paper. We refer to a setting in which the source schema is relational and the ontology is expressed in a basic Description Logic language [3], which actually corresponds to $DL\text{-}Lite_{core}$ [6]. Since S is relational, queries in the body of mapping assertions are encoded in SQL.

Then, consider the following schema S of the database used in a zoo for handling information about the animals and the area of the zoo they live. In the schema, the underlined attributes represent the keys of the tables, and we also assume that a foreing key is specified between the attribute AREA of ANM_TAB and the table AREA_TAB.

> ANM_TAB(<u>ANM_CODE</u>,NAME,BREED,AREA)
> AREA_TAB(<u>AREA_CODE</u>,SIZE)

An ontology $\mathcal{O}$ modeling a very small portion of the zoo domain is as follow.

$\mathcal{O} = \{$ Lion $\sqsubseteq$ Animal, Monkey $\sqsubseteq$ Animal, Lion $\sqsubseteq \neg$Monkey, Animal $\sqsubseteq \exists$name, Animal $\sqsubseteq \exists$locatedIn, $\exists$locatedIn$^-$ $\sqsubseteq$ Area, Area $\sqsubseteq \exists$size$\}$

In words, $\mathcal{O}$ specifies that both lions (Lion) and monkeys (Monkey) are animals (Animal), a lion cannot be a monkey, and every animal has a name (name) and is located in (locatedIn) an area (Area). Moreover, every area has a size (size).

An example of mapping $\mathcal{M}$ between $\mathcal{O}$ and S is as follows:

$m1$: `SELECT ANM_CODE AS X, NAME AS Y` $\rightsquigarrow$ Animal$(X) \wedge$ name(X,Y)
       `FROM ANM_TAB`
$m2$: `SELECT ANM_CODE AS X, AREA AS Y` $\rightsquigarrow$ Lion$(X) \wedge$ locatedIn(X,Y)
       `FROM ANM_TAB WHERE BREED = 'Lion'`
$m3$: `SELECT ANM_CODE AS X, AREA AS Y` $\rightsquigarrow$ Monkey$(X) \wedge$ locatedIn(X,Y)
       `FROM ANM_TAB WHERE BREED = 'Monkey'`
$m4$: `SELECT ANM_CODE AS X, AREA AS Y` $\rightsquigarrow$ locatedIn(X,Y)
       `FROM ANM_TAB`
$m5$: `SELECT AREA_CODE AS X, SIZE AS Y` $\rightsquigarrow$ Area$(X) \wedge$ size(X,Y)
       `FROM AREA_TAB`

$\square$

The semantics of an OBDA specification $\mathcal{J}$ is defined with respect to a source instance that is legal for S. More precisely, a source instance D is a set of facts over Σ_S. Given such a D, we denote by $\mathcal{I}_D$ the interpretation over Σ_S that is isomorphic to D. Then, we say that D *is legal for* $\mathcal{S}$ if $\mathcal{I}_D \models \mathcal{S}$. For example, if $\mathcal{S}$ is relational, we consider as legal only the instances that satisfy the integrity constraints on $\mathcal{S}$. We assume that for each $\mathcal{S}$ a legal instance always exists. Then, for each mapping assertion $m \in \mathcal{M}$ we denote with $\pi(m)$ the FOL formula

$$\forall \boldsymbol{x}.\phi(\boldsymbol{x}) \rightarrow \exists \boldsymbol{z}.\psi(\boldsymbol{y}, \boldsymbol{z})$$

where $\boldsymbol{z}$ denotes the existential variables in $head(m)$, and we pose $\pi(\mathcal{M}) = \{\pi(m) \mid m \in \mathcal{M}\}$. Then, the *models of* $\mathcal{J}$ *w.r.t.* D are the models of the FOL theory $\mathcal{O} \cup \pi(\mathcal{M}) \cup D$ that are isomorphic to D on the interpretation of the predicates in $\mathcal{S}$. We denote with $Models(\mathcal{J}, D)$ the set of models of $\mathcal{J}$ w.r.t. D.

Mapping Languages. In this paper we study specific cases of OBDA specifications where we fix the fragment of FOL adopted for the queries in the head and in the body of mapping assertions. In particular, we mainly focus on the following mapping languages:

- *FO2DCQ*, where, for each $m \in \mathcal{M}$, $body(m)$ is a FOL query over $\mathcal{S}$ and $head(m)$ is a conjunctive query over $\mathcal{O}$ without existential variables;
- *CQ2DCQ*, where, for each $m \in \mathcal{M}$, $body(m)$ is a conjunctive query over $\mathcal{S}$ and $head(m)$ is a conjunctive query over $\mathcal{O}$ without existential variables.

Obviously, *FO2DCQ* subsumes *CQ2DCQ*, and thus all definitions we give in the following for *FO2DCQ* mappings also apply to *CQ2DCQ* mappings.

Both languages above are extended forms of the so-called GAV mapping, which, differently from the LAV mapping, does not allow for non-free variables in the head of assertions [17,9]. On the other-hand, GAV is the only kind of mapping that has been used in practical OBDA and data integration applications [13,1]. An example of *CQ2DCQ* language is given in Example 1.

We notice that classical GAV mapping only allows for single atom queries in the head of assertions (instead of conjunctions of atoms). However, it is easy to see that each *FO2DCQ* assertion can be rephrased into a logically equivalent set of classic GAV assertions. More precisely, let

$$m : \exists \boldsymbol{w}.\phi(\boldsymbol{x}, \boldsymbol{w}) \rightsquigarrow \psi(\boldsymbol{x})$$

be one such assertion, where we have explicited the existential variables in the body query, then, we can rephrase m into the following set of mapping assertions

$$\{\exists \overline{\boldsymbol{x}}_i, \boldsymbol{w}.\phi(\boldsymbol{x}_i, \overline{\boldsymbol{x}}_i, \boldsymbol{w}) \rightsquigarrow \psi_i(\boldsymbol{x}_i) \mid \text{ for each atom } \psi_i(\boldsymbol{x}_i) \text{ in } body(m)\}$$

where $\overline{\boldsymbol{x}}_i$ denotes the free variables of $\boldsymbol{x}$ that do not occur in $\boldsymbol{x}_i$. Given a *FO2DCQ* mapping $\mathcal{M}$, we denote with $\mathsf{Split}(\mathcal{M})$ the above set of mappings.

Example 2. Consider the $m1$ mapping assertion of Example 1. The set $\mathsf{Split}(m1)$ contains the following mapping assertions:

$m1'$: `SELECT ANM_CODE AS X FROM ANM_TAB` $\qquad\rightsquigarrow$ Animal(X)
$m1''$: `SELECT ANM_CODE AS X, NAME AS Y FROM ANM_TAB` $\rightsquigarrow$ name(X, Y)

$\square$

Let m be a *FO2DCQ* mapping assertion of arity n and let t be an n-tuple of constants. We denote by $m(t)$ the mapping assertion obtained from m by replacing the frontier variables of m with the constants in t. Then, let D be a source instance, we define the *facts retrieved by m on D*, denoted by $Retr(m, D)$, as the set of ground atoms

$$\{\alpha \mid t \text{ is a tuple of constants and } \mathcal{I}_D \models body(m(t)) \text{ and } \alpha \text{ occurs in } head(m(t))\}$$

Moreover, given a *FO2DCQ* mapping $\mathcal{M}$ and a source instance D, we define the *facts retrieved by $\mathcal{M}$ on D*, denoted by $Retr(\mathcal{M}, D)$, as the set of ground atoms

$$\bigcup_{m \in \mathcal{M}} Retr(m, D)$$

Finally, given an ontology predicate A, we define the *extension of A retrieved by $\mathcal{M}$ on D*, denoted by $Retr(A, \mathcal{M}, D)$, as the set $\{t \mid A(t) \in Retr(\mathcal{M}, D)\}$.

From now on, without loss of generality we assume that different mapping assertions use different sets of variable symbols.

3   Mapping Analysis Tasks

In this section we provide the formal definitions that constitute the basis of the mapping analysis functionalities that will be studied in Section 4. We first deal with mapping consistency, then we turn our attention to mapping redundancy and subsumption. If not otherwise specified, definitions and properties given in this section apply to mappings that contain general assertions of the form (1).

3.1   Consistency

We start by providing some notions of inconsistency relative to a single mapping assertion. Informally, with such notions we characterize the anomalous situations in which either the query in the head of an assertion has certainly an empty evaluation in every model for the ontology $\mathcal{O}$ (we call this situation head-inconsistency), or the query in the body of an assertion has certainly an empty evaluation in every model for the source schema $\mathcal{S}$ (we call this situation body-inconsistency).

Definition 1. *(mapping head-inconsistency) Let $\langle \mathcal{O}, \mathcal{S}, \mathcal{M} \rangle$ be an OBDA specification and $m : \phi(\boldsymbol{x}) \rightsquigarrow \psi(\boldsymbol{x})$ be a mapping assertion in $\mathcal{M}$. We say that m is head-inconsistent for $\langle \mathcal{O}, \mathcal{S} \rangle$ if $\mathcal{O} \models \forall \boldsymbol{x}.(\neg \psi(\boldsymbol{x}))$.*

Example 3. Let $\langle \mathcal{O}, \mathcal{S}, \mathcal{M} \rangle$ be an OBDA specification where $\mathcal{O}$ and $\mathcal{S}$ are as in Example 1. Suppose that the mapping $\mathcal{M}$ contains the following assertion:

$$m : \texttt{SELECT ANM\_CODE AS X} \rightsquigarrow \textsf{Lion}(X) \wedge \textsf{Monkey}(X)$$
$$\texttt{FROM ANM\_TAB}$$

Then, m is head-inconsistent for $\langle \mathcal{O}, \mathcal{S} \rangle$, since we have that $\mathcal{O} \models \textsf{Lion} \sqsubseteq \neg\textsf{Monkey}$. □

Definition 2. *(mapping body-inconsistency) Let $\langle \mathcal{O}, \mathcal{S}, \mathcal{M} \rangle$ be an OBDA specification and $m : \phi(\boldsymbol{x}) \rightsquigarrow \psi(\boldsymbol{x})$ be a mapping assertion in $\mathcal{M}$. We say that m is body-inconsistent for $\langle \mathcal{O}, \mathcal{S} \rangle$ if $\mathcal{S} \models \forall \boldsymbol{x}.(\neg\phi(\boldsymbol{x}))$.*

Example 4. Let $\langle \mathcal{O}, \mathcal{S}, \mathcal{M} \rangle$ be an OBDA specification where $\mathcal{O}$ and $\mathcal{S}$ are as in Example 1. Suppose that the mapping $\mathcal{M}$ contains the following mapping assertion:

$$m : \texttt{SELECT ANM\_CODE AS X} \qquad \rightsquigarrow \quad \text{Animal}(X)$$
$$\texttt{FROM ANM\_TAB}$$
$$\texttt{WHERE BREED = 'Lion' AND}$$
$$\texttt{BREED = 'Monkey'}$$

Since, obviously, for every tuple in ANM_TAB the attribute BREED can assume only a single value, we can easily conclude that m is body-inconsistent for $\langle \mathcal{O}, \mathcal{S} \rangle$. □

We compose the above two notions into the following notion of inconsistency of a single mapping assertion.

Definition 3. *(mapping inconsistency) Let $\langle \mathcal{O}, \mathcal{S}, \mathcal{M} \rangle$ be an OBDA specification and m be a mapping assertion in $\mathcal{M}$. We say that m is inconsistent for $\langle \mathcal{O}, \mathcal{S} \rangle$ if m is head-inconsistent or body-inconsistent for $\langle \mathcal{O}, \mathcal{S} \rangle$.*

Then, we provide a "global" notion of inconsistency, that is, inconsistency relative to a whole mapping specification. To this aim, we first need to define when a mapping is active on a source instance.

We say that a mapping $\mathcal{M}$ *is active on* a source instance D if, for every mapping assertion $m : \phi(\boldsymbol{x}) \rightsquigarrow \psi(\boldsymbol{x})$ in $\mathcal{M}$, $\mathcal{I}_D \models \exists \boldsymbol{x}.\phi(\boldsymbol{x})$ (in other words, every mapping assertion is "activated" by D and retrieves at least one tuple from D).

Definition 4. *(global mapping inconsistency) Let $\mathcal{J} = \langle \mathcal{O}, \mathcal{S}, \mathcal{M} \rangle$ be an OBDA specification. We say that $\mathcal{M}$ is globally inconsistent for $\langle \mathcal{O}, \mathcal{S} \rangle$ if there does not exist a source instance D legal for $\mathcal{S}$ such that $\mathcal{M}$ is active on D and $Models(\mathcal{J}, D) \neq \emptyset$.*

Intuitively, if a mapping is globally inconsistent, then it is not possible to simultaneously activate all its mapping assertions without causing inconsistency of the whole specification. This is certainly an anomalous situation, as shown by the following example.

Example 5. Let $\langle \mathcal{O}, \mathcal{S}, \mathcal{M} \rangle$ be an OBDA specification where $\mathcal{O}$ and $\mathcal{S}$ are as in Example 1. Suppose that $\mathcal{M}$ contains the following mapping assertions:

$$m1 : \texttt{SELECT ANM\_CODE AS X FROM ANM\_TAB} \rightsquigarrow \text{Lion}(X)$$
$$m2 : \texttt{SELECT ANM\_CODE AS X FROM ANM\_TAB} \rightsquigarrow \text{Monkey}(X)$$

It is easy to see that $\mathcal{M}$ is globally inconsistent for $\langle \mathcal{O}, \mathcal{S} \rangle$.   □

The following property relates the two notions of mapping inconsistency and global mapping inconsistency.

Proposition 1. *Let $\langle \mathcal{O}, \mathcal{S}, \{m\} \rangle$ be a OBDA specification. If the mapping assertion m is inconsistent for $\langle \mathcal{O}, \mathcal{S} \rangle$, then every mapping $\mathcal{M}$ that contains m is globally inconsistent for $\langle \mathcal{O}, \mathcal{S} \rangle$.*

Note that a mapping $\mathcal{M}$ that is globally inconsistent for some $\langle \mathcal{O}, \mathcal{S} \rangle$ may not contain any mapping assertion m that is inconsistent for $\langle \mathcal{O}, \mathcal{S} \rangle$, which is actually the case shown in Example 5. In other terms, inconsistency of a mapping assertion is a sufficient but not necessary condition for global inconsistency.

3.2   Redundancy and Subsumption

We now deal with mapping redundancy and subsumption. First, given an ODBA specification $\mathcal{J} = \langle \mathcal{O}, \mathcal{S}, \mathcal{M} \rangle$ where $\mathcal{M} = \{m\}$, we consider a mapping assertion m' to be redundant for m, if adding m' to $\mathcal{M}$ produces a specification equivalent to $\mathcal{J}$. This is formalized below.

Definition 5. *(mapping redundancy) Let $\mathcal{O}$ be an ontology, let $\mathcal{S}$ be a source schema, and let m, m' be mapping assertions of the same arity. We say that m' is redundant for m under $\langle \mathcal{O}, \mathcal{S} \rangle$ if, for every source instance D that is legal for $\mathcal{S}$, $Models(\langle \mathcal{O}, \mathcal{S}, \{m\} \rangle, D) = Models(\langle \mathcal{O}, \mathcal{S}, \{m, m'\} \rangle, D)$.*

Our aim now is to characterize the above notion of redundancy in terms of composition of separate entailment checks on the source schema level and the ontology level of the OBDA specification. We thus define the notions of head-subsumption and body-subsumption for a pair of mapping assertions.

Definition 6. *(mapping body-subsumption, mapping head-subsumption) Let $\mathcal{S}$ be a source schema, let m_1, m_2 be mapping assertions of the same arity, let $FR(m_2) = \{x_1, \ldots, x_n\}$, and let μ be a bijective mapping from $FR(m_1)$ to $FR(m_2)$. We say that m_1 body-subsumes m_2 under $\mathcal{S}$ and μ if the schema $\mathcal{S}$ entails the sentence $\forall x_1, \ldots, x_n (body(m_2) \rightarrow \mu(body(m_1)))$. Moreover, we say that m_1 head-subsumes m_2 under $\mathcal{O}$ and μ if the ontology $\mathcal{O}$ entails the sentence $\forall x_1, \ldots, x_n (head(m_2) \rightarrow \mu(head(m_1)))$.*

Informally, body-subsumption characterizes the case when the body of the mapping assertion m_2 entails the body of m_1 under the schema $\mathcal{S}$ and under a mapping μ of the frontier variables of m_1 and m_2. Head-subsumption is defined in an analogous way.

Example 6. Consider again the ontology $\mathcal{O}$ and the schema $\mathcal{S}$ of Example 1 and the following mapping assertions:

$m1$: `SELECT AREA_CODE AS X, SIZE AS Y` $\rightsquigarrow$ $\mathsf{size}(X, Y)$
          `FROM AREA_TAB`
$m2$: `SELECT AREA_CODE AS X, SIZE AS Y` $\rightsquigarrow$ $\mathsf{Area}(X) \wedge \mathsf{size}(X, Y)$
          `FROM AREA_TAB WHERE SIZE > 10`
$m3$: `SELECT ANM_CODE AS X` $\rightsquigarrow$ $\mathsf{Animal}(X) \wedge \mathsf{name}(X, Y)$
          `FROM ANM_TAB WHERE BREED = 'Monkey'`
$m4$: `SELECT ANM_CODE AS X, NAME AS Y` $\rightsquigarrow$ $\mathsf{Lion}(X) \wedge \mathsf{name}(X, Y)$
          `FROM ANM_TAB WHERE BREED = 'Lion'`

It is easy to see that $m1$ body-subsumes $m2$. Moreover, since the ontology $\mathcal{O}$ entails that a lion is an animal, we have that $m3$ head-subsumes $m4$.   □

The relationship between the notion of redundancy and the notions of head- and body-subsumption is stated by the following proposition.

Proposition 2. *Let $\mathcal{O}$ be an ontology, let S be a source schema, and let m, m' be mapping assertions of the same arity. Then, m' is redundant for m under $\langle \mathcal{O}, S \rangle$ iff there exists a bijective mapping $\mu : FR(m) \to FR(m')$ such that m body-subsumes m' under S and μ and m' head-subsumes m under $\mathcal{O}$ and μ.*

Notice that, for m' to be redundant for m, we require that (under the same bijective mapping of the frontier variables) m body-subsumes m', whereas m' head-subsumes m. This indeed reflects the "semantic flow" of the data: m' is redundant since it retrieves from the sources less data than m, and at the same time the instantiation of ontology predicates that m' realizes with these data is less specific than the instantiation due to m, but implies it under $\mathcal{O}$.

Example 7. Let $\mathcal{O}$ and S be respectively the ontology and the source schema of Example 1. Consider the following mapping assertions:

$m1$: `SELECT ANM_CODE AS X, NAME AS Y` $\rightsquigarrow$ $\mathsf{name}(X, Y)$
          `FROM ANM_TAB WHERE BREED = 'Monkey'`
$m2$: `SELECT ANM_CODE AS X, NAME AS Y` $\rightsquigarrow$ $\mathsf{Animal}(X) \wedge \mathsf{name}(X, Y)$
          `FROM ANM_TAB`

We have that $m1$ is redundant for $m2$ under $\langle \mathcal{O}, \mathcal{S} \rangle$. Indeed, it easy to see that $m2$ body-subsumes $m1$ under S and that $m1$ head-subsumes $m2$ under $\mathcal{O}$. Notice that, if we add the atom $\mathsf{Monkey}(X)$ in the head of $m1$, the redundancy does no longer hold, since in that case $m2$ head-subsumes $m1$.   □

Then, we define a more general, global notion of mapping redundancy which is relative to a whole mapping specification.

Definition 7. *(global mapping redundancy) Let $\mathcal{O}$ be an ontology, let S be a source schema, and let $\mathcal{M}, \mathcal{M}'$ be mappings. We say that $\mathcal{M}'$ is globally redundant for $\mathcal{M}$ under $\langle \mathcal{O}, S \rangle$ if, for every source instance D that is legal for S, $Models(\langle \mathcal{O}, S, \mathcal{M} \rangle, D) = Models(\langle \mathcal{O}, S, \mathcal{M} \cup \mathcal{M}' \rangle, D)$.*

Notice that global redundancy of a mapping $\mathcal{M}'$ for a mapping $\mathcal{M}$ under $\langle \mathcal{O}, \mathcal{S} \rangle$ does not imply that there exists an assertion m' in $\mathcal{M}'$ and an assertion m in $\mathcal{M}$ such that m' is redundant for m under $\langle \mathcal{O}, \mathcal{S} \rangle$, as shown below.

Example 8. Consider the ontology $\mathcal{O} = \{A_1 \sqsubseteq A, B_1 \sqsubseteq B\}$, the source schema composed by the only unary predicate Q, and the following mapping assertions:

$$
\begin{aligned}
m_1 &: Q(X) \rightsquigarrow A_1(X) \\
m_2 &: Q(X) \rightsquigarrow B_1(X) \\
m_3 &: Q(X) \rightsquigarrow A(X) \wedge B(X)
\end{aligned}
$$

Then, $\mathcal{M}' = \{m_3\}$ is globally redundant for $\mathcal{M} = \{m_1, m_2\}$ under $\langle \mathcal{O}, \mathcal{S} \rangle$, but m_3 is not redundant under $\langle \mathcal{O}, \mathcal{S} \rangle$ for any mapping assertion in $\mathcal{M}$.   □

Conversely, it is easy to see the if a mapping $\mathcal{M}'$ contains only assertions that, taken one by one, are redundant under $\langle \mathcal{O}, \mathcal{S} \rangle$ for some assertion contained in a mapping $\mathcal{M}$, then $\mathcal{M}'$ is globally redundant for $\mathcal{M}$ under $\langle \mathcal{O}, \mathcal{S} \rangle$.

Finally, we define extensional predicate subsumption, a mapping-based notion of subsumption between ontology predicates. Differently from all the other definitions and propositions given in this section, such notion applies only to GAV mappings, and thus we give it for *FO2DCQ* mappings, which subsume all GAV mappings considered in this paper.

Definition 8. *(extensional predicate subsumption and emptiness) Let $\mathcal{S}$ be a source schema, let $\mathcal{M}$ be a FO2DCQ mapping, and let A, A' be ontology predicates having the same arity. We say that A extensionally subsumes A' under $\langle \mathcal{S}, \mathcal{M} \rangle$ if, for every source instance D that is legal for $\mathcal{S}$, $Retr(A, \mathcal{M}, D) \supseteq Retr(A', \mathcal{M}, D)$. Moreover, given a predicate A, we say that A is extensionally empty under $\langle \mathcal{S}, \mathcal{M} \rangle$ if, for every source instance D that is legal for $\mathcal{S}$, $Retr(A, \mathcal{M}, D) = \emptyset$.*

Informally, the above notion of extensional predicate subsumption checks containment of the instances of the predicates retrieved by the mapping on every legal source instance.

4 Verification

In this section we study the problem of decidability of the verification of the formal properties of mappings defined in Section 3. It can be immediately observed that verification for mappings expressed in the language *FO2DCQ* poses a serious decidability issue independently of the ontology language $\mathcal{L}_\mathcal{O}$ and the source schema language $\mathcal{L}_\mathcal{S}$, since arbitrary FOL expressions can appear in the body of such mapping assertions. Therefore, our first analysis focuses on identifying sufficient conditions for the decidability of the verification of the properties under examination.

For ease of exposition, in the rest of this section we assume that the ontology language $\mathcal{L}_\mathcal{O}$ has a predefined empty predicate $\bot$. More precisely, we assume the existence of an ontology predicate $\bot$ of arity 0 that is false in every interpretation.

4.1   Head-Subsumption and Head-Inconsistency

Let us consider mapping head-subsumption. In the following, let m_1, m_2 be either $FO2DCQ$ or $CQ2DCQ$ mapping assertions of the same arity, let μ be a bijective mapping from $FR(m_1)$ to $FR(m_2)$, and let $q_1 = head(m_1)$, $q_2 = head(m_2)$.

The following algorithm checks whether m_2 head-subsumes m_1 under $\mathcal{S}$ and μ:

1. freeze query $\mu(q_1)$, i.e., generate a source instance (set of ground atoms) $D_{\mu(q_1)}$ from q_1 by replacing every occurrence of a variable x with a constant symbol c_x;
2. let q_2' be the formula obtained from q_2 by replacing every occurrence of a variable x with a constant symbol c_x. Notice that q_2' is a conjunction of ground atoms;
3. if, for every ground atom α in q_2', $\mathcal{O} \cup D_{\mu(q_1)} \models \alpha$ (ground atom entailment problem in $\mathcal{L_O}$), then return true, otherwise return false.

It can be shown that the above algorithm is correct. This implies that mapping head-subsumption is decidable as soon as ground atom entailment in $\mathcal{L_O}$ is decidable. Conversely, undecidability of head-subsumption when ground atom entailment in $\mathcal{L_O}$ is undecidable can be shown by an easy reduction of ground atom entailment in $\mathcal{L_O}$ to mapping head-subsumption. Consequently, the following property holds.

Theorem 1. *For both FO2DCQ mappings and CQ2DCQ mappings, mapping head-subsumption is decidable iff ground atom entailment in $\mathcal{L_O}$ is decidable.*

Mapping head-inconsistency can be immediately reduced to mapping head-subsumption, since $\mathcal{L_O}$ allows for the empty predicate $\perp$. Then, m is head-inconsistent for $\langle \mathcal{O}, \mathcal{S} \rangle$ iff m is head-subsumed by m' under $\langle \mathcal{O}, \mathcal{S} \rangle$ and μ, where m' is the mapping obtained from m by adding the atom $\perp$ in the head of m, and μ is the identity mapping on $FR(m)$.

Moreover, it can be shown that ground atom entailment can be reduced to mapping head-inconsistency, under some assumptions on the ontology language $\mathcal{L_O}$. In particular, we say that $\mathcal{L_O}$ allows for binary denial formulas if, for every pair of predicate names p, p' in Σ_O of the same arity, the formula $\forall x\, (p(x) \wedge p'(x) \to \perp)$ belongs to $\mathcal{L_O}$. The above assumption is a sufficient condition for reducing ground atom entailment to head-inconsistency.

Theorem 2. *For both FO2DCQ mappings and CQ2DCQ mappings, mapping head-inconsistency is decidable if ground atom entailment in $\mathcal{L_O}$ is decidable. Moreover, if $\mathcal{L_O}$ allows for binary denial formulas, then ground atom entailment in $\mathcal{L_O}$ is decidable if mapping head-inconsistency is decidable.*

4.2   Body-Subsumption and Body-Inconsistency

Body-subsumption and body-inconsistency are undecidable for $FO2DCQ$ mappings (due to the undecidability of the validity problem in FOL).

Concerning $CQ2DCQ$ mappings, the following property immediately follows from the definitions of mapping body-subsumption.

Theorem 3. *For CQ2DCQ mappings, mapping body-subsumption is decidable iff conjunctive query containment is decidable in $\mathcal{L_S}$.*

Notice that several schema languages are known to satisfy the hypothesis of the above theorem. E.g., conjunctive query containment is decidable in the language of non-key-conflicting keys and inclusion dependencies studied in [5], as well as in several classes of TGDs [4,15].

For mapping body-inconsistency, a similar property holds under some sufficient assumptions on the language $\mathcal{L_S}$. In particular, we say that $\mathcal{L_S}$ allows for CQ-denial formulas if, for every conjunctive query $q(\boldsymbol{x})$ over $\Sigma_\mathcal{S}$, the formula $\forall \boldsymbol{x}(q(\boldsymbol{x}) \to \bot)$ belongs to $\mathcal{L_S}$.

Theorem 4. *For CQ2DCQ mappings, mapping body-inconsistency is decidable iff conjunctive query containment is decidable in $\mathcal{L_S}$. Moreover, if $\mathcal{L_S}$ allows for CQ-denial formulas, then conjunctive query containment in $\mathcal{L_S}$ is decidable if mapping body-inconsistency is decidable.*

4.3   Redundancy and Inconsistency

Given the above undecidability results for head- and body-subsumption, it obviously follows that both redundancy and inconsistency of mapping assertions are undecidable properties for *FO2DCQ* mappings.

However, the situation is different for *CQ2DCQ* mappings, In fact, it is immediate to see that Proposition 2, Definition 3, Theorem 1, and Theorem 3, imply the following properties.

Theorem 5. *For CQ2DCQ mappings, mapping redundancy is decidable iff ground atom entailment is decidable in $\mathcal{L_O}$ and conjunctive query containment is decidable in $\mathcal{L_S}$.*

Theorem 6. *For CQ2DCQ mappings, mapping inconsistency is decidable if ground atom entailment is decidable in $\mathcal{L_O}$ and conjunctive query containment is decidable in $\mathcal{L_S}$. Moreover, if $\mathcal{L_S}$ allows for CQ-denial formulas and $\mathcal{L_O}$ allows for binary denial formulas, then ground atom entailment is decidable in $\mathcal{L_O}$ and conjunctive query containment is decidable in $\mathcal{L_S}$ if mapping inconsistency is decidable.*

4.4   Extensional Subsumption and Emptiness

We start by relating the notion of extensional predicate subsumption with the notion of global mapping redundancy.

Let $\mathcal{M}$ be a mapping and let A be a predicate name. We define $\mathcal{M}_A$ as the mapping obtained from $\mathsf{Split}(\mathcal{M})$ by considering only the mapping assertions in which A occurs in the head. Moreover, given such a mapping $\mathcal{M}_A$ and a predicate name B of the same arity as A, we define $\mathcal{M}_A(B)$ as the mapping obtained from $\mathcal{M}_A$ by replacing every occurrence of the predicate A with B.

The relationship between global mapping redundancy and extensional predicate subsumption is stated by the following property.

Theorem 7. *Let A, A' be ontology predicates of the same arity. Then, A extensionally subsumes A' under $\langle S, \mathcal{M} \rangle$ iff $\mathcal{M}_{A'}(A)$ is redundant for $\mathcal{M}_A$ under $\langle \emptyset, S \rangle$.*

From the above theorem, it can be easily verified that extensional predicate subsumption is a generalization of the the notion of concept (and role) inclusion in *extensional constraints* (also known as *ABox dependencies*) studied in Description Logics [21,23,8].

As shown by the previous theorem, extensional predicate subsumption reduces to a special case of global mapping redundancy, in which the ontology is empty. Under this simplification, it can be easily verified that this task can be reduced to a containment check between two unions of conjunctive queries (UCQs) in the language $\mathcal{L}_S$. Consequently, the following property holds.

Theorem 8. *For CQ2DCQ mappings, extensional subsumption is decidable iff UCQ containment is decidable in $\mathcal{L}_S$.*

Furthermore, it is immediate to verify that, for *FO2DCQ* mappings, extensional subsumption (as well as extensional emptiness) is undecidable.

4.5   Global Mapping Inconsistency

Given the above undecidability results, it immediately follows that, for *FO2DCQ* mappings, verifying global mapping inconsistency is undecidable.

For *CQ2DCQ* mappings, we present a technique that is able to decide global inconsistency in the case when the source schema S is empty.

Let $\mathcal{M}$ be a *CQ2DCQ* mapping and let $\mathcal{C}_{\mathcal{M}}$ be any set of constant symbols whose arity is the same as the number of variable symbols occurring in the bodies of the mapping assertions in $\mathcal{M}$. We call *grounding of $\mathcal{M}$ over $\mathcal{C}_{\mathcal{M}}$* any mapping obtained from $\mathcal{M}$ by replacing, in every mapping assertion, every variable symbol with a constant from $\mathcal{C}_{\mathcal{M}}$.

Given such a grounding $\mathcal{M}_G$ of $\mathcal{M}$, let $D(\mathcal{M}_G)$ be the source instance containing all the ground atoms that occur in the bodies of the mapping assertions of $\mathcal{M}_G$.

Theorem 9. *Given an OBDA specification $\langle \mathcal{O}, \emptyset, \mathcal{M} \rangle$ where $\mathcal{M}$ is a CQ2DCQ mapping, $\mathcal{M}$ is globally inconsistent for $\langle \mathcal{O}, \emptyset \rangle$ iff there exists no grounding $\mathcal{M}_G$ of $\mathcal{M}$ over $\mathcal{C}_{\mathcal{M}}$ such that $\mathcal{O} \cup \{Retr(\mathcal{M}, D(\mathcal{M}_G))\}$ is satisfiable.*

Notice that checking satisfiability of $\mathcal{O} \cup \{Retr(\mathcal{M}, D(\mathcal{M}_G))\}$ can be reduced to ground atom entailment, in particular, entailment of the ground atom $\bot$ with respect to the theory $\mathcal{O} \cup \{Retr(\mathcal{M}, D(\mathcal{M}_G))\}$. Therefore, from Theorem 9 it follows that decidability of global inconsistency is implied by decidability of ground atom entailment in $\mathcal{L}_{\mathcal{O}}$. For the other direction, the proof easily follows from Theorem 6 and from the fact that mapping inconsistency can be obviously reduced to global mapping inconsistency.

	$\mathcal{L}_S \in$ UCQ-dec, arbitrary $\mathcal{L}_O$	arbitrary $\mathcal{L}_S$, $\mathcal{L}_O \in$ GAE-dec	$\mathcal{L}_S \in$ UCQ-dec, $\mathcal{L}_O \in$ GAE-dec
head-subsumption/inconsistency	U	D	D
body-subsumption/inconsistency	U	U	U
redundancy/inconsistency	U	U	U
ext. subsumption/emptiness	U	U	U
global inconsistency	U	U	U

Results for *FO2DCQ* mappings (D=decidable, U=undecidable).

	$\mathcal{L}_S \in$ UCQ-dec, arbitrary $\mathcal{L}_O$	arbitrary $\mathcal{L}_S$, $\mathcal{L}_O \in$ GAE-dec	$\mathcal{L}_S \in$ UCQ-dec, $\mathcal{L}_O \in$ GAE-dec
head-subsumption/inconsistency	U	D	D
body-subsumption/inconsistency	D	U	D
redundancy/inconsistency	U	U	D
ext. subsumption	D	U	D
global inconsistency*	U	D	D

Results for *CQ2DCQ* mappings
(D=decidable, U=undecidable, *=The result holds only when S is empty).

Fig. 1. Summary of decidability/undecidabilty results

Theorem 10. *For CQ2DCQ mappings and empty source schemas, global mapping inconsistency is decidable if ground atom entailment is decidable in $\mathcal{L}_O$. Moreover, if $\mathcal{L}_S$ allows for CQ-denial formulas and $\mathcal{L}_O$ allows for binary denial formulas, then ground atom entailment is decidable in $\mathcal{L}_O$ and conjunctive query containment is decidable in $\mathcal{L}_S$ if global mapping inconsistency is decidable.*

The results shown in this section are summarized in Figure 1. The figure reports two tables: the first one is relative to the *FO2DCQ* mappping language, while the second one is relative to the *CQ2DCQ* mapping language. In the two tables, we denote by *UCQ-dec* the class of FO languages for which UCQ containment is decidable, and denote by *GAE-dec* the class of FO languages for which entailment of ground atoms is decidable. We remark that the undecidability results for head-inconsistency hold under the assumption that $\mathcal{L}_O$ allows for binary denial formulas (Theorem 2); moreover, for *CQ2DCQ* mappings, the undecidability results for body-inconsistency hold under the assumption that $\mathcal{L}_S$ allows for CQ-denial formulas (Theorem 4), and the undecidability results for mapping inconsistency and global mapping inconsistency hold under the assumption that $\mathcal{L}_O$ allows for binary denial formulas and $\mathcal{L}_S$ allows for CQ-denial formulas (Theorem 6 and Theorem 9).

5   Beyond GAV Mappings

In this section we draw some initial considerations on extending our analysis towards mapping languages beyond GAV. As we have seen, GAV-like mappings enjoy useful properties when it comes to mappings analysis. However, there are cases in OBDA systems where GAV-like mappings are insufficient. For example, consider the simple case of relating the answers of a database query Q to the existential restriction of a role R in a Description Logic ontology. With GAV mappings, the only way to do so is to map Q to a new concept A in the ontology, and add to the ontology the concept inclusion axiom $A \sqsubseteq \exists R$. This may not be desirable, as it clutters the ontology with concepts that may have little relation to the domain being described.

In this section, we therefore consider the languages obtained from $CQ2DCQ$ and $FO2DCQ$ by allowing existential variables to occur in the heads of mapping assertions. Doing so without restriction gives us the languages $FO2CQ$ and $CQ2CQ$, that is, the head of a mapping assertion is simply a conjunctive query over $\mathcal{O}$.

Unfortunately, such an increased expressiveness causes computational complications: for instance, for these two mapping languages the task of query unfolding is equivalent to query answering using views [17], which is much harder than unfolding with GAV-like mappings [14]. Furthermore, given such a mapping $\mathcal{M}$, it cannot be rephrased into a set of mapping assertions with single atoms in the head, as existentially quantified variables may occur in multiple atoms. Thus, Split($\mathcal{M}$) does not yield an equivalent mapping. To address these issues, we consider the languages $FO2CQE$ and $CQ2CQE$, where for each $m \in \mathcal{M}$, $head(m)$ is a conjunctive query over O, and every existential variable in $head(m)$ occurs in exactly one atom. These languages allow us to map queries to existential restrictions of roles, but avoid the difficulties discussed. For example, it is easy to verify that for a mapping $\mathcal{M}$ in either of these two languages, Split($\mathcal{M}$) produces an equivalent mapping.

For all four languages, all the definitions of inconsistency, subsumption and redundancy (with the exception of extensional predicate subsumption) provided by Section 3 apply, as well as Proposition 1 and Proposition 2. For $FO2CQ$ and $CQ2CQ$, we have the following analogue of Theorem 1.

Theorem 11. *For both FO2CQ and CQ2CQ mappings, mapping head-subsumption is decidable iff conjunctive query containment is decidable in $\mathcal{L}_{\mathcal{O}}$.*

For $FO2CQE$ and $CQ2CQE$ we can do better. Since Split($\mathcal{M}$) produces an equivalent mapping in these languages, checking head-subsumption can be done atom by atom. As such, for these languages mapping head-subsumption is decidable iff containment of positive single-atom queries is decidable.

By a similar argument, it is possible to show that, in order to check global mapping inconsistency for these languages over empty source schemas, we likewise need entailment of ground atoms in $\mathcal{L}_{\mathcal{O}}$.

6   Conclusions

In this paper we have formally defined some properties of interest for the mapping component of an OBDA specification, and we have provided several (un)decidability results concerning the task of verifying such properties for some typical mapping languages.

Our study is still in its initial stage, and several further issues need to be investigated. In particular, we left as future work the study of global redundancy and of global inconsistency for OBDA specifications with a non-empty source schema. Furthermore, we intend to extend our analysis to forms of mapping that go beyond the GAV setting (e.g., consider LAV and GLAV [17]), for which we have only provided some preliminary discussion and results in Section 5. Also, we want to study verification of the various forms of subsumption, redundancy, and inconsistency introduced in this paper for concrete instantiations of both the $\mathcal{L}_{\mathcal{O}}$ and the $\mathcal{L}_{\mathcal{S}}$ languages, and characterize its computational complexity.

Finally, we notice that the analysis conducted in this work is based on a "classical" notion of equivalence and subsumption, i.e., equality/containment between the sets of models of two specifications. Recent work in the data exchange area [10,11,19] has studied alternative notions of equivalence for schema mappings. One possible extension of the present work is applying these alternative semantic approaches to the case of OBDA mappings.

Acknowledgments. We thank the anonymous reviewers for precious suggestions. This research has been partially supported by the EU under FP7 project Optique (grant n. FP7-318338).

References

1. Antonioli, N., Castanò, F., Civili, C., Coletta, S., Grossi, S., Lembo, D., Lenzerini, M., Poggi, A., Savo, D.F., Virardi, E.: Ontology-based data access: the experience at the Italian Department of Treasury. In: Proc. of the Industrial Track of CAiSE. CEUR, vol. 1017, pp. 9–16 (2013), ceur-ws.org

2. Arenas, M., Barceló, P., Libkin, L., Murlak, F.: Foundations of Data Exchange. Cambridge University Press (2014)

3. Baader, F., Calvanese, D., McGuinness, D., Nardi, D., Patel-Schneider, P.F. (eds.): The Description Logic Handbook: Theory, Implementation and Applications, 2nd edn. Cambridge University Press (2007)

4. Calì, A., Gottlob, G., Lukasiewicz, T.: A general Datalog-based framework for tractable query answering over ontologies. J. of Web Semantics 14, 57–83 (2012)

5. Calì, A., Lembo, D., Rosati, R.: On the decidability and complexity of query answering over inconsistent and incomplete databases. In: Proc. of PODS, pp. 260–271 (2003)

6. Calvanese, D., De Giacomo, G., Lembo, D., Lenzerini, M., Rosati, R.: Tractable reasoning and efficient query answering in description logics: The *DL-Lite* family. JAR 39(3), 385–429 (2007)

7. Civili, C., Console, M., De Giacomo, G., Lembo, D., Lenzerini, M., Lepore, L., Mancini, R., Poggi, A., Rosati, R., Ruzzi, M., Santarelli, V., Savo, D.F.: MASTRO STUDIO: Managing ontology-based data access applications. PVLDB 6, 1314–1317 (2013)
8. Console, M., Lenzerini, M., Mancini, R., Rosati, R., Ruzzi, M.: Synthesizing extensional constraints in Ontology-Based Data Access. In: Proc. of DL. CEUR, vol. 1014, pp. 628–639 (2013), ceur-ws.org
9. Doan, A., Halevy, A.Y., Ives, Z.G.: Principles of Data Integration. Morgan Kaufmann (2012)
10. Fagin, R., Kolaitis, P.G., Nash, A., Popa, L.: Towards a theory of schema-mapping optimization. In: Proc. of PODS, pp. 33–42 (2008)
11. Gottlob, G., Pichler, R., Savenkov, V.: Normalization and optimization of schema mappings. VLDBJ 20(2), 277–302 (2011)
12. Haarslev, V., Hidde, K., Möller, R., Wessel, M.: The RacerPro knowledge representation and reasoning system. Semantic Web J. 3(3), 267–277 (2012)
13. Haase, P., Horrocks, I., Hovland, D., Hubauer, T., Jiménez-Ruiz, E., Kharlamov, E., Klüwer, J.W., Pinkel, C., Rosati, R., Santarelli, V., Soylu, A., Zheleznyakov, D.: Optique system: Towards ontology and mapping management in OBDA solutions. In: Proc. of WoDOOM, pp. 21–32 (2013)
14. Halevy, A.Y.: Answering queries using views: A survey. VLDBJ 10(4), 270–294 (2001)
15. König, M., Leclère, M., Mugnier, M.-L., Thomazo, M.: On the exploration of the query rewriting space with existential rules. In: Faber, W., Lembo, D. (eds.) RR 2013. LNCS, vol. 7994, pp. 123–137. Springer, Heidelberg (2013)
16. Kontchakov, R., Lutz, C., Toman, D., Wolter, F., Zakharyaschev, M.: The combined approach to ontology-based data access. In: Proc. of IJCAI, pp. 2656–2661 (2011)
17. Lenzerini, M.: Data integration: A theoretical perspective. In: Proc. of PODS, pp. 233–246 (2002)
18. Lenzerini, M.: Ontology-based data management. In: Proc. of CIKM, pp. 5–6 (2011)
19. Pichler, R., Sallinger, E., Savenkov, V.: Relaxed notions of schema mapping equivalence revisited. Theory of Computing Systems 52(3), 483–541 (2013)
20. Poggi, A., Lembo, D., Calvanese, D., De Giacomo, G., Lenzerini, M., Rosati, R.: Linking data to ontologies. J. on Data Semantics X, 133–173 (2008)
21. Rodriguez-Muro, M., Calvanese, D.: High performance query answering over DL-Lite ontologies. In: Proc. of KR, pp. 308–318 (2012)
22. Rodríguez-Muro, M., Kontchakov, R., Zakharyaschev, M.: Ontology-based data access: Ontop of databases. In: Alani, H., et al. (eds.) ISWC 2013, Part I. LNCS, vol. 8218, pp. 558–573. Springer, Heidelberg (2013)
23. Rosati, R.: Prexto: Query rewriting under extensional constraints in DL-Lite. In: Simperl, E., Cimiano, P., Polleres, A., Corcho, O., Presutti, V. (eds.) ESWC 2012. LNCS, vol. 7295, pp. 360–374. Springer, Heidelberg (2012)
24. Shearer, R., Motik, B., Horrocks, I.: HermiT: A highly-efficient OWL reasoner. In: Proc. of OWLED. CEUR, vol. 432 (2008), ceur-ws.org
25. Sirin, E., Parsia, B., Cuenca Grau, B., Kalyanpur, A., Katz, Y.: Pellet: A practical OWL-DL reasoner. J. of Web Semantics 5(2), 51–53 (2007)
26. Tsarkov, D., Horrocks, I.: FaCT++ description logic reasoner: System description. In: Proc. of IJCAR, pp. 292–297 (2006)
27. Venetis, T., Stoilos, G., Stamou, G.B.: Query extensions and incremental query rewriting for OWL 2 QL ontologies. J. on Data Semantics 3(1), 1–23 (2014)

Conjunctive Query Answering in Finitely-Valued Fuzzy Description Logics*

Theofilos Mailis[1], Rafael Peñaloza[1,2], and Anni-Yasmin Turhan[1]

[1] Chair for Automata Theory, Theoretical Computer Science, TU Dresden, Germany
[2] Center for Advancing Electronics Dresden, Germany
{mailis,penaloza,turhan}@tcs.inf.tu-dresden.de

Abstract. Fuzzy Description Logics (DLs) generalize crisp ones by providing membership degree semantics for concepts and roles. A popular technique for reasoning in fuzzy DL ontologies is by providing a reduction to crisp DLs and then employ reasoning in the crisp DL. In this paper we adopt this approach to solve conjunctive query (CQ) answering problems for fuzzy DLs. We give reductions for Gödel, and Łukasiewicz variants of fuzzy $\mathcal{SROIQ}$ and two kinds of fuzzy CQs. The correctness of the proposed reduction is proved and its complexity is studied for different fuzzy variants of $\mathcal{SROIQ}$.

1 Introduction

Description Logics (DLs) are a class of knowledge representation languages with well-defined semantics that are widely used to represent the conceptual knowledge of an application domain in a structured and formally well-understood way. DLs have been successfully employed to formulate ontologies for several knowledge domains such as bio-medical applications. DLs provide the formal foundation for the standard web ontology language OWL, a milestone for the Semantic Web. In this paper we focus on the DL $\mathcal{SROIQ}$, the DL underlying (full) OWL 2.

DLs represent knowledge by means of concepts that correspond to sets of objects, and roles that relate pairs of objects. Ontology axioms are used to restrict the possible interpretations of our domain of interest. For example, we can express the fact that a CPU cpuA is overutilized and that a server that has a part that is overutilized is a server with limited resources by stating:

$$(\mathsf{CPU} \sqcap \mathsf{Overutilized})(\mathsf{cpuA}) \tag{1}$$

$$\mathsf{Server} \sqcap \exists\mathsf{hasPart}.\mathsf{Overutilized} \sqsubseteq \mathsf{ServerWithLimitedResources} \tag{2}$$

Some applications require to describe sets for which there exists no sharp, unambiguous distinction between the members and nonmembers. In our running example Overutilized is such a notion. We can say that cpuA is overutilized to

* Partially supported by DFG SFB 912 (HAEC) and the Cluster of Excellence 'cfAED'.

R. Kontchakov and M.-L. Mugnier (Eds.): RR 2014, LNCS 8741, pp. 124–139, 2014.
© Springer International Publishing Switzerland 2014

Table 1. Families of fuzzy logic operators

Family	t-norm $a \otimes b$	t-conorm $a \oplus b$	negation $\ominus a$	implication $\alpha \Rightarrow b$
Gödel	$\min(a, b)$	$\max(a, b)$	$\begin{cases} 1, & a = 0 \\ 0, & a > 0 \end{cases}$	$\begin{cases} 1, & a \leqslant b \\ b, & a > b \end{cases}$
Łukasiewicz	$\max(a + b - 1, 0)$	$\min(a + b, 1)$	$1 - a$	$\min(1 - a + b, 1)$

a certain degree Overutilized(cpuA) $\geqslant 0.8$. To represent this kind of information faithfully, fuzzy variants of DLs were introduced. Fuzzy DLs generalize crisp DLs by providing membership degree semantics for their concepts and roles by fuzzy sets. The membership degree of an individual to a fuzzy concept can be understood as a weight extending the logic with the possibility of expressing imprecision. Likewise, axioms describing the domain knowledge are equipped with a weight that gives additional flexibility in the restrictions of the membership degrees used. In fuzzy DLs, all crisp set operations are extended to the fuzzy case. The intersection, union, complement and implication set operations are performed by a t-norm function $\otimes$, a t-conorm function $\oplus$, a negation function $\ominus$, and an implication function $\rightarrow$, respectively. These functions or fuzzy operators are grouped in families, also simply called fuzzy logics. It is well known that different families of fuzzy operators lead to fuzzy DLs with different properties. In this paper we concentrate on the families of fuzzy logic operators displayed in Table 1. We use the prefixes f_G and $f_{Ł_n}$ to distinguish between Gödel and Łukasiewicz based semantics. We investigate the reasoning task of conjunctive query answering in these settings. Conjunctive queries are a very powerful way to access the facts in the ontology and it has been widely studied in the recent years for crisp DLs. We are considering finitely-valued fuzzy DLs, since unrestricted fuzzy DLs easily turn undecidable [1,8]. An alternative to implementing a fuzzy DL reasoner from scratch is to reduce reasoning within fuzzy DLs to reasoning in crisp DLs, which allows for the use of existing DL reasoners and to benefit from new optimizations implemented in these systems.

Although there has been a significant amount of work regarding the reduction from fuzzy to crisp DLs, this body of work concentrates mainly on the following problems: instance checking or concept satisfiability [4,23]. In this paper we extend these reductions to the interesting problem of conjunctive query answering. By which we can answer queries that ask for all pairs of servers and CPUs such that the CPU is a part of the server and also is over utilized to an at least 0.6 degree:

$$\text{Server}(\mathsf{x}) \geqslant 1 \land \text{hasPart}(x, y) \geqslant 1 \land \text{CPU}(y) \geqslant 1 \land \text{Overutilized}(y) \geqslant 0.6.$$

The contributions made in this paper are the following:

- We give a reduction from fuzzy $\mathcal{SROIQ}$ under Gödel and Łukasiewicz semantics to $\mathcal{SROIQ}$ for answering conjunctive queries in the finitely-valued setting and prove its correctness. The presented proof builds on the reductions presented in [2,3,4].

- We prove that, if there exists a reduction from the fuzzy DL f-$\mathcal{L}$ to the corresponding crisp DL $\mathcal{L}$ and there exists an algorithm for conjunctive query answering w.r.t. $\mathcal{L}$, then it can also be applied to answer conjunctive queries w.r.t. f-$\mathcal{L}$ in the finitely-valued setting.
- We assess the complexity of the presented conjunctive query answering technique for different fuzzy extensions of the DL $\mathcal{SHIQ}$. $\mathcal{SHIQ}$ is a sublanguage of $\mathcal{SROIQ}$ for which the query answering problem has been studied and solved [12].
- Finally, in order to ensure the correctness of our approach, we have extended the correctness proof sketched in [4] for the Łukasiewicz based extension of $\mathcal{SROIQ}$ (for a detailed proof see the technical report accompanying this paper [16]).

The rest of the paper is structured as follows: Section 2 presents the syntax and semantics of classic and fuzzy DLs based on the DL $\mathcal{SROIQ}$, along with the reduction procedure from the fuzzy to the crisp DL. Section 3 defines the different types of conjunctive queries in the fuzzy setting, while Section 4 presents the actual reduction from fuzzy to crisp conjunctive query answering, along with a proof of its correctness. Finally, Section 5 presents the current literature on reduction techniques and conjunctive query answering for fuzzy DLs, while Section 6 gives an overview of the paper and refers to future work and implementations.

2   Preliminaries

We start with a brief introduction to DL syntax and semantics and present the DL $\mathcal{SROIQ}$ [13]. This specific DL was chosen since: it is one of the most expressive decidable DLs, it provides the direct model-theoretic semantics of OWL 2, and there exists a reduction technique from fuzzy to classic $\mathcal{SROIQ}$ ontologies [2,4,23]. DL ontologies are constructed from countable, and pairwise disjoint sets of individual names N_I, of concept names N_C, and of role names N_R. Individuals correspond to elements of the domain, concept names are used to describe sets of elements, and role names describe binary relations between elements. The set N_S is the subset of N_R containing only *simple* roles. Based on these, concept and role descriptions can be built using different constructors.

In the first and the second column of Table 2 we see most of the concept and role constructors of the highly expressive DL $\mathcal{SROIQ}$ –for a more detailed presentation of the crisp $\mathcal{SROIQ}$ language the reader may refer to [13], while the Gödel and Łukasiewicz fuzzy variants of the language are thoroughly presented in [4,23]–. In Tables 2–4 we have the following notation: $o_1, \ldots, o_m, a, b \in N_I$; $r, r_1, \ldots, r_n \in N_R$; $s \in N_S$; $d_1, \ldots, d_m, d \in (0, 1]_\mathcal{O}$, C, D correspond to concept descriptions, while $\triangleright \in \{\geqslant, >\}$ and $\bowtie \in \{\leqslant, <, \geqslant, >\}$. As usual the simplest form of a concept description is an element $A \in N_C$.

Table 2. Concept constructors from $\mathcal{SROIQ}$

	Syntax	Crisp Semantics	Fuzzy Semantics
Conjunction	$C \sqcap D$	$C^{\mathcal{I}} \cap D^{\mathcal{I}}$	$C^{\mathcal{I}}(x) \otimes D^{\mathcal{I}}(x)$
Disjunction	$C \sqcup D$	$C^{\mathcal{I}} \cup D^{\mathcal{I}}$	$C^{\mathcal{I}}(x) \oplus D^{\mathcal{I}}(x)$
Negation	$\neg C$	$\Delta^{\mathcal{I}} \setminus C^{\mathcal{I}}$	$\ominus C^{\mathcal{I}}(x)$
Value restriction	$\forall r.C$	$\{x \mid \forall y, (x,y) \notin R^{\mathcal{I}} \text{ or } y \in C^{\mathcal{I}}\}$	$\inf_{y \in \Delta^{\mathcal{I}}}\{R^{\mathcal{I}}(x,y) \Rightarrow C^{\mathcal{I}}(y)\}$
Existential restr.	$\exists r.C$	$\{x \mid \exists y, (x,y) \in R^{\mathcal{I}} \text{ and } y \in C^{\mathcal{I}}\}$	$\sup_{y \in \Delta^{\mathcal{I}}}\{R^{\mathcal{I}}(x,y) \otimes C^{\mathcal{I}}(y)\}$
Nominals	$\{o\}$	$\{o^{\mathcal{I}}\}$	1 if $x \in \{o\}$, 0 otherwise
fuzzy	$\{d/o\}$	—	$\sup\{d \mid x = o^{\mathcal{I}}\}$
At-least restr.	$\geq n\; s.C$	$\{x \mid \#\{y : (x,y) \in s^{\mathcal{I}} \text{ and } y \in C^{\mathcal{I}}\} \geq n\}$	$\sup_{y_1,\dots y_n \in \Delta^{\mathcal{I}}} (\min_{i=1}^{n}\{s^{\mathcal{I}}(x,y_i) \otimes C^{\mathcal{I}}(y_i)\}) \otimes \left(\bigotimes_{1 \leq j < k \leq n} \{y_i \neq y_k\}\right)$
At-most restr.	$\leq n\; s.C$	$\{x \mid \#\{y : (x,y) \in s^{\mathcal{I}} \text{ and } y \in C^{\mathcal{I}}\} \leq n\}$	$\inf_{y_1,\dots y_{n+1} \in \Delta^{\mathcal{I}}} (\min_{i=1}^{n+1}\{s^{\mathcal{I}}(x,y_i) \otimes C^{\mathcal{I}}(y_i)\}) \Rightarrow \left(\bigoplus_{1 \leq j < k \leq n+1} \{y_i = y_k\}\right)$

Table 3. $\mathcal{SROIQ}$ TBox axioms

	Syntax	Crisp Semantics	Fuzzy Semantics
GCI	$C \sqsubseteq D$	$C^{\mathcal{I}} \subseteq D^{\mathcal{I}}$	$\inf_{x \in \Delta^{\mathcal{I}}}\{C^{\mathcal{I}}(x) \Rightarrow D^{\mathcal{I}}(x)\} = 1$
fuzzy	$\langle C \sqsubseteq D, \rhd d\rangle$	—	$\inf_{x \in \Delta^{\mathcal{I}}}\{C^{\mathcal{I}}(x) \Rightarrow D^{\mathcal{I}}(x)\} \rhd d$
RI	$r_1 \dots r_n \sqsubseteq r$	$r_1^{\mathcal{I}} \circ \dots \circ r_n^{\mathcal{I}} \subseteq r^{\mathcal{I}}$	$\inf_{x,y \in \Delta^{\mathcal{I}}}\{[r_1^{\mathcal{I}} \circ^{\otimes} \dots \circ^{\otimes} r_n^{\mathcal{I}}](x,y) \Rightarrow r^{\mathcal{I}}(x,y)\} = 1$
fuzzy	$\langle r_1 \dots r_n \sqsubseteq r \rhd d\rangle$	—	$\inf_{x,y \in \Delta^{\mathcal{I}}}\{[r_1^{\mathcal{I}} \circ^{\otimes} \dots \circ^{\otimes} r_n^{\mathcal{I}}](x,y) \Rightarrow r^{\mathcal{I}}(x,y)\} \rhd d$
Inverse role	r^-	$\{(y,x) \mid (x,y) \in r^{\mathcal{I}}\}$	$(r^-)^{\mathcal{I}}(x,y) = r^{\mathcal{I}}(y,x)$
Transitive role	$\text{trans}(r)$	$r^{\mathcal{I}} \circ r^{\mathcal{I}} \subseteq r^{\mathcal{I}}$	

An ontology $\mathcal{O}$ comprises of the intentional and extensional knowledge related to an application domain. The intensional knowledge, i.e. general knowledge about an application domain, is expressed via the Terminological Box (TBox) $\mathcal{T}$ and the Role Box (RBox) $\mathcal{R}$. The extensional knowledge, i.e. particular knowledge about specific situations, is expressed via an Assertional Box (ABox) $\mathcal{A}$ containing statements about individuals. Table 3 presents the syntax of statements for TBoxes and Table 4 that of ABoxes for the crisp and fuzzy variants of $\mathcal{SROIQ}$. As depicted in Tables 3,4, fuzzy ABoxes and TBoxes have the same syntax as their crisp counterparts, while they may also contain fuzzy assertions, fuzzy General Concept Inclusions (GCIs), and fuzzy Role Inclusions (RIs). In order to ensure decidability of the crisp DL, a set of restrictions regarding the use of roles and simple roles in GCIs and RIs is imposed, e.g. a simple role cannot subsume any transitive role, for more details see [13]. The same restrictions are also adopted for the fuzzy versions of $\mathcal{SROIQ}$ [2,3,4].

Table 4. $\mathcal{SROIQ}$ ABox axioms

	Syntax	Crisp Semantics	Fuzzy Semantics
Concept a.	$C(a)$	$a^{\mathcal{I}} \in C^{\mathcal{I}}$	$C^{\mathcal{I}}(a^{\mathcal{I}}) = 1$
fuzzy	$C(a) \bowtie d$	—	$C^{\mathcal{I}}(a^{\mathcal{I}}) \bowtie d$
Role a.	$r(a,b)$	$(a^{\mathcal{I}}, b^{\mathcal{I}}) \in r^{\mathcal{I}}$	$r^{\mathcal{I}}(a^{\mathcal{I}}, b^{\mathcal{I}}) = 1$
fuzzy	$r(a,b) \bowtie d$	—	$r^{\mathcal{I}}(a^{\mathcal{I}}, b^{\mathcal{I}}) \bowtie d$
Negated role a.	$\neg r(a,b)$	$(a^{\mathcal{I}}, b^{\mathcal{I}}) \notin r^{\mathcal{I}}$	$r^{\mathcal{I}}(a^{\mathcal{I}}, b^{\mathcal{I}}) = 0$
fuzzy	$\neg r(a,b) \bowtie d$	—	$\ominus r^{\mathcal{I}}(a^{\mathcal{I}}, b^{\mathcal{I}}) \bowtie d$
Inequality a.	$a \neq b$	$a^{\mathcal{I}} \neq b^{\mathcal{I}}$	$a^{\mathcal{I}} \neq b^{\mathcal{I}}$
Equality a.	$a = b$	$a^{\mathcal{I}} = b^{\mathcal{I}}$	$a^{\mathcal{I}} = b^{\mathcal{I}}$

Example 1. Based on the concept assertion and inclusion axioms presented and explained in equation 2 we can create the following crisp ABox and TBox:

$$\mathcal{A} := \{\mathsf{CPU}(\mathsf{cpuA}), \mathsf{Overutilized}(\mathsf{cpuA})\}$$

$$\mathcal{T} := \{\mathsf{Server} \sqcap \exists \mathsf{hasPart}.\mathsf{Overutilized} \sqsubseteq \mathsf{ServerWithLimitedResources}\}$$

where $\mathsf{cpuA} \in N_I$; $\mathsf{CPU}, \mathsf{Overutilized}, \mathsf{ServerWithLimitedResources}, \mathsf{Server} \in N_C$; and $\mathsf{hasPart} \in N_R$. As expected $\mathsf{Server} \sqcap \exists \mathsf{hasPart}.\mathsf{Overutilized}$ corresponds to a complex concept description. A fuzzy version of the previous ABox can occur if for example we add a degree of truth to the concept assertion $\mathsf{Overutilized}(\mathsf{cpuA})$. The fuzzy assertion $\mathsf{Overutilized}(\mathsf{cpuA}) \geqslant 0.8$ states that cpuA is overutilized with a degree of at least 0.8.

The semantics of crisp $\mathcal{SROIQ}$ are given via an interpretation $\mathcal{I}$ that is a pair $(\Delta^{\mathcal{I}}, \cdot^{\mathcal{I}})$ consisting of a non empty set $\Delta^{\mathcal{I}}$ and an interpretation function $\cdot^{\mathcal{I}}$ mapping every individual $a \in N_I$ onto an element $a^{\mathcal{I}} \in \Delta^{\mathcal{I}}$, every concept name $A \in N_C$ to a set $A^{\mathcal{I}} \subseteq \Delta^{\mathcal{I}}$, every atomic role $r \in N_R$ onto a relation $r^{\mathcal{I}} \subseteq \Delta^{\mathcal{I}} \times \Delta^{\mathcal{I}}$. The interpretations of complex concepts, GCIs and assertions are presented on the third column of Tables 2,3, and 4.

In a fuzzy extension of $\mathcal{SROIQ}$, concepts denote fuzzy sets of individuals and roles denote fuzzy binary relations. Likewise fuzzy axioms may hold to some degree. The semantics of $f\text{-}\mathcal{SROIQ}$ is given via interpretations $\mathcal{I}$ that are pairs $(\Delta^{\mathcal{I}}, \cdot^{\mathcal{I}})$ consisting of a non empty set $\Delta^{\mathcal{I}}$ and an interpretation function $\cdot^{\mathcal{I}}$ mapping every individual $a \in N_I$ to an element $a^{\mathcal{I}} \in \Delta^{\mathcal{I}}$, every concept name $A \in N_C$ onto a membership function $A^{\mathcal{I}} : \Delta^{\mathcal{I}} \to [0,1]$, every atomic role $r \in N_R$ onto a membership function $r^{\mathcal{I}} : \Delta^{\mathcal{I}} \times \Delta^{\mathcal{I}} \to [0,1]$. In the finitely-valued setting, which we consider here, the membership function mapps to a finite subset of $[0,1]$. The interpretations of complex concepts, TBox axioms, and ABox assertions are presented on the fourth column of Tables 3,4, for the different families of fuzzy logic operators ($\otimes, \oplus, \ominus, \Rightarrow$) presented in Table 1. Based on the semantics reasoning services can be defined. In this paper we are interested in conjunctive query answering.

Definition 2 (Conjunctive Query for Classic DLs - CQ [12]). *Let N_V be a countably infinite set of variables disjoint from N_C, N_R, and N_I. An atom* *is an expression* $A(x)$ *(concept atom) or* $r(x,y)$ *(role atom), where* $A \in N_C$, $r \in N_R$, *and* $x, y \in N_V \cup N_I$. *A* conjunctive query q *is a non-empty set of atoms. Intuitively, such a set represents the conjunction of its elements. We use* Var(q) *to denote the set of variables occurring in the query* q. *Let* $\mathcal{I}$ *be an interpretation,* q *a conjunctive query, and* $\pi : Var(q) \to \Delta^{\mathcal{I}}$ *a total function, s.t.* $\pi(a) = a^{\mathcal{I}}$ *for all* $a \in N_I$. *We write:* $\mathcal{I} \models^{\pi} C(x)$ *if* $\pi(x) \in C^{\mathcal{I}}$ *and* $\mathcal{I} \models^{\pi} r(x,y)$ *if* $(\pi(x), \pi(y)) \in r^{\mathcal{I}}$. *If* $\mathcal{I} \models^{\pi}$ *at for all atoms* $at \in q$, *we write* $\mathcal{I} \models^{\pi} q$ *and call* π *a match for* $\mathcal{I}$ *and* q. *We say that* $\mathcal{I}$ *satisfies* q *and write* $\mathcal{I} \models q$ *if there is a match* π *for* $\mathcal{I}$ *and* q. *If* $\mathcal{I} \models q$ *for all models* $\mathcal{I}$ *of an ontology* $\mathcal{O}$, *we write* $\mathcal{O} \models q$ *and say that* $\mathcal{O}$ *entails* q.

Finally, a *union of conjunctive queries* q_{UCQ} is a set of conjunctive queries. We write $\mathcal{O} \models q_{UCQ}$ and say that $\mathcal{O}$ entails q_{UCQ} if for every model $\mathcal{I}$ of $\mathcal{O}$ we have that $\mathcal{I} \models q$ for some conjunctive query $q \in q_{UCQ}$.

Reduction to the Crisp Case

The goal is to devise a reduction of answering UCQs over a fuzzy ontology $\mathcal{O}$ to answer UCQs over a crisp ontology $\mathcal{O}_C$. The basic idea is that each concept and role in $\mathcal{O}$ is mapped onto a set of concepts and roles corresponding to their α-cuts, which is the crisp set containing all elements that belong to a fuzzy set up to a given degree. For example, if the concept Overutilized in $\mathcal{O}$ maps each CPU to the degree to which it is overutilized, then the concept Overutilized$_{\geq 0.6}$ in $\mathcal{O}_C$ represents the set of CPUs that are overutilized to a degree of at least 0.6.

We present the reduction algorithm for the fuzzy versions of $\mathcal{SROIQ}$ corresponding to the Gödel, and Łukasiewicz based semantics. We employ the notation $[0,1]_{\mathcal{O}}$ in order to represent the *finite* set of degrees that appear in our ontology. We also use the notation $(a,b)_{\mathcal{O}}$ to represent the $(a,b) \cap [0,1]_{\mathcal{O}}$ subset of $[0,1]_{\mathcal{O}}$.

It has been proved for fuzzy ontologies under Gödel logics that the set of degrees of truth that must be considered for any reasoning task is the set $[0,1]_{\mathcal{O}} \cup \{0,1\}$ [4]. In order to ensure that the reduction technique can be applied for f-$\mathcal{SROIQ}$ with Łukasiewicz based semantics, we need restrict to a finite number of degrees that have the form of $\{0, \frac{1}{n}, \ldots, \frac{n-1}{n}, 1\}$ where n is a natural number [4]. From now on when using the notation $[0,1]_{\mathcal{O}}$ we consider that the corresponding set satisfies this restriction when referring to a Łukasiewicz based fuzzy DL.

A compact form of the reduction rules from fuzzy to crisp $\mathcal{SROIQ}$ is displayed in Table 5. It should be noted that the uppercase bold letters in this Table correspond to the conditions illustrated in Table 6. For a detailed description of the reduction rules the reader may refer to [3,4]. The reduced ontology $\mathcal{O}_C$ has the following form:

- In order to preserve the semantics of α-cuts of atomic concepts and roles the following axioms are added to $\mathcal{T}_C$ for every $A \in N_C$, $r \in N_R$:

$$A_{\geq d_{i+1}} \sqsubseteq A_{> d_i} \quad A_{> d_i} \sqsubseteq A_{\geq d_i}$$
$$r_{\geq d_{i+1}} \sqsubseteq r_{> d_i} \quad r_{> d_i} \sqsubseteq r_{\geq d_i} \tag{3}$$

Table 5. Mapping of concept and role expressions in fuzzy $\mathcal{SROIQ}$

Reduction of concepts and axioms	Gödel / Lukasiewicz
$\rho(A, \geqslant d)$	$A_{\geqslant d}$
$\rho(A, \leqslant d)$	$\neg A_{> d}$
$\rho(\neg C, \geqslant d)$	$\rho(C, \leqslant 0) \ / \ \neg\rho(C, > 1 - d)$
$\rho(\neg C, \leqslant d)$	$\rho(C, > 0) \ / \ \rho(C, \geqslant 1 - d)$
$\rho(C \sqcap D, \geqslant d)$	$\rho(C, \geqslant d) \sqcap \rho(D, \geqslant d) \ / \ \bigsqcup_{\mathbf{A}} \left(\rho(C, \geqslant d_1) \sqcap \rho(D, \geqslant d_2) \right)$
$\rho(C \sqcap D, \leqslant d)$	$\rho(C, \leqslant d) \sqcup \rho(D, \leqslant d) \ / \ \rho(\neg C \sqcup \neg D, \geqslant 1 - d)$
$\rho(C \sqcup D, \geqslant d)$	$\rho(C, \geqslant d) \sqcup \rho(D, \geqslant d) \ /$
	$\rho(C, \geqslant d) \sqcup \rho(D, \geqslant d) \sqcup \bigsqcup_{\mathbf{B}} \left(\rho(C, \geqslant d_1) \sqcap \rho(D, \geqslant d_2) \right)$
$\rho(C \sqcup D, \leqslant d)$	$\rho(C, \leqslant d) \sqcap \rho(D, \leqslant d) \ / \ \rho(\neg C \sqcap \neg D, \geqslant 1 - d)$
$\rho(\exists r.C, \geqslant d)$	$\exists\rho(r, \geqslant d).\rho(C, \geqslant d) \ / \ \bigsqcup_{\mathbf{A}} \left(\exists\rho(r, \geqslant d_1).\rho(C, \geqslant d_2) \right)$
$\rho(\exists r.C, \leqslant d)$	$\forall\rho(r, > d).\rho(C \leqslant d) \ / \ \rho(\forall r.\neg C, \geqslant 1 - d)$
$\rho(\forall r.C, \geqslant d)$	$\bigsqcap_{\mathbf{C}} (\forall\rho(r, \geqslant d').\rho(C, \geqslant d')) \sqcap \bigsqcap_{\mathbf{D}} (\forall\rho(r, > d').\rho(C, > d')) \ /$
	$\bigsqcap_{\mathbf{E}} (\forall\rho(r, \geqslant d_1).\rho(C, \geqslant d_2))$
$\rho(\forall r.C, \leqslant d)$	$\bigsqcup_{\mathbf{F}} (\exists\rho(r, > d).\rho(C, \leqslant d)) \ / \ \rho(\exists r.\neg C, \geqslant 1 - d)$
$\rho(\cup_{i=1}^{m}\{d_i/o_i\}, \bowtie d)$	$\{o_i \mid d_i \bowtie d, 1 \leqslant i \leqslant m\}$
$\rho(\geqslant m \ s.C, \geqslant d)$	$\geqslant m \ \rho(s, \geqslant d).\rho(C, \geqslant d) \ /$
	$\bigsqcup_{\mathbf{G}} (\exists\rho(s, \geqslant d_1).\rho(B_1 \sqcap \rho(C, \geqslant e_1)) \sqcap \quad \cdots$
	$\sqcap \exists\rho(s, \geqslant d_m).(B_m \sqcap \rho(C, \geqslant e_m)))$
$\rho(\geqslant m \ s.C, \leqslant d)$	$\leqslant m - 1 \ \rho(s, > d).\rho(C, > d) \ /$
	$\neg \Big(\bigsqcup_{\mathbf{H}} (\exists\rho(s, \geqslant d_1).\rho(B_1 \sqcap \rho(C, \geqslant e_1)) \sqcap \quad \cdots$
	$\sqcap \exists\rho(s, \geqslant d_m).(B_m \sqcap \rho(C, \geqslant e_m))) \Big)$
$\rho(\leqslant n \ s.C, \geqslant d)$	$\leqslant n \ \rho(s, > 0).\rho(C, > 0) \ / \ \rho(\neg(\geqslant n + 1 \ s.C), \geqslant d)$
$\rho(\leqslant n \ s.C, \leqslant d)$	$\geqslant n + 1 \ \rho(s, > 0).\rho(C, > 0) \ / \ \rho(\neg(\geqslant n + 1 \ s.C), \leqslant d)$
$\rho(r, \geqslant d)$	$r_{\geqslant d}$
$\rho(r, \leqslant d)$	$\neg r_{> d}$
$\rho(r^{-}, \geqslant d)$	$r_{\geqslant d}^{-}$
$\rho(r^{-}, \leqslant d)$	$\neg r_{> d}^{-}$
$\kappa(C(a) \bowtie d)$	$\rho(C, \bowtie d)(a)$
$\kappa(r(a, b) \bowtie d)$	$\rho(r, \bowtie d)(a, b)$
$\kappa(\langle C \sqsubseteq D \geqslant d \rangle)$	$\bigcup_{\mathbf{C}} (\rho(C, \geqslant d') \sqsubseteq \rho(D, \geqslant d')) \cup \bigcup_{\mathbf{D}} (\rho(C, > d') \sqsubseteq \rho(D, > d')) \ /$
	$\bigcup_{\mathbf{I}} (\rho(C, \geqslant d_1 \sqsubseteq \rho(D, \geqslant d_2))$
$\kappa(\langle r_1 \ldots r_n \sqsubseteq r \geqslant d \rangle)$	$\bigcup_{\mathbf{C}} (\rho(r_1, \geqslant d') \ldots \rho(r_n, \geqslant d') \sqsubseteq \rho(r, \geqslant d')) \cup$
	$\bigcup_{\mathbf{D}} (\rho(r_1, \geqslant d') \ldots \rho(r_n, \geqslant d') \sqsubseteq \rho(r, \geqslant d')) \ /$
	$\bigcup_{\mathbf{J}} (\rho(r_1, \geqslant d_1) \ldots \rho(r_n, \geqslant d_n) \sqsubseteq \rho(r, \geqslant d_{n+1}))$

where d_i, d_{i+1} correspond to every pair of degrees d_i, d_{i+1} such that (i) $d_{i+1} > d_i$, (ii) there exists no element $e \in [0, 1]_{\mathcal{O}}$ such that $d_{i+1} > e > d_i$, and (iii) the subscript $_{>1}$ is not considered in any of the GCIs. For the f_{L_n} variant of $\mathcal{SROIQ}$, since admitting for a finite truth space, we must add to our ontology that $A_{> d_i} \equiv A_{\geqslant d_{i+1}}$ and $r_{> d_i} \equiv r_{\geqslant d_{i+1}}$.

- For each complex concept C appearing in $\mathcal{O}$ the complex concept $\rho(C, \bowtie d)$ in $\mathcal{O}_C$ represents its corresponding α-cut. These complex concepts are inductively defined according to the set of reduction rules presented in the first part of Table 5.
- Each ABox axiom in $\mathcal{A}$ is represented by its corresponding axiom in $\mathcal{A}_C$ presented in the second part of Table 5.

Table 6. Conditions corresponding to the uppercase letters of Table 5

A. for every pair $d_1, d_2 \in (0,1]_\mathcal{O}$ such that $d_1 + d_2 = 1 + d$.
B. for every pair $d_1, d_2 \in (0,1]_\mathcal{O}$ such that $d_1 + d_2 = d$.
C. for every $d' \in (0,1]_\mathcal{O}$ such that $d' \leqslant d$.
D. for every $d' \in [0,1]_\mathcal{O}$ such that $d' < d$.
E. for every pair $d_1, d_2 \in (0,1]_\mathcal{O}$ such that $d_1 = d_2 + 1 - d$.
F. for every $d' \in [0,1]_\mathcal{O}$ such that $d' \leqslant d$.
G. for every combination of $d_1, e_1, \ldots d_m, e_m \in (0,1]_\mathcal{O}$ such that $d_i + e_i = 1 + d$, for $i = \{1, \ldots, m\}$.
H. for every combination of $d_1, e_1, \ldots d_m, e_m \in (0,1]_\mathcal{O}$ such that (i) $d_i + e_i > 1 + d$, for $i = \{1, \ldots, m\}$, (ii) $\nexists d' \in (0,1]_\mathcal{O}$ such that $d' < d_i$ and $d' + e_i > 1 + d$, (iii) $\nexists d' \in (0,1]_\mathcal{O}$ such that $d' < e_i$ and $d' + d_i > 1 + d$.
I. for every pair $d_1, d_2 \in (0,1]_\mathcal{O}$ such that $d_1 = d_2 + 1 - d$.
J. for every combination of $d_1, \ldots d_{n+1} \in (0,1]_\mathcal{O}$ such that: $d_1 + \ldots + d_n = d_{n+1} + n - d$.

- Each TBox axiom in $\mathcal{T}$ is represented by its corresponding axiom or set of axioms in $\mathcal{T_C}$ according to the set of reduction rules presented in the third part of Table 5.

3   Conjunctive Queries for Fuzzy DLs

Our main objective is to find an algorithm for answering to conjunctive queries for fuzzy DLs based on a reduction procedure to classic ones. Different forms of conjunctive queries for fuzzy DLs have been proposed in the literature. According to [19], these are classified to queries of two different types, namely threshold conjunctive queries and general fuzzy queries. With respect to the example provided on the introduction a threshold query of the form:

$$\mathsf{Server}(x) \geqslant 1 \wedge \mathsf{hasPart}(x, y) \geqslant 1 \wedge \mathsf{CPU}(y) \geqslant 1 \wedge \mathsf{Overutilized}(y) \geqslant 0.6 \quad (4)$$

searches for all pairs of servers and CPUs such that the CPU is a part of the server and is also overutilized to a degree of at least 0.6. In contrast, a fuzzy query of the form:

$$\mathsf{Server}(x) \wedge \mathsf{hasPart}(x, y) \wedge \mathsf{CPU}(y) \wedge \mathsf{Overutilized}(y) \quad (5)$$

searches for the pairs of elements that satisfy it along with the degree of satisfaction (provided that this degree is greater than 0).

Definition 3 (Threshold Conjunctive Query - $\mathbf{CQ_\theta}$). *Let N_V be a countably infinite set of variables disjoint from N_C, N_R, and N_I. A degree atom is an expression $P(\overline{X}) \rhd d$ where $P \in N_C \cup N_R$, $\overline{X}$ is an ordered tuple of elements* of $N_I \cup N_V$ having an arity of 1 if $P \in N_C$ and 2 if $P \in N_R$, $\rhd \in \{\geqslant, >\}$, and $d \in (0, 1]$. A Threshold Conjunctive Query has the form:

$$\bigwedge_{i=1}^{\lambda} P_i(\overline{X}_i) \rhd_i d_i$$

We use $VarIndivs(q_\theta)$ to denote the set of variables and individuals occurring in a CQ_θ named q_θ. Let $\mathcal{I}$ be an interpretation and $\pi : VarIndivs(q) \to \Delta^{\mathcal{I}}$ a total function that maps each element $a \in N_I$ to $a^{\mathcal{I}}$. If $P_i^{\mathcal{I}}(\pi(\overline{X}_i)) \rhd_i d_i$ for all degree atoms in q_θ, we write $\mathcal{I} \models^\pi q_\theta$ and call π a match for $\mathcal{I}$ and q_θ. We say that $\mathcal{I}$ satisfies q_θ and write $\mathcal{I} \models q_\theta$ if there is a match π for $\mathcal{I}$ and q_θ. If $\mathcal{I} \models q_\theta$ for all models $\mathcal{I}$ of an ontology $\mathcal{O}$, we write $\mathcal{O} \models q_\theta$ and say that $\mathcal{O}$ entails q_θ.

Definition 4 (Fuzzy Conjunctive Query - CQ$_\phi$). *A plain atom is an expression $P(\overline{X})$. A Fuzzy Conjunctive Query with plain atoms has the form:*

$$\bigwedge_{i=1}^{\lambda} P_i(\overline{X}_i)$$

Let $\mathcal{I}$ be an interpretation, q_ϕ a CQ_ϕ, π a mapping, and $\otimes$ a fuzzy logic t-norm –we assume that the t-norms of the query and the DL are the same–. If $P_i^{\mathcal{I}}(\pi(\overline{X}_i)) = d_i$ for all atoms in q_ϕ and $\otimes_{i=1}^{\kappa} d_i \geqslant d$ we write $\mathcal{I} \models^\pi q_\phi \geqslant d$ and call π a match for $\mathcal{I}$ and q_ϕ with a degree of at least d. We say that $\mathcal{I}$ satisfies q_ϕ with a degree of at least d and write $\mathcal{I} \models q_\phi \geqslant d$ if there is a corresponding match. If $\mathcal{I} \models q_\phi \geqslant d$ for all models $\mathcal{I}$ of a an ontology $\mathcal{O}$, we write $\mathcal{O} \models q_\phi \geqslant d$ and say that $\mathcal{O}$ entails q_ϕ with a degree of at least d. The problem of determining whether $\mathcal{O} \models q_\phi > d$ is defined analogously.

The query entailment problem for a CQ_θ is to decide whether $\mathcal{O} \models q_\theta$ for a given assignment of the variables. For CQ_ϕs we may consider two variants of the query entailment problem, namely to decide whether $\mathcal{O} \models q_\phi \geqslant d$ for some degree $d \in (0, 1]$, and to find the degree $\sup\{d \mid \mathcal{O} \models q_\phi \geqslant d\}$. Since the f_G, $f_{Ł_n}$ variants of $\mathcal{SROIQ}$ admit for the finite truth space $[0, 1]_\mathcal{O}$ we can assume without loss of generality that the two problems can be reduced to each other. The query answering problem requests for the specific assignments that satisfy the query, thus the reduction can be achieved by testing all assignments, which give an exponential blow-up. It is well-known from crisp DLs that query entailment and query answering can be mutually reduced and that decidability and complexity results carry over [7] modulo the mentioned blow-up.

Example 5. Suppose that we have the queries described in equations 4,5, the ABox

$$\mathcal{A} = \{\mathsf{Server}(s_1) \geqslant 1, \mathsf{hasPart}(s_1, cpu_1) \geqslant 1,$$
$$\mathsf{CPU}(cpu_1) \geqslant 1, \mathsf{Overutilized}(cpu_1) \geqslant 0.7\}$$

and an empty TBox $\mathcal{T}$. Then the answer to equation 4 would be the pair (s_1, cpu_1), while the answer to equation 5 would be (s_1, cpu_1) with a degree of at least 0.7.

A *union of* $CQ_\theta s$ is a set of $CQ_\theta s$. An ontology $\mathcal{O}$ entails such a union U_{q_θ}, i.e. $\mathcal{O} \models U_{q_\theta}$, when for every model $\mathcal{I} \models \mathcal{O}$ there exists some $q_\theta \in U_{q_\theta}$ such that $\mathcal{I} \models q_\theta$. Another type of union is one consisting of a set of $CQ_\phi s$. An ontology $\mathcal{O}$ entails such a union U_{q_ϕ} to a degree of at least $d \in (0, 1]$, i.e. $\mathcal{O} \models U_{q_\phi} \geqslant d$, when for every model $\mathcal{I} \models \mathcal{O}$ there exists some $q_\phi \in U_{q_\phi}$ such that $\mathcal{I} \models q_\phi \geqslant d$.

Remark 6. In the context of the reduction algorithms, we focus on $\geqslant, >$ inequalities appearing in threshold/fuzzy conjunctive queries. A threshold conjunctive query with $\leqslant, <$ inequalities would be reduced to a crisp conjunctive query containing negated role atoms. Moreover, the reduction of a fuzzy conjunctive query q_ϕ with a less or equal degree $\kappa(q_\phi, \leqslant d)$ would be reduced to a disjunction of negated atoms. Since the problems of negated atoms and disjunctive queries have not been studied for expressive classic DLs, we focus on $\geqslant, >$ inequalities.

4   Conjunctive Query Answering by Reduction

In this section we provide the corresponding steps so as to solve the problem of conjunctive query answering for fuzzy DLs by taking advantage of existing crisp DL algorithms for the same problem. The solution we provide operates on the DLs and is based on the reduction techniques presented in [4,23]. We denote with κ the reduction process from $CQ_\theta s$ and $CQ_\phi s$ queries to crisp CQs and UCQs. The reduction process operates differently for each query type.

For the CQ_θ described in Definition 3 the reduction process takes the following form:

$$\kappa \left(\bigwedge_{i=1}^{\lambda} P_i(\overline{X}_i) \rhd d_i \right) = \bigwedge_{i=1}^{\lambda} \rho(P_i, \rhd d_i)(\overline{X}_i) \qquad (6)$$

Since P_i is either a concept name $A_i \in N_C$ or a role name $r_i \in N_R$ we have that $\rho(A_i, \rhd d_i) = A_{i \rhd d_i}$ or $\rho(r_i, \rhd d_i) = r_{i \rhd d_i}$ as presented on Table 5.

The reduction process for a CQ_ϕ has two inputs, the first input is the query itself and the second input is the degree that we want to examine. In addition, for $CQ_\phi s$ the reduction process depends on the t-norm operator that has been adopted to provide semantics for conjunction. For the CQ_ϕ described in Definition 4 the reduction process takes the form presented in equation 7 when the CQ_ϕ refers to an f_G-$\mathcal{SROIQ}$ ontology. When the CQ_ϕ refers to an f_{L_n}-$\mathcal{SROIQ}$ ontology the corresponding reduction is the union of conjunctive queries presented in equation 8 ($\otimes$ in equation 8 stands for the Łukasiewicz t-norm operator).

$$\kappa \left(\bigwedge_{i=1}^{\lambda} P_i(\overline{X}_i), \geqslant d \right) = \bigwedge_{i=1}^{\lambda} \rho(P_i, \geqslant d)(\overline{X}_i) \qquad (7)$$

$$\kappa \left(\bigwedge_{i=1}^{\lambda} P_i(\overline{X}_i), \geqslant d \right) = \bigcup_{\otimes_{i=1}^{\lambda} d_i = d \text{ and } d_i \in [0,1]_{\mathcal{O}}} \left\{ \bigwedge_{i=1}^{\lambda} \rho(P_i, \geqslant d_i)(\overline{X}_i) \right\} \qquad (8)$$

Example 7. The reduced form of the CQ_θ presented in equation 4 follows in equation 9. The reduced form of the CQ_ϕ in equation 5 for the degree of at least 0.75 for the f_G-$\mathcal{SROIQ}$ logic follows in equation 10. Finally if we consider the f_{L_4}-$\mathcal{SROIQ}$ logic we have that $[0,1]_\mathcal{O} = \{0, 0.25, 0.5, 0.75, 1\}$ and the reduced form of equation 5 for the degree of at least 0.75 is the UCQ presented in equation 11:

$$\mathsf{Server}_{\geqslant 1}(x) \wedge \mathsf{hasPart}_{\geqslant 1}(x,y) \wedge \mathsf{CPU}_{\geqslant 1}(y) \wedge \mathsf{Overutilized}_{\geqslant 0.6}(y) \qquad (9)$$

$$\mathsf{Server}_{\geqslant 0.75}(x) \wedge \mathsf{hasPart}_{\geqslant 0.75}(x,y) \wedge \mathsf{CPU}_{\geqslant 0.75}(y) \wedge \mathsf{Overutilized}_{\geqslant 0.75}(y) \qquad (10)$$

$$\begin{aligned}
&\{\mathsf{Server}_{\geqslant 0.75}(x) \wedge \mathsf{hasPart}_{\geqslant 1}(x,y) \wedge \mathsf{CPU}_{\geqslant 1}(y) \wedge \mathsf{Overutilized}_{\geqslant 1}(y)\} \cup \\
&\{\mathsf{Server}_{\geqslant 1}(x) \wedge \mathsf{hasPart}_{\geqslant 0.75}(x,y) \wedge \mathsf{CPU}_{\geqslant 1}(y) \wedge \mathsf{Overutilized}_{\geqslant 1}(y)\} \cup \\
&\{\mathsf{Server}_{\geqslant 1}(x) \wedge \mathsf{hasPart}_{\geqslant 1}(x,y) \wedge \mathsf{CPU}_{\geqslant 0.75}(y) \wedge \mathsf{Overutilized}_{\geqslant 1}(y)\} \cup \\
&\{\mathsf{Server}_{\geqslant 1}(x) \wedge \mathsf{hasPart}_{\geqslant 1}(x,y) \wedge \mathsf{CPU}_{\geqslant 1}(y) \wedge \mathsf{Overutilized}_{\geqslant 0.75}(y)\}
\end{aligned} \qquad (11)$$

The following Theorem states that our query reduction algorithm is sound and complete. Since we consider the f_G, f_{L_n} variants of $\mathcal{SROIQ}$ the theorem applies for these DLs and only. A generalization of the theorem follows in Corollaries 9,10.

Theorem 8. *Let $\mathcal{O}_C$ be the crisp version of the fuzzy Ontology $\mathcal{O}$ such that $\kappa(\mathcal{O}) = \mathcal{O}_C$, q_θ be a CQ_θ and $\kappa(q_\theta)$ its form obtained by the reduction, q_ϕ is a CQ_ϕ and $\kappa(q_\phi, \geqslant d)$ its reduced form for the degree $d \in [0,1]_\mathcal{O}$. Then the following equivalences apply:*

1. $\mathcal{O} \models q_\theta \Leftrightarrow \mathcal{O}_C \models \kappa(q_\theta)$
2. $\mathcal{O} \models q_\phi \geqslant d \Leftrightarrow \mathcal{O}_C \models \kappa(q_\phi, \geqslant d)$.

Proof (Sketch). In order to prove that $\mathcal{O}_C \models \kappa(q_\theta) \Rightarrow \mathcal{O} \models q_\theta$, we build for every model $\mathcal{I}$ of $\mathcal{O}$ a non fuzzy interpretation $\mathcal{I}_C = \{\Delta^{\mathcal{I}_C}, \cdot^{\mathcal{I}_C}\}$ as follows:

$$\Delta^{\mathcal{I}_C} = \Delta^{\mathcal{I}} \qquad A^{\mathcal{I}_C}_{\rhd d} = \{\beta \mid A^{\mathcal{I}}(\beta) \rhd d\}$$

$$a^{\mathcal{I}_C} = a^{\mathcal{I}} \qquad r^{\mathcal{I}_C}_{\rhd d} = \{(\beta, \gamma) \mid r^{\mathcal{I}}(\beta, \gamma) \rhd d\}. \qquad (12)$$

It is shown in [3,4] that $\mathcal{I}_C$ is a model of the crisp ontology $\mathcal{O}_C$. Since $\mathcal{I}_C \models \mathcal{O}_C$ and $\mathcal{O}_C \models \kappa(q_\theta)$ it applies that $\mathcal{I}_C \models \kappa(q_\theta)$. Based on the construction of $\mathcal{I}_C$ and the form of $\kappa(q_\theta)$ (equation 6), it can be verified that $\mathcal{I} \models q_\theta$ must also apply. It can be shown in a similar way that $\mathcal{O} \models q_\phi \geqslant d \Rightarrow \mathcal{O}_C \models \kappa(q_\phi, \geqslant d)$.

The proof of the opposite direction is performed by building a fuzzy interpretation $\mathcal{I}$ for each model $\mathcal{I}_C$ of $\mathcal{O}_C$ as follows:

$$\Delta^{\mathcal{I}} = \Delta^{\mathcal{I}_C} \qquad A^{\mathcal{I}}(\beta) = \sup\left\{d \mid \beta \in A^{\mathcal{I}_C}_{\geqslant d}\right\} \cup \left\{d_+ \mid \beta \in A^{\mathcal{I}_C}_{>d}\right\}$$

$$a^{\mathcal{I}} = a^{\mathcal{I}_C} \qquad r^{\mathcal{I}}(\beta, \gamma) = \sup\left\{d \mid (\beta, \gamma) \in r^{\mathcal{I}_C}_{\geqslant d}\right\} \cup \left\{d_+ \mid (\beta, \gamma) \in r^{\mathcal{I}_C}_{>d}\right\} \qquad (13)$$

where the degree d_+ for the language of f_G-$\mathcal{SROIQ}$ is defined to be some degree in $[0,1]$ such that $d < d_+$ and there exists no $d' \in [0,1]_\mathcal{O}$ with $d < d' < d_+$. For the language of f_{L_n}-$\mathcal{SROIQ}$ the degree d_+ is defined in a similar way with the main difference that it has to belong to $[0,1]_\mathcal{O}$. □

Corollary 9. *If (i) there is a reduction technique from a fuzzy DL f-$\mathcal{L}$ to a crisp DL $\mathcal{L}$, (ii) for each model $\mathcal{I}$ of an ontology $\mathcal{O}$ in the DL of f-$\mathcal{L}$ there exists a corresponding model $\mathcal{I}_C$ for the reduced ontology $\mathcal{O}_C$ that can be built based on equation 12, (iii) for each model $\mathcal{I}_C$ of the reduced ontology $\mathcal{O}_C$ there exists a corresponding model $\mathcal{I}$ for the initial ontology $\mathcal{O}$ that can be built based on equation 13, (iv) there exists a query answering algorithm for the DL of f-$\mathcal{L}$ then: the reduction technique can be applied in order to answer to threshold queries for the DL of f-$\mathcal{L}$.*

Corollary 10 (Generalization of Corollary 9). *If f-$\mathcal{L}$ and $\mathcal{L}$ satisfy criteria (i), (ii), (iii) presented in Corollary 9 and (iv) f-$\mathcal{L}_{sub}$ is a sub-language of f-$\mathcal{L}$ (v) f-$\mathcal{L}_{sub}$ can be reduced to a sub-language $\mathcal{L}_{sub}$ of $\mathcal{L}$ for which there exists a query answering algorithm, then the reduction technique can be applied in order to answer to threshold queries for the language of f-$\mathcal{L}_{sub}$.*

Since there are algorithms for conjunctive query answering for the DLs $\mathcal{SHIQ}$ [12] and Horn fragments of $\mathcal{SROIQ}$ [18] (both are sub-languages of $\mathcal{SROIQ}$), we can apply the reduction technique for conjunctive query answering for the language of f-$\mathcal{SHIQ}$ and Horn fragments of f-$\mathcal{SROIQ}$ (where f correspond to one of $f_G, f_Ł$ fuzzy logics).

Complexity Results

Complexity in the size of $\mathcal{O}$. According to [3], the reduction process for the DL f_G-$\mathcal{SROIQ}$ creates an ontology $\mathcal{O}_C$ that has size $O(|\mathcal{O}|^2)$ compared to the initial ontology $\mathcal{O}$. If we combine the latter with the facts that: i) *"conjunctive query entailment in the crisp $\mathcal{SHIQ}$ can be decided in time exponential in the size of the ontology [12]"*(†) ii) the language $\mathcal{SHIQ}$ is a sublanguage of $\mathcal{SROIQ}$, we get an exponential complexity with respect to the size of the initial ontology. Regarding the $f_{Ł_n}$-$\mathcal{SROIQ}$ DL, the size of the resulting ontology $\mathcal{O}_C$ is $O(|\mathcal{O}|\,|[0,1]_\mathcal{O}|^k)$ in case no number restrictions occur in $\mathcal{O}$, where k is the maximal depth of the concepts appearing in $\mathcal{O}$ (proof in [4]). Intuitively the depth of some $A \in N_C$ is 1 while the depth of $\exists r.(\forall r.A)$ is 3. The latter results are discouraging, with the absence of number restrictions the size of $\mathcal{O}_C$ may become exponential w.r.t. the size of $\mathcal{O}$. If we combine these results with fact (†) we get a double exponential upper bound for threshold query answering w.r.t. a $f_{Ł_n}$-$\mathcal{SHIQ}$ ontology, even with the absence of number restrictions.

Complexity in the size of the query. We examine the complexity w.r.t. the size of the examined threshold/fuzzy conjunctive query. Suppose that $\kappa(q_\theta)$ is the reduced form of a threshold conjunctive query denoted with q_θ. We have that the size of $\kappa(q_\theta)$ is linear to the size of q_θ. The size of the reduced form $\kappa(q_\phi, \geq d)$ of a fuzzy conjunctive query also remains linear w.r.t. the size of the initial fuzzy query q_ϕ if we consider the f_G-$\mathcal{SHIQ}$ semantics. For fuzzy conjunctive queries under the Łukasiewicz semantics, the size of $\kappa(q_\phi, \geq d)$ belongs to the complexity class $O(|[0,1]_\mathcal{O}|^{k-1})$ where k is the number of conjuncts in q_ϕ. Therefore it is exponential compared to the size of q_ϕ. If we combine the latter results with the fact that *"conjunctive query entailment in $\mathcal{SHIQ}$ can be decided in time double exponential in the size of the query [12]"* we get a double exponential complexity for threshold query answering and fuzzy conjunctive query answering w.r.t. f_G-$\mathcal{SHIQ}$ ontologies. Otherwise, we get a triple exponential upper bound for fuzzy query answering w.r.t. a $f_{\text{Ł}_n}$-$\mathcal{SHIQ}$ ontology.

Generalizing the Query Component

So far we have examined the reduction technique for answering threshold and fuzzy CQs. These two types of queries are immediate extensions of the classic CQ problem. Nevertheless, the existence of degrees may lead to more general forms of fuzzy CQs in which the score of a query is computed via a monotone scoring function:

Example 11. Lets extend the query in Equation 5 by asking for servers that have overutilized CPU and RAM memory, while the utilization of the CPU is more important than that of the RAM memory. The resulting query will take the form:

$$\text{Server}(x) \wedge \text{hasPart}(x,y) \wedge \text{CPU}(y) \wedge \text{hasPart}(x,z) \wedge$$
$$\text{RAM}(z) \wedge \frac{0.6 \cdot \text{Overutilized}(y) + 0.4 \cdot \text{Overutilized}(z)}{2} \quad (14)$$

where the fraction corresponds to an aggregation scoring function that takes into account the degree of overutilization of a CPU and the degree of overutilization of a RAM memory with weights 0.6 and 0.4 respectively.

Such kind of queries have already been defined in the literature [19,26,27] and the question is if the reduction technique can be applied to answer them. By taking account the fact that the reduction technique works on finite valued fuzzy DLs, these problems can be solved by considering for all possible combinations of degrees in $[0,1]_\mathcal{O}$. We consider the previous example for the $f_{\text{Ł}_n}$-$\mathcal{SROIQ}$ with $[0,1]_\mathcal{O} = \{0, 0.25, 0.5, 0.75, 1\}$, where the concepts Server, CPU, RAM and the role hasPart are essentially crisp. The (crisp) conjunctive query

$$\text{Server}_{\geqslant 1}(x) \wedge \text{hasPart}_{\geqslant 1}(x,y) \wedge \text{CPU}_{\geqslant 1}(y) \wedge \text{hasPart}_{\geqslant 1}(x,z) \wedge$$
$$\text{RAM}_{\geqslant 1}(z) \wedge \text{Overutilized}_{\geqslant 0.25}(y) \wedge \text{Overutilized}_{\geqslant 0.75}(z)$$

if applied on the reduced ontology will return the triples of Server, CPU, and RAM that satisfy the query in Equation 14 with a degree greater or equal than 0.45 (i.e. $0.6 \cdot 0.25 + 0.4 \cdot 0.75$).

Another interesting problem, specific to fuzzy DLs, is the top-k query answering problem presented in [25,26,27]. This variation of the fuzzy query answering problem focuses on the k answers with the highest degrees of satisfaction. In a naive approach to solve this problem, the reduction technique for CQ_ϕs can be iteratively applied starting from the highest to the lowest degrees in $[0,1]_\mathcal{O}$ until the limit of k answers is reached. It has to be investigated if a more sophisticated approach can be adopted to solve this problem.

5   Related Work

Non-fuzzy representations of fuzzy DLs have been extensively studied for several families of DLs that can be classified based on their fuzzy and DL parts. Reduction techniques have been proposed in [23,22,2] for the DLs of f-$\mathcal{ALCH}$, f-$\mathcal{SHOIN}$, and f-$\mathcal{SROIQ}$, that are based on the Zadeh fuzzy logic semantics. An experimental evaluation of the reduction technique for the DL of f_{KD}-$\mathcal{SHIN}$ is presented in [11]. A reduction procedure for the $\mathcal{SROIQ}$ DL under Gödel semantics is considered in [3], while in [4] the reduction technique for the finitely many valued Łukasiewicz fuzzy Description Logic f_{L_n}-$\mathcal{SROIQ}$ is studied. Based on a different approach, a family of fuzzy DLs using α-cuts as atomic concepts and roles is considered in [15], while a generalization of existing approaches where a finite totally ordered set of linguistic terms or labels is assumed is presented in [5].

Conjunctive query answering for fuzzy DLs has been mostly studied for the fuzzy DL-Lite family of DLs. In [24,25] the problem of evaluating ranked top-k queries in the Description logic fuzzy DL-Lite is considered, while a variety of query languages by which a fuzzy DL-Lite knowledge base can be queried is presented in [19]. Tableaux based approaches for conjunctive query answering have also been studied. A tableaux algorithm for conjunctive query answering for the language of fuzzy CARIN, a knowledge representation language combining the DL f_Z-$\mathcal{ALCNR}$ with Horn rules, is provided in [17] . An algorithm for answering expressive fuzzy conjunctive queries is presented in [10,9]. The algorithm allows the occurrence of both lower bound and the upper bound of thresholds in a query atom, over the DLs f_Z-$\mathcal{ALCN}$, and f_Z-$\mathcal{SHIN}$. Finally, practical approach for storing and querying fuzzy knowledge in the semantic web have been investigated in [21].

6   Conclusions and Future Work

This paper describes how non fuzzy representation of fuzzy DLs can be adopted in order to solve the threshold and fuzzy conjunctive query answering problems. Specifically, the previously mentioned problems on fuzzy DLs are reduced to their equisatisfiable conjunctive query (or union of conjunctive queries) answering problems on crisp DLs. The correctness of the suggested techniques is proved and their complexity is studied for different fuzzy variants of the $\mathcal{SROIQ}$ DL. As far as we know no similar theoretical results have been presented. The proofs rely on the fact that each model of a fuzzy ontology $\mathcal{O}$ can be mapped to a model of its reduced crisp form $\mathcal{O}_C$ and vice versa (soundness and completeness of the reduction technique), while they are based on the structure of the two constructed models. To verify the correctness of our approach we have extended the correctness proofs sketched in [4]. Therefore this paper can be considered complementary to the existing literature on non fuzzy representation of fuzzy DLs.

Our current line of works involves implementing the reduction techniques for the f_Z, f_G, and f_{L_n} variants of $\mathcal{SROIQ}$. The upcoming implementation is based on the HermiT OWL Reasoner [20] extended with the OWL BGP SPARQL wrapper [14] that is used for conjunctive query answering. Future work involves evaluating the proposed reduction techniques on real data, studying their performance, and examining if available optimizations techniques for fuzzy and crisp DLs can be applied to improve the performance of these algorithms. Another interesting line of work involves applying these reduction based threshold and fuzzy query answering algorithms for the more general family of finite lattice based fuzzy DLs presented in [6].

References

1. Baader, F., Borgwardt, S., Peñaloza, R.: On the decidability status of fuzzy $\mathcal{ALC}$ with general concept inclusions. Journal of Philosophical Logic (2014)
2. Bobillo, F., Delgado, M., Gómez-Romero, J.: Optimizing the Crisp Representation of the Fuzzy Description Logic $\mathcal{SROIQ}$. In: da Costa, P.C.G., d'Amato, C., Fanizzi, N., Laskey, K.B., Laskey, K.J., Lukasiewicz, T., Nickles, M., Pool, M. (eds.) URSW 2005 - 2007. LNCS (LNAI), vol. 5327, pp. 189–206. Springer, Heidelberg (2008)
3. Bobillo, F., Delgado, M., Gómez-Romero, J., Straccia, U.: Fuzzy Description Logics under Gödel semantics. International Journal of Approximate Reasoning 50(3), 494–514 (2009)
4. Bobillo, F., Straccia, U.: Reasoning with the finitely many-valued Łukasiewicz fuzzy Description Logic $\mathcal{SROIQ}$. Information Sciences 181(4), 758–778 (2011)
5. Bobillo, F., Straccia, U.: Finite fuzzy Description Logics and crisp representations. In: Bobillo, F., Costa, P.C.G., d'Amato, C., Fanizzi, N., Laskey, K.B., Laskey, K.J., Lukasiewicz, T., Nickles, M., Pool, M. (eds.) URSW 2008-2010/UniDL 2010. LNCS, vol. 7123, pp. 99–118. Springer, Heidelberg (2013)
6. Borgwardt, S., Peñaloza, R.: Consistency reasoning in lattice-based fuzzy Description Logics. International Journal of Approximate Reasoning (2013)
7. Calvanese, D., De Giacomo, G., Lenzerini, M.: On the decidability of query containment under constraints. In: Proceedings of the Seventeenth ACM SIGACT-SIGMOD-SIGART Symposium on Principles of Database Systems, pp. 149–158. ACM (1998)
8. Cerami, M., Straccia, U.: On the (un)decidability of fuzzy Description Logics under Łukasiewicz t-norm. Information Sciences 227, 1–21 (2013)
9. Cheng, J., Ma, Z.M., Zhang, F., Wang, X.: Deciding query entailment for fuzzy $\mathcal{SHIN}$ ontologies. In: Gómez-Pérez, A., Yu, Y., Ding, Y. (eds.) ASWC 2009. LNCS, vol. 5926, pp. 120–134. Springer, Heidelberg (2009)
10. Cheng, J., Ma, Z.M., Zhang, F., Wang, X.: Deciding query entailment in fuzzy Description Logic knowledge bases. In: Bhowmick, S.S., Küng, J., Wagner, R. (eds.) DEXA 2009. LNCS, vol. 5690, pp. 830–837. Springer, Heidelberg (2009)
11. Cimiano, P., Haase, P., Ji, Q., Mailis, T., Stamou, G., Stoilos, G., Tran, D.T., Tzouvaras, V.: Reasoning with large A-Boxes in fuzzy Description Logics using DL reasoners: an experimental evaluation. In: Proceedings of the ESWC Workshop on Advancing Reasoning on the Web: Scalability and Commonsense (2008)
12. Glimm, B., Horrocks, I., Lutz, C., Sattler, U.: Conjunctive query answering for the Description Logic $\mathcal{SHIQ}$. Journal of Artificial Intelligence Research 31, 157–204 (2008)

13. Horrocks, I., Kutz, O., Sattler, U.: The even more irresistible $\mathcal{SROIQ}$. In: Proc. of the 10th International Conference of Knowledge Representation and Reasoning (KR 2006), vol. 6, pp. 57–67 (2006)

14. Kollia, I., Glimm, B.: Optimizing SPARQL query answering over OWL ontologies. Journal of Artificial Intelligence Research 48, 253–303 (2013)

15. Li, Y., Xu, B., Lu, J., Kang, D., Wang, P.: A family of extended fuzzy Description Logics. In: Computer Software and Applications Conference, vol. 1, pp. 221–226. IEEE (2005)

16. Mailis, T., Peñaloza, R., Turhan, A.Y.: Conjunctive query answering in finitely-valued fuzzy Description Logics. Tech. rep., Chair of Automata Theory, Institute of Theoretical Computer Science, Technische Universität Dresden, Dresden, Germany (2014)

17. Mailis, T., Stoilos, G., Stamou, G.: Expressive reasoning with Horn rules and fuzzy Description Logics. Knowledge and Information Systems 25(1), 105–136 (2010)

18. Ortiz, M., Rudolph, S., Šimkus, M.: Query answering in the Horn fragments of the description logics $\mathcal{SHOIQ}$ and $\mathcal{SROIQ}$. In: Proceedings of the Twenty-Second International Joint Conference on Artificial Intelligence, vol. 2, pp. 1039–1044. AAAI Press (2011)

19. Pan, J.Z., Stamou, G.B., Stoilos, G., Thomas, E.: Expressive querying over fuzzy DL-Lite ontologies. In: 20th International Workshop on Description Logics (2007)

20. Shearer, R., Motik, B., Horrocks, I.: Hermit: A highly-efficient OWL reasoner. In: Proc. of the 5th Int. Workshop on OWL: Experiences and Directions (OWLED), Karlsruhe, Germany (2008)

21. Simou, N., Stoilos, G., Tzouvaras, V., Stamou, G.B., Kollias, S.D.: Storing and querying fuzzy knowledge in the Semantic Web. In: 7th International Workshop on Uncertainty Reasoning For the Semantic Web, Karlsruhe, Germany (2008)

22. Stoilos, G., Stamou, G.B.: Extending fuzzy Description Logics for the Semantic Web. In: 3rd International Workshop of OWL: Experiences and Directions, Innsbruck (2007)

23. Straccia, U.: Transforming fuzzy Description Logics into classical Description Logics. In: Alferes, J.J., Leite, J. (eds.) JELIA 2004. LNCS (LNAI), vol. 3229, pp. 385–399. Springer, Heidelberg (2004)

24. Straccia, U.: Answering vague queries in fuzzy DL-Lite. In: Proceedings of the 11th International Conference on Information Processing and Management of Uncertainty in Knowledge-Based Systems (IPMU 2006), pp. 2238–2245 (2006)

25. Straccia, U.: Towards top-k query answering in Description Logics: the case of DL-Lite. In: Fisher, M., van der Hoek, W., Konev, B., Lisitsa, A. (eds.) JELIA 2006. LNCS (LNAI), vol. 4160, pp. 439–451. Springer, Heidelberg (2006)

26. Straccia, U.: Foundations of Fuzzy Logic and Semantic Web Languages. CRC Press (2013)

27. Straccia, U.: On the top-k retrieval problem for ontology-based access to databases. In: Dehne, F., Fiala, F., Koczkodaj, W.W. (eds.) ICCI 1991. LNCS, vol. 497, pp. 95–114. Springer, Heidelberg (1991)

Exchange-Repairs: Managing Inconsistency in Data Exchange

Balder ten Cate[1,2], Richard L. Halpert[1], and Phokion G. Kolaitis[1,3]

[1] University of California Santa Cruz, USA
[2] LogicBlox, Inc., USA
[3] IBM Research - Almaden, USA
{btencate,rhalpert,kolaitis}@ucsc.edu

Abstract. In a data exchange setting with target constraints, it is often the case that a given source instance has no solutions. Intuitively, this happens when data sources contain inconsistent or conflicting information that is exposed by the target constraints at hand. In such cases, the semantics of target queries trivialize, because the certain answers of every target query over the given source instance evaluate to "true". The aim of this paper is to introduce and explore a new framework that gives meaningful semantics in such cases by using the notion of exchange-repairs. Informally, an exchange-repair of a source instance is another source instance that differs minimally from the first, but has a solution. In turn, exchange-repairs give rise to a natural notion of exchange-repair certain answers (in short, XR-certain answers) for target queries in the context of data exchange with target constraints.

After exploring the structural properties of exchange-repairs, we focus on the problem of computing the XR-certain answers of conjunctive queries. We show that for schema mappings specified by source-to-target GAV dependencies and target equality-generating dependencies (egds), the XR-certain answers of a target conjunctive query can be rewritten as the consistent answers (in the sense of standard database repairs) of a union of source conjunctive queries over the source schema with respect to a set of egds over the source schema, thus making it possible to use a consistent query-answering system to compute XR-certain answers in data exchange. In contrast, we show that this type of rewriting is not possible for schema mappings specified by source-to-target LAV dependencies and target egds. We then examine the general case of schema mappings specified by source-to-target GLAV constraints, a weakly acyclic set of target tgds and a set of target egds. The main result asserts that, for such settings, the XR-certain answers of conjunctive queries can be rewritten as the certain answers of a union of conjunctive queries with respect to the stable models of a disjunctive logic program over a suitable expansion of the source schema.

Keywords: data exchange, certain answers, database repairs, consistent query answering, disjunctive logic programming, stable models.

R. Kontchakov and M.-L. Mugnier (Eds.): RR 2014, LNCS 8741, pp. 140–156, 2014.

1   Introduction and Summary of Contributions

Data exchange is the problem of transforming data structured under one schema, called the source schema, into data structured under a different schema, called the target schema, in such a way that pre-specified constraints on these two schemas are satisfied. Data exchange is a ubiquitous data inter-operability task that has been explored in depth during the past decade (see [3]). This task is formalized with the aid of schema mappings $\mathcal{M} = (\mathbf{S}, \mathbf{T}, \Sigma_{st}, \Sigma_t)$, where $\mathbf{S}$ is the source schema, $\mathbf{T}$ is the target schema, Σ_{st} is a set of constraints between $\mathbf{S}$ and $\mathbf{T}$, and Σ_t is a set of constraints on $\mathbf{T}$. The most thoroughly investigated schema mappings are the ones in which Σ_{st} is a set of source-to-target tuple-generating dependencies (s-t tgds) and Σ_t is a set of target tuple-generating dependencies (target tgds) and target equality-generating dependencies (target egds) [11].

Every schema mapping $\mathcal{M} = (\mathbf{S}, \mathbf{T}, \Sigma_{st}, \Sigma_t)$ gives rise to two distinct algo- rithmic problems. The first is the existence and construction of solutions: given a source instance I, determine whether a *solution* for I exists (i.e., a target in- stance J so that (I, J) satisfies $\Sigma_{st} \cup \Sigma_t$) and, if it does, construct such a "good" solution. The second is to compute the *certain answers* of target queries, where if q is a target query and I is a source instance, then certain$(q, I, \mathcal{M})$ is the inter- section of the sets $q(J)$, as J varies over all solutions for I. For arbitrary schema mappings specified by s-t tgds and target tgds and egds, both these problems can be undecidable [16]. However, as shown in [11], if the set Σ_t of target tgds obeys a mild structural condition, called *weak acyclicity*, then both these prob- lems can be solved in polynomial time using the *chase procedure*. This procedure, given a source instance I, attempts to build a "most general" solution J for I by generating facts that satisfy each s-t tgd and each target tgd as needed, and by equating two nulls or equating a null to a constant, as dictated by the egds. If the chase procedure encounters an egd that equates two distinct constants, then it terminates and reports that no solution for I exists. Otherwise, it constructs a *universal* solution J for I, which can also be used to compute the certain answers of conjunctive queries in time bounded by a polynomial in the size of I.

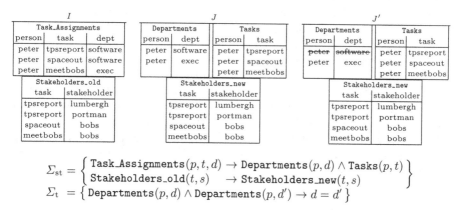

$$\Sigma_{st} = \left\{ \begin{array}{l} \texttt{Task\_Assignments}(p, t, d) \rightarrow \texttt{Departments}(p, d) \land \texttt{Tasks}(p, t) \\ \texttt{Stakeholders\_old}(t, s) \quad \rightarrow \texttt{Stakeholders\_new}(t, s) \end{array} \right\}$$
$$\Sigma_t = \left\{ \texttt{Departments}(p, d) \land \texttt{Departments}(p, d') \rightarrow d = d' \right\}$$

Fig. 1. An instance I, an inconsistent result of a chase J, and a subset-repair J' of J

(I_1, J_1)

Task_Assignments

person	task	dept
~~peter~~	~~tpsreport~~	~~software~~
~~peter~~	~~spaceout~~	~~software~~
peter	meetbobs	exec

Departments		Tasks	
person	dept	person	task
peter	exec		
		peter	meetbobs

(I_2, J_2)

Task_Assignments

person	task	dept
peter	tpsreport	software
peter	spaceout	software
~~peter~~	~~meetbobs~~	~~exec~~

Departments		Tasks	
person	dept	person	task
peter	software	peter	tpsreport
		peter	spaceout

(I_3, J_3)

Task_Assignments

person	task	dept
peter	tpsreport	software
~~peter~~	~~spaceout~~	~~software~~
~~peter~~	~~meetbobs~~	~~exec~~

Departments		Tasks	
person	dept	person	task
peter	software	peter	tpsreport

Fig. 2. Three repairs of $(I, \emptyset)$ w.r.t. $\Sigma_{st} \cup \Sigma_t$ (**Stakeholders** tables omitted)

Consider the situation in which the chase terminates and reports that no solution exists. In such cases, for every boolean target query q, the certain answers certain$(q, I, \mathcal{M})$ evaluate to "true". Even though the certain answers have become the standard semantics of queries in the data exchange context, there is clearly something unsatisfactory about this state of affairs, since the certain answers trivialize when no solutions exist. Intuitively, the root cause for the lack of solutions is that the source instance contains inconsistent or conflicting information that is exposed by the target constraints of the schema mapping at hand. In turn, this suggests that alternative semantics for target queries could be obtained by adopting the notions of database *repairs* and *consistent answers* from the study of inconsistent databases (see [5] for an overview). One concrete possibility is to perform all chase steps involving s-t tgds and target tgds, and then treat the result as an inconsistent target instance in need of repair. Unfortunately, this *exchange-then-repair* approach fails to take into account how the data in the target instance was derived, and, as a result, may give unreasonable answers to some queries. Figure 1 gives an example in which a company, Initech, migrates their database instance I by splitting the **Task_Assignments** table and applying a key constraint to the resulting **Departments** table requiring that each employee belongs to no more than one department. Notice that the subset repair J' places peter in the exec department, yet still has him performing tasks for the software department – the fact that tpsreport and spaceout are derived from a tuple placing peter in the software department has been lost. The only other repair of J similarly fails to reflect the shared origin of tuples in the **Tasks** and **Departments** tables, and this disconnect manifests in the consistent answers to queries over J. For example, in the exchange-then-repair approach, $q(p, s) = \exists t\, \text{Tasks}(p, t) \wedge \text{Stakeholders\_new}(t, s)$ gives (peter,bobs), (peter,portman), and (peter,lumbergh). However, the last two tuples are derived from facts placing peter in the software department, which in J' he is not.

The situation is no better if we treat data exchange as just one part of a repair problem involving the union $\Sigma_{st} \cup \Sigma_t$ of all constraints and the pair $(I, \emptyset)$ consisting of the source instance I and the empty target instance (in the spirit of [14]). While the first two symmetric-difference[1] repairs in Figure 2 seem reasonable, in the third symmetric-difference repair we have eliminated Task_Assignments(peter, spaceout, software), even though our key constraint is already satisfied by the removal of Task_Assignments(peter, meetbobs, exec) alone. In symmetric-difference repairs, it is equally valid to satisfy a violated tgd by removing a tuple as by adding a tuple. However, in a data exchange setting, the target instance is initially empty, so it is natural to satisfy violated tgds by deriving new tuples rather than by deleting existing tuples.

Our aim in this paper is to introduce and explore a new framework that gives meaningful and non-trivial semantics to queries in data exchange, including cases in which no solutions exist for a given source instance. At the conceptual level, the main contribution is the introduction of the notion of an *exchange-repair*. Informally, an exchange-repair of a source instance is another source instance that differs minimally from the first, but has a solution. In turn, exchange-repairs give rise to a natural notion of *exchange-repair certain answers* (in short, *XR-certain answers*) for target queries in the context of data exchange. Note that if a source instance I has a solution, then the XR-certain answers of target queries on I coincide with the certain answers of the queries on I. If I has no solutions, then, unlike the certain answers, the XR-certain answers are non-trivial.

The idea of using a repair-based semantics in cases where no solutions exist was also considered by Calì et al. in the context of data integration [7]; moreover, the semantics used in [7] was motivated by similar semantics for repairs of inconsistent databases in [6]. However, the approach in [7] suffers from the same problem as other exchange-then-repair approaches; for example, the target instance J' is a possible world in their setting, but not in ours.

After exploring the structural properties of exchange-repairs, we focus on the problem of computing the XR-certain answers of conjunctive queries. We show that for schema mappings specified by source-to-target GAV (global-as-view) dependencies and target egds, the XR-certain answers of a target conjunctive query can be rewritten as the consistent answers (in the sense of standard database repairs) of a union of source conjunctive queries over the source schema with respect to a set of egds over the source schema, thus making it possible to use a consistent query-answering system to compute XR-certain answers in data exchange. In contrast, we show that this type of rewriting is not possible for schema mappings specified by source-to-target LAV (local-as-view) dependencies and target egds. We then examine the general case of schema mappings specified by s-t tgds, a weakly acyclic set of target tgds and a set of target egds. The main result asserts that, for such settings, the XR-certain answers of conjunctive queries can be rewritten as the certain answers of a union of conjunctive queries with respect to the stable models of a disjunctive logic program over a suitable expansion of the source schema.

Due to space limitations, we will restrict ourselves to giving hints of proofs.

---

[1] Subset-repairs would always leave the target instance empty. Superset-repairs would fail whenever there is an egd violation equating two constants.

2 Preliminaries

This section contains definitions of basic notions and a minimum amount of background material. Detailed information about schema mappings and certain answers can be found in [3, 11], and about repairs and consistent answers in [4, 5].

Schema Mappings and Certain Answers. A *tuple-generating dependency (tgd)* is an expression of the form $\forall \mathbf{x}(\phi(\mathbf{x}) \rightarrow \exists \mathbf{y} \psi(\mathbf{x}, \mathbf{y}))$, where $\phi(\mathbf{x})$ and $\psi(\mathbf{x}, \mathbf{y})$ are conjunctions of atoms over some relational schema. Assume that we have two disjoint relational schemas $\mathbf{S}$ and $\mathbf{T}$, called the *source schema* and the *target schema*. A *source-to-target tgd (s-t tgd)* is a tgd as above such that $\phi(\mathbf{x})$ is a conjunction of atoms over $\mathbf{S}$ and $\psi(\mathbf{x}, \mathbf{y}))$ is a conjunction of atoms over $\mathbf{T}$.

Tgds are also known as GLAV (global-and-local-as-view) constraints. Two important special cases are the GAV constraints and the LAV constraints: the former are the tgds of the form $\forall \mathbf{x}(\phi(\mathbf{x}) \rightarrow P(\mathbf{x}))$ and the latter are the tgds of the form $\forall \mathbf{x}(R(\mathbf{x}) \rightarrow \exists \mathbf{y} \psi(\mathbf{x}, \mathbf{y}))$, where P and R are individual relation symbols.

An *equality-generating dependency (egd)* is an expression of the form $\forall \mathbf{x}(\phi(\mathbf{x}) \rightarrow x_i = x_j)$ with $\phi(\mathbf{x})$ a conjunction of atoms over a relational schema.

For the sake of readability, we will frequently drop the universal quantifiers when writing tgds and egds.

A *schema mapping* is a quadruple $\mathcal{M} = (\mathbf{S}, \mathbf{T}, \Sigma_{\mathrm{st}}, \Sigma_{\mathrm{t}})$, where $\mathbf{S}$ is a source schema, $\mathbf{T}$ is a target schema, Σ_{st} is a finite set of s-t tgds, and Σ_{t} is a finite set of target tgds and target egds (i.e., tgds and egds over the target schema).

We will use the notation GLAV, GAV, LAV, EGD to denote the classes consisting of finite sets of, respectively, GLAV constraints, GAV constraints, LAV constraints, and egds. If C is a class of sets of source-to-target dependencies and D is a class of sets of target dependencies, then the notation $C+D$ denotes the class of all schema mappings $\mathcal{M} = (\mathbf{S}, \mathbf{T}, \Sigma_{\mathrm{st}}, \Sigma_{\mathrm{t}})$ such that Σ_{st} is a member of C and Σ_{t} is a member of D. For example, GLAV+EGD denotes the class of all schema mappings $\mathcal{M} = (\mathbf{S}, \mathbf{T}, \Sigma_{\mathrm{st}}, \Sigma_{\mathrm{t}})$ such that Σ_{st} is a finite set of s-t tgds and Σ_{t} is a finite set of egds. Moreover, we will use the notation (D_1, D_2) to denote the union of two classes D_1 and D_2 of sets of target dependencies. For example, GAV+(GAV, EGD) denotes the class of all schema mappings $\mathcal{M} = (\mathbf{S}, \mathbf{T}, \Sigma_{\mathrm{st}}, \Sigma_{\mathrm{t}})$ such that Σ_{st} is a set of GAV s-t tgds and Σ_{t} is the union of a finite set of GAV target tgds with a finite set of target egds.

Let $\mathcal{M} = (\mathbf{S}, \mathbf{T}, \Sigma_{\mathrm{st}}, \Sigma_{\mathrm{t}})$ be a schema mapping. A target instance J is a *solution* for a source instance I w.r.t. $\mathcal{M}$ if the pair (I, J) satisfies $\mathcal{M}$, i.e., (I, J) satisfies Σ_{st}, and J satisfies Σ_{t}. A *universal* solution for I is a solution J for I such that if J' is a solution for I, then there is a homomomorphism h from J to J' that is the identity on the active domain of I. If $\mathcal{M} = (\mathbf{S}, \mathbf{T}, \Sigma_{\mathrm{st}}, \Sigma_{\mathrm{t}})$ is an arbitrary schema mapping, then a given source instance may have no solution or it may have a solution, but no universal solution. However, if Σ_{t} is the union of a *weakly acyclic* set of target tgds and a set of egds, then a solution exists if and only if a universal solution exists. Moreover, the *chase procedure* can be used to determine if, given a source instance I, a solution for I exists and, if it does, to actually construct a universal solution $chase(I, \mathcal{M})$ for I in time polynomial in the size of I (see [11] for details). The definition of weak acyclicity is as follows.

Definition 1. Let Σ be a set of tgds over a schema $\mathbf{T}$. Construct a directed graph, called the *dependency graph*, as follows:
- Nodes: For every pair (R, A) with R a relation symbol of the schema and A an attribute of R, there is a distinct node; call such a pair (R, A) a *position*.
- Edges: For every tgd $\forall \mathbf{x}(\phi(\mathbf{x}) \rightarrow \exists \mathbf{y}\, \psi(\mathbf{x}, \mathbf{y}))$ in Σ and for every x in $\mathbf{x}$ that occurs in ψ, and for every occurrence of x in ϕ in position (R, A_i):
  1. For every occurrence of x in ψ in position (S, B_j), add an edge $(R, A_i) \rightarrow (S, B_j)$ (if it does not already exist).
  2. For every existentially quantified variable y and for every occurrence of y in ψ in position (T, C_k), add a *special edge* $(R, A_i) \rightarrow (T, C_k)$ (if it does not already exists).

We say that Σ is *weakly acyclic* if the dependency graph has no cycle going through a special edge. We say that a tgd θ is *weakly acyclic* if the singleton set $\{\theta\}$ is weakly acyclic.

WA-GLAV denotes the class of all finite weakly acyclic sets of target tgds.

The tgd $\forall x \forall y (E(x, y) \rightarrow \exists z\, E(x, z))$ is weakly acyclic; in contrast, the tgd $\forall x \forall y (E(x, y) \rightarrow \exists z\, E(y, z))$ is not, because the dependency graph contains a special self-loop. Moreover, every set of GAV tgds is weakly acyclic, since the position graph contains no special edges in this case.

We will also make heavy use of the notion of *rank*. Let Σ be a finite weakly acyclic set of tgds. For every node (R, A) in the dependency graph of Σ, define an *incoming path* to be any (finite or infinite) path ending in (R, A). Define the *rank* of (R, A), denoted by $rank(R, A)$, as the maximum number of special edges on any such incoming path. Since Σ is weakly acyclic, there are no cycles going through special edges; hence, $rank(R, A)$ is finite. The *rank* of Σ, denoted $rank(\Sigma)$ is the maximum of $rank(R, A)$ over all positions (R, A) in the dependency graph of Σ.

If q is query over the target schema $\mathbf{T}$ and I is a source instance, then the *certain answers* of q with respect to $\mathcal{M}$ are defined as

$$\text{certain}(q, I, \mathcal{M}) = \bigcap \{q(J) : \ J \text{ is a solution for } I \text{ w.r.t. } \mathcal{M}\}$$

Repairs and Consistent Answers. Let Σ be a set of constraints over some relational schema. An *inconsistent* database is a database that violates at least one constraint in Σ. Informally, a *repair* of an inconsistent database I is a consistent database I' that differs from I in a "minimal" way. This notion can be formalized in several different ways.

1. A *symmetric-difference-repair* of I, denoted $\oplus$-repair of I, is an instance I' that satisfies Σ and where there is no instance I'' such that $I \oplus I'' \subset I \oplus I'$ and I'' satisfies Σ. Here, $I \oplus I'$ denotes the set of facts that form the symmetric difference of the instances I and I'.

2. A *subset-repair* of I is an instance I' that satisfies Σ and where there is no instance I'' such that $I' \subset I'' \subseteq I$ and I'' satisfies Σ.
3. A *superset-repair* of I is an instance I' that satisfies Σ and where there is no instance I'' such that $I \supset I'' \supseteq I$ and I'' satisfies Σ.

Clearly, subset-repair and superset-repairs are also $\oplus$-repairs; however, a $\oplus$-repair need not be a subset-repair or a superset-repair.

The *consistent answers* of a query q on I with respect to Σ are defined as:

$$\mathrm{CQA}(q, I, \Sigma) = \bigcap \{q(I') :\ I' \text{ is a } \oplus\text{-repair of } I \text{ w.r.t. } \Sigma\}$$

3   The Exchange-Repair Framework

In this section, we introduce the exchange-repair framework and discuss the structural and algorithmic properties of exchange-repairs.

Definition 2. Let $\mathcal{M} = (\mathbf{S}, \mathbf{T}, \Sigma_{\mathrm{st}}, \Sigma_{\mathrm{t}})$ be a schema mapping, I a source instance, and (I', J') a pair of a source instance and a target instance.
1. We say that (I', J') is a *symmetric-difference exchange-repair solution* (in short, a $\oplus$-XR-solution) for I w.r.t. $\mathcal{M}$ if (I', J') satisfies $\mathcal{M}$ and there is no pair of instances (I'', J'') such that $I \oplus I'' \subset I \oplus I'$ and (I'', J'') satisfies $\mathcal{M}$.
2. We say that (I', J') is a *subset exchange-repair solution* (in short, a subset-XR-solution) for I with respect to $\mathcal{M}$ if (I', J') satisfies $\mathcal{M}$ and there is no pair of instances (I'', J'') such that $I' \subset I'' \subseteq I$ and (I'', J'') satisfies $\mathcal{M}$.

Note that the minimality condition in the preceding definitions applies to the source instance I', but not to the target instance J' of the pair (I', J'). The source instance I' of a $\oplus$-XR-solution (subset-XR-solution) for I is called a $\oplus$-*source-repair* (respectively, subset source-repair) of I.

Source-repairs constitute a new notion that, in general, has different properties from those of the standard database repairs. Indeed, as mentioned earlier, a $\oplus$-repair need not be a subset repair. In contrast, the first result in this section asserts that the state of affairs is different for source-repairs. Recall that, according to the notation introduced earlier, $\mathrm{GLAV}+(\mathrm{WA\text{-}GLAV}, \mathrm{EGD})$ denotes the collection of all schema mappings $\mathcal{M} = (\mathbf{S}, \mathbf{T}, \Sigma_{\mathrm{st}}, \Sigma_{\mathrm{t}})$ such that Σ_{st} is a finite set of s-t tgds and Σ_{t} is the union of a finite weakly acyclic set of target tgds with a finite set of target egds.

Theorem 1. *Let $\mathcal{M}$ be a $\mathrm{GLAV}+(\mathrm{WA\text{-}GLAV}, \mathrm{EGD})$ schema mapping. If I is a source instance and (I', J') is a $\oplus$-XR-solution of I w.r.t. $\mathcal{M}$, then (I', J') is actually a subset-XR-solution of I w.r.t. $\mathcal{M}$. Consequently, every $\oplus$-source-repair of I is also a subset source-repair of I.*

Proof. (Hint) We use the following property of the chase procedure: if I and I' are source instances such that $I \subseteq I'$, then we can chase I and I' with the dependencies of $\mathcal{M}$ in such a way that $chase(I, \mathcal{M}) \subseteq chase(I', \mathcal{M})$. It follows that if I is a source instance that has no solution w.r.t. $\mathcal{M}$, then adding tuples to I yields a source instance I' that also has no solution w.r.t. $\mathcal{M}$. Using this fact, it can be shown that every $\oplus$-XR-solution is actually a subset-XR-solution.

From here on and in view of Theorem 1, we will use the term XR-solution to mean subset-XR-solution; similarly, source-repair will mean subset source-repair.

Note that if $\mathcal{M}$ is a GLAV+(WA-GLAV, EGD) schema mapping, then source-repairs always exist. The reason is that, since the pair $(\emptyset, \emptyset)$ trivially satisfies $\mathcal{M}$, then for every source instance I, there must exist a maximal subinstance I' of I for which a solution J' w.r.t. $\mathcal{M}$ exists; hence, (I', J') is a source repair for I w.r.t. $\mathcal{M}$.

We now claim that the following statements are true.

1. Repairs of an inconsistent result of the chase procedure are not necessarily XR-solutions.
2. Repairs of $(I, \emptyset)$ are not necessarily XR-solutions.

For the first statement, consider the pair (I, J') in Figure 1, where J' is a subset-repair of the inconsistent result J of the chase of I. Clearly, (I, J') is not an XR-solution, because J' is not a solution for I. For the second statement, consider the pairs (I_1, J_1), (I_2, J_2), (I_3, J_3) in Figure 2, all of which are $\oplus$-repairs of $(I, \emptyset)$. The first two are also XR-solutions of I, but the third one is not.

It can also be shown that XR-solutions are not necessarily $\oplus$-repairs of $(I, \emptyset)$. We now describe an important case in which XR-solutions are $\oplus$-repairs of $(I, \emptyset)$. For this, we recall the notion of a *core universal solution* from [11]. By definition, a *core universal solution* is a universal solution that has no homomorphism to a proper subinstance. If a universal solution exists, then a core universal solution also exists. Moreover, core universal solutions are unique up to isomorphism.

Proposition 1. *Let $\mathcal{M}$ be a GLAV+(WA-GLAV, EGD) schema mapping. If I is source instance and (I', J') is an XR-solution for I w.r.t. $\mathcal{M}$ such that J' is a core universal solution for I' w.r.t. $\mathcal{M}$, then (I', J') is a $\oplus$-repair of $(I, \emptyset)$ w.r.t. $\Sigma_{st} \cup \Sigma_t$.*

Next, we present the second key notion in the exchange-repair framework.

Definition 3. Let $\mathcal{M} = (\mathbf{S}, \mathbf{T}, \Sigma_{st}, \Sigma_t)$ be a schema mapping and q a query over the target schema $\mathbf{T}$. If I is a source instance, then the *XR-certain answers* of q on I w.r.t. $\mathcal{M}$ is the set

$$\text{XR-certain}(q, I, \mathcal{M}) = \bigcap \{q(J') : (I', J') \text{ is an XR-solution for } I\}.$$

The next result provides a comparison of the XR-certain answers with the consistent answers.

Proposition 2. *Let $\mathcal{M} = (\mathbf{S}, \mathbf{T}, \Sigma_{st}, \Sigma_t)$ be a GLAV+(WA-GLAV, EGD) schema mapping and q a conjunctive query over the target schema $\mathbf{T}$. If I is a source instance, then $\text{XR-certain}(q, I, \mathcal{M}) \supseteq \text{CQA}(q, (I, \emptyset), \Sigma_{st} \cup \Sigma_t)$. Moreover, this containment may be a proper one.*

The containment of the consistent answers in the XR-certain answers follows from Proposition 1 and the properties of core universal solutions. To see that this containment may be a proper one, consider the schema mapping $\mathcal{M}$ in Figure 1, the repairs of $(I, \emptyset)$ in Figure 2, and the conjunctive query $q(x, y) : \exists t\mathsf{Tasks}(x, t) \wedge$ $\mathsf{Stakeholders\_new}(t, y)$. It is easy to verify that $\mathrm{CQA}(q, (I, \emptyset), \Sigma_{st} \cup \Sigma_t) = \emptyset$, while $\mathrm{XR\text{-}certain}(q, I, \mathcal{M}) = \{(\mathtt{peter}, \mathtt{bobs})\}$.

Let $\mathcal{M} = (\mathbf{S}, \mathbf{T}, \Sigma_{st}, \Sigma_t)$ be a schema mapping and q a Boolean query over $\mathbf{T}$. We consider two natural decision problems in the exchange-repair framework, and give upper bounds for their computational complexity.

- **Source-Repair Checking**: Given a source instance I and a source instance $I' \subseteq I$, is I' a source-repair of I w.r.t. $\mathcal{M}$?
- **XR-certain Query Answering**: Given a source instance I, does $\mathrm{XR\text{-}certain}(q, I, \mathcal{M})$ evaluate to true? In other words, is $q(J')$ true on every target instance J' for which there is a source instance I' such that (I', J') is an XR-solution for I?

Theorem 2. *Let $\mathcal{M}$ be a* GLAV+(WA-GLAV, EGD) *schema mapping.*
1. *The source-repair checking problem is in* PTIME.
2. *Let q be a conjunctive query over the target schema. The* XR-certain *query answering problem for q is in* coNP.

Moreover, there is a schema mapping specified by copy s-t tgds and target egds, and a Boolean conjunctive query for which the XR-certain *query answering problem is* coNP-complete. *Thus, the data complexity of the* XR-certain *answers for Boolean conjunctive queries is* coNP-complete.

Proof. (Hint) For the first part, given I and I' with $I' \subseteq I$, we use repeatedly the chase to check that there is a solution for I', as well as for every extension $I' \cup \{t\}$ of I' with a fact t in $I \setminus I'$.

For the second part, the complement of the XR-certain answers is in NP, because we can guess a source-repair I' of I, use the chase procedure to compute $chase(I, \mathcal{M})$, and then verify in polynomial time that $q(chase(I, \mathcal{M}))$ is false. For the matching lower bound, we use well known results about the consistent answers of conjunctive queries w.r.t. key constraints [9, 12] and the easily checked fact that if $\mathcal{M}$ is specified by copy s-t tgds and target egds, then the XR-certain answers of a target query coincide with the consistent answers of the query on the copy of the given source instance.

The preceding Theorem 2 implies that the algorithmic properties of exchange-repairs are quite different from those of $\oplus$-repairs. Indeed, as shown in [1, 8], for GLAV+(WA-GLAV, EGD) schema mappings, the $\oplus$-repair problem is in coNP (and can be coNP-complete), while the data complexity of the consistent answers of Boolean conjunctive queries is Π_2^p-complete.

4   CQA-Rewritability

In this section, we show that, for GAV+EGD schema mappings $\mathcal{M} = (\mathbf{S}, \mathbf{T}, \Sigma_{st}, \Sigma_t)$, it is possible to construct a set of egds Σ_s over $\mathbf{S}$ such that an $\mathbf{S}$-instance I is consistent with Σ_s if and only if I has a solution w.r.t. $\mathcal{M}$.

We use this to show that XR-certain$(q, I, \mathcal{M})$, for a conjunctive query q, co-incides with subset-CQA(q_s, I, Σ_s), where q_s is a union of conjunctive queries. Thus, we can employ tools for consistent query answering with respect to egds, in order to compute XR-certain answers for GAV+EGD schema mappings.

We will use the well-known technique of *GAV unfolding* (see, e.g., [17]). Let $\mathcal{M} = (\mathbf{S}, \mathbf{T}, \Sigma_{st}, \Sigma_t)$ be a GAV+EGD schema mapping. For each k-ary target relation $T \in \mathbf{T}$, let q_T be the set of all conjunctive queries $q(x_1, \ldots, x_k) = \exists \mathbf{y}(\phi(\mathbf{y}) \wedge x_1 = y_{i_1} \wedge \cdots \wedge x_k = y_{i_k})$, for $\phi(\mathbf{y}) \rightarrow T(y_{i_1}, \ldots, y_{i_k})$ a GAV tgd belonging to Σ_{st} (recall that we frequently omit universal quantifiers in our notation, for the sake of readability).

A *GAV unfolding* of a conjunctive query $q(\mathbf{z})$ over $\mathbf{T}$ w.r.t. Σ_{st} is a conjunctive query over $\mathbf{S}$ obtained by replacing each occurrence of a target atom $T(\mathbf{z}')$ in $q(\mathbf{z})$ with one of the conjunctive queries in q_T (substituting variables from $\mathbf{z}'$ for $x_1, \ldots, x_k$, and pulling existential quantifiers out to the front of the formula).

Similarly, we define a *GAV unfolding* of an egd $\phi(\mathbf{x}) \rightarrow x_k = x_l$ over $\mathbf{T}$ w.r.t. Σ_{st} to be an egd over $\mathbf{S}$ obtained by replacing each occurrence of a target atom $T(\mathbf{z}')$ in $\phi(\mathbf{x})$ by one of the conjunctive queries in q_T (substituting variables from $\mathbf{z}'$ for $x_1, \ldots, x_k$, and pulling existential quantifiers out to the front of the formula as needed, where they become universal quantifiers).

Theorem 3. *Let $\mathcal{M} = (\mathbf{S}, \mathbf{T}, \Sigma_{st}, \Sigma_t)$ be a GAV+EGD schema mapping, and let Σ_s be the set of all GAV unfoldings of egds in Σ_t w.r.t. Σ_{st}. Let I be an $\mathbf{S}$-instance.*

1. *I satisfies Σ_s if and only if I has a solution w.r.t. $\mathcal{M}$.*
2. *The subset-repairs of I w.r.t. Σ_s are the source repairs of I w.r.t. $\mathcal{M}$.*
3. *For each conjunctive query q over $\mathbf{T}$, we have that XR-certain$(q, I, \mathcal{M}) =$ CQA(q_s, I, Σ_s), where q_s is the union of GAV-unfoldings of q w.r.t. Σ_{st}.*

The following result tells us that Theorem 3 cannot be extended to schema mappings containing LAV s-t tgds.

Theorem 4. *Consider the LAV+EGD schema mapping $\mathcal{M} = (\mathbf{S}, \mathbf{T}, \Sigma_{st}, \Sigma_t)$, where*

- *$\mathbf{S} = \{R\}$ and $\mathbf{T} = \{T\}$,*
- *$\Sigma_{st} = \{R(x, y) \rightarrow \exists u\, T(x, u) \wedge T(y, u)\}$, and*
- *$\Sigma_t = \{T(x, y) \wedge T(x, z) \rightarrow y = z\}$.*

Consider the query $q(x, y) = \exists z.\ T(x, z) \wedge T(y, z)$ over $\mathbf{T}$. There does not exist a UCQ q_s over $\mathbf{S}$ and a set of universal first-order sentences (in particular, egds) Σ_s such that, for every instance I, we have that XR-certain$(q, I, \mathcal{M}) =$ CQA(q_s, I, Σ_s).

The proof makes use of the fact that XR-certain$(q, I, \mathcal{M})$ defines a reachability query: $(a, b) \in$ certain$(q, I, \mathcal{M})$ if and only if b is reachable from a along an undirected R-path in I.

It is worth noting that the schema mapping $\mathcal{M}$ in the statement of Theorem 4 is such that every source instance has a solution, and hence "XR-certain" could be replaced by "certain" in the statement.

5   DLP-Rewritability

We saw in the previous section that the applicability of the CQA-rewriting approach is limited to GAV+EGD schema mappings. In this section, we consider another approach to computing XR-certain answers, based on a reduction to the problem of computing certain answers over the stable models of a disjunctive logic program. Our reduction is applicable to GLAV+(WA-GLAV, EGD) schema mappings. First, we reduce the case of GLAV+(WA-GLAV, EGD) schema mappings to the case of GAV+(GAV, EGD) schema mappings.

Theorem 5. *From a* GLAV+(WA-GLAV, EGD) *schema mapping* $\mathcal{M}$ *we can construct a* GAV+(GAV, EGD) *schema mapping* $\hat{\mathcal{M}}$ *such that, from a conjunctive query* q, *we can construct a union of conjunctive queries* $\hat{q}$ *with* XR-certain$(q, I, \mathcal{M}) = $ XR-certain$(\hat{q}, I, \hat{\mathcal{M}})$.

The proof of Theorem 5 is given in Section 6. Theorem 5 and Theorem 4 together imply that the CQA-rewriting approach studied in Section 4 is, in general, not applicable to GAV+(GAV, EGD) schema mappings and unions of conjunctive queries. In fact, a direct argument can be used to show that the same holds even for conjunctive queries. To address this problem, we will now consider a different approach to computing XR-certain answers, using disjunctive logic programs.

Stable models of disjunctive logic programs have been well-studied as a way to compute database repairs ([19] provides thorough treatment). In [7], Calì et al. give an encoding of their loosely-sound semantics for data integration as a disjunctive logic program. Their encoding is applicable for *non-key-conflicting* sets of constraints, a structural condition which is orthogonal to *weak acyclicity*, and which eliminates the utility of named nulls. Although their semantics involves a similar notion of minimality to the one in exchange-repairs, their setting differs sufficiently from ours that we consider this result to be complementary.

Fix a domain *Const*. A *disjunctive logic program* (DLP) Π over a schema $\mathbf{R}$ is a finite collection of rules of the form

$$\alpha_1 \vee \ldots \vee \alpha_n \leftarrow \beta_1, \ldots, \beta_m, \neg\gamma_1, \ldots, \neg\gamma_k.$$

where $n, m, k \geq 0$ and $\alpha_1, \ldots, \alpha_n, \beta_1, \ldots, \beta_m, \gamma_1, \ldots, \gamma_k$ are atoms formed from the relations in $\mathbf{R} \cup \{=\}$, using the constants in *Const* and first-order variables. A DLP is said to be *positive* if it consists of rules that do not contain negated atoms except possibly for inequalities. A DLP is said to be *ground* if it consists of rules that do not contain any first-order variables. A *model* of Π is an $\mathbf{R}$-instance I over domain *Const* that satisfies all rules of Π (viewed as universally quantified first-order sentences). A *minimal model* of Π is a model M of Π such that there does not exist a model M' of Π where the facts of M' form a strict subset of the facts of M. More generally, for subsets $\mathbf{R}_M, \mathbf{R}_F \subseteq \mathbf{R}$, an $\langle \mathbf{R}_M, \mathbf{R}_F \rangle$-*minimal model* of Π is a model M of Π such that there does not exist a model M' of Π where the facts of M' involving relations from $\mathbf{R}_M$ form a strict subset of the facts of M involving relations from $\mathbf{R}_M$, and the set of facts of M' involving relations from $\mathbf{R}_F$ is equal to the set of facts of M involving relations from $\mathbf{R}_F$ [15]. Although minimal models are a well-behaved semantics for positive DLPs, it is not well suited for programs with negations. The *stable model* semantics is a widely used semantics of DLPs that are not necessarily positive. For positive DLPs, it coincides with the minimal model semantics. For a ground DLP Π over a schema $\mathbf{R}$ and an $\mathbf{R}$-instance M over the domain *Const*, the *reduct* Π^M of Π with respect to M is the DLP containing, for each rule $\alpha_1 \vee \ldots \vee \alpha_n \leftarrow \beta_1, \ldots, \beta_m, \neg\gamma_1, \ldots, \neg\gamma_k$, with $M \not\models \gamma_i$ for all $i \leq k$, the rule $\alpha_1 \vee \ldots \vee \alpha_n \leftarrow \beta_1, \ldots, \beta_m$. A *stable model* of a ground DLP Π is an $\mathbf{R}$-instance M over the domain *Const* such that M is a minimal model of the reduct Π^M. See [13] for more details.

In this section, we will construct positive DLP programs whose $\langle \mathbf{R}_M, \mathbf{R}_F \rangle$-minimal models correspond to XR-solutions. In the light of Theorem 5, we may restrict our attention to GAV+(GAV, EGD) schema mappings.

Theorem 6. *Given a GAV+(GAV, EGD) schema mapping $\mathcal{M} = (\mathbf{S}, \mathbf{T}, \Sigma_{st}, \Sigma_t)$, we can construct in linear time a positive DLP Π over a schema $\mathbf{R}$ that contains $\mathbf{S} \cup \mathbf{T}$, and subsets $\mathbf{R}_M, \mathbf{R}_F \subseteq \mathbf{R}$, such that, for every union q of conjunctive queries over $\mathbf{T}$, and for every $\mathbf{S}$-instance I, we have that* XR-certain$(q, I, \mathcal{M}) = \bigcap \{q(M) \mid M$ *is an $\langle \mathbf{R}_M, \mathbf{R}_F \rangle$-minimal model of $\Pi \cup I\}$.*

In [15] it was shown that a positive ground DLP Π over a schema $\mathbf{R}$, together with subset $\mathbf{R}_M, \mathbf{R}_F \subseteq \mathbf{R}$, can be translated in linear time to a (not necessarily positive) DLP Π' over a possibly larger schema that includes $\mathbf{R}$, such that there is a bijection between the $\langle \mathbf{R}_M, \mathbf{R}_F \rangle$-minimal models of Π and the stable models of Π', where every pair of instances that stand in the bijection agree on all facts over the schema $\mathbf{R}$. This shows that DLP reasoners based on the stable model semantics, such as DLV [18, 2], can be used to evaluate positive ground disjunctive logic programs under the $\langle \mathbf{R}_M, \mathbf{R}_F \rangle$-minimal model semantics. Although stated only for ground programs in [15], this technique can be used for arbitrary positive DLPs, through grounding. Note that, when a program is grounded, inequalities are reduced to $\top$ or $\bot$.

Proof (Hint). We construct a disjunctive logic program $\Pi_{\mathrm{XRc}}(\mathcal{M})$ for a GAV+(GAV, EGD) schema mapping $\mathcal{M} = (\mathbf{S}, \mathbf{T}, \Sigma_{st}, \Sigma_t)$ as follows:
1. For each source relation S with arity n, add the rules

$$S_k(x_1, \ldots, x_n) \vee S_d(x_1, \ldots, x_n) \leftarrow S(x_1, \ldots, x_n)$$
$$\bot \leftarrow S_k(x_1, \ldots, x_n), S_d(x_1, \ldots, x_n)$$
$$S(x_1, \ldots, x_n) \leftarrow S_k(x_1, \ldots, x_n)$$

where S_k and S_d represent the *kept* and *deleted* atoms of S, respectively.
2. For each st-tgd $\phi(\mathbf{x}) \rightarrow T(\mathbf{x}')$ in Σ_{st}, add the rule

$$T(\mathbf{x}') \leftarrow \alpha_1, \ldots, \alpha_m$$

where $\alpha_1, \ldots, \alpha_m$ are the atoms in $\phi(\mathbf{x})$, in which each relation S has been uniformly replaced by S_k.

3. For each tgd $\phi(\mathbf{x}) \rightarrow T(\mathbf{x}')$ in Σ_t, add the rule

$$T(\mathbf{x}) \leftarrow \alpha_1, \ldots, \alpha_m$$

where $\alpha_1, \ldots, \alpha_m$ are the atoms in $\phi(\mathbf{x})$.
4. For each egd $\phi(\mathbf{x}) \rightarrow x_1 = x_2$, where $x_1, x_2 \in \mathbf{x}$, add the rule

$$\perp \leftarrow \alpha_1, \ldots, \alpha_m, x_1 \neq x_2,$$

where $\alpha_1, \ldots, \alpha_m$ are the atoms in $\phi(\mathbf{x})$.

We minimize w.r.t. $\mathbf{R}_M = \{S_d \mid S \in \mathbf{S}\}$, and fix $\mathbf{R}_F = \{S \mid S \in \mathbf{S}\}$. The disjunctive logic program for $\mathcal{M}$, denoted $\Pi_{\mathrm{XRc}}(\mathcal{M})$, is a straightforward encoding of the constraints in Σ_{st} and Σ_t as disjunctive logic rules over an indefinite view of the source instance. Since the source instance is fixed, the rules of the form $S(x_1, \ldots, x_n) \leftarrow S_k(x_1, \ldots, x_n)$ in $\Pi_{\mathrm{XRc}}(\mathcal{M})$ force the kept atoms to be a sub-instance of the source instance. Notice that egds are encoded as denial constraints, and that disjunction is used only to non-deterministically choose a subset of the source instance.

We first show that the restriction of every $\langle \mathbf{R}_M, \mathbf{R}_F \rangle$-minimal model of $\Pi_{\mathrm{XRc}}(\mathcal{M}) \cup I$ to the schema $\{S_k \mid S \in \mathbf{S}\} \cup \mathbf{T}$ constitutes an exchange-repair solution. We then show that for every exchange-repair solution, we can build a corresponding $\langle \mathbf{R}_M, \mathbf{R}_F \rangle$-minimal model of $\Pi_{\mathrm{XRc}}(\mathcal{M}) \cup I$.

6   From GLAV+(WA-GLAV, EGD) to GAV+(GAV, EGD)

We will now proceed with the proof of Theorem 5, showing how to translate GLAV+(WA-GLAV, EGD) schema mappings to GAV+(GAV, EGD) schema mappings. The translation involves the concepts of *skolemization* and *skeletons*, a first-order representation of second-order terms, similar to [10].

Let Θ be a collection of function symbols, each having a designated arity. By a Θ-*term*, we mean an expression built up from variables and/or constants using the function symbols in Θ, such that the arity of the function symbols is respected. We will omit Θ from the notation, when it is understood from context. The *depth* of a term is the maximal nesting of function symbols, with $depth(d) = 0$ for d a constant or variable. The *skeleton* of a term is the expression obtained by replacing all constants and variables by $\bullet$, where $\bullet$ is a fixed symbol that does not belong to Θ. Thus, for example, the skeleton of $f(g(x,y),z)$ is $f(g(\bullet,\bullet),\bullet)$. The *arity* of a skeleton s, denoted by $arity(s)$, is the number of occurrences of $\bullet$, and the depth of a skeleton is defined in the same way as for terms. If $s, s_1', \ldots, s_k'$ are skeletons with $arity(s) = k$, then we denote by $s(s_1', \ldots, s_k')$ the skeleton of arity $arity(s_1') + \cdots + arity(s_k')$ obtained by replacing, for each $i \leq k$, the i-th occurrence of $\bullet$ in s by s_i'.

Consider now a tgd of the form $\forall \mathbf{x}(\phi(\mathbf{x}) \rightarrow \exists \mathbf{y} \psi(\mathbf{x}, \mathbf{y}))$ with $\mathbf{x} = x_1, \ldots, x_n$ and $\mathbf{y} = y_1, \ldots, y_m$, and for each $i \leq m$, let f_i be a corresponding fresh n-ary function symbol. By the *skolemizations* of this tgd, we will mean the formulas of the form $\forall \mathbf{x}(\phi(\mathbf{x}) \rightarrow \alpha[f_1(\mathbf{x})/y_1, \ldots, f_m(\mathbf{x})/y_m])$ where α is a conjunct of ψ. Here, $\alpha[f_1(\mathbf{x})/y_1, \ldots, f_m(\mathbf{x})/y_m]$ refers to the result of replacing, in α, each variable y_i by the term $f_i(\mathbf{x})$. Thus, for example, the skolemizations of $\forall xy(R(x,y) \rightarrow \exists z(S(x,z) \wedge T(y,z)))$ are $\forall xy(R(x,y) \rightarrow S(x,f(x,y)))$ and $\forall xy(R(x,y) \rightarrow T(y,f(x,y)))$.

Definition 4. Let $\mathcal{M} = (\mathbf{S}, \mathbf{T}, \Sigma_{st}, \Sigma_t)$ be a GLAV+(WA-GLAV, EGD) schema mapping with rank r, and let Σ_{st}^{Sko} and Σ_t^{Sko} be the sets of skolemizations of the tgds in Σ_{st} and Σ_t, respectively, using a finite set of function symbols Θ. The *skeleton rewriting* of $\mathcal{M}$ is the GAV+(GAV, EGD) schema mapping $\hat{\mathcal{M}} = (\mathbf{S}, \hat{\mathbf{T}}, \hat{\Sigma}_{st}, \hat{\Sigma}_t^1 \cup \hat{\Sigma}_t^2 \cup \hat{\Sigma}_t^3 \cup \hat{\Sigma}_t^4)$ of $\mathcal{M}$ given by:

- $\hat{\mathbf{T}} = \{R_{s_1,\ldots,s_n} \mid R \in \mathbf{T}$ and each s_i is a Θ-skeleton of depth at most r$\}$, where the arity of R is n and the arity of $R_{s_1,\ldots,s_n}$ is $arity(s_1) + \cdots + arity(s_n)$.
- $\hat{\Sigma}_{st} = \{\phi(\mathbf{x}) \rightarrow T_{t_1,\ldots,t_n}(\mathbf{x}_1,\ldots,\mathbf{x}_n) \mid \phi(\mathbf{x}) \rightarrow T(t_1(\mathbf{x}_1),\ldots,t_n(\mathbf{x}_n)) \in \Sigma_{st}^{Sko}\}$.
- $\hat{\Sigma}_t^1$ consists of all tgds of the form

$$\phi_{s_1,\ldots,s_m}(\mathbf{y}_1,\ldots,\mathbf{y}_m) \rightarrow T_{s_1',\ldots,s_n'}(\bar{\mathbf{y}}_1,\ldots,\bar{\mathbf{y}}_n)$$

for $\phi(\mathbf{x}) \rightarrow T(t_1(\mathbf{x}_1),\ldots t_n(\mathbf{x}_n)) \in \Sigma_t^{Sko}$ with $\mathbf{x} = x_1,\ldots,x_m$, where $s_1,\ldots,s_m$ are Θ-skeletons of depth at most r; each $\mathbf{y}_i$ is a sequence of $arity(s_i)$ fresh variables; $\phi_{s_1,\ldots,s_m}(\mathbf{y}_1,\ldots,\mathbf{y}_m)$ is obtained from ϕ by replacing each atom $R(x_{i_1},\ldots,x_{i_k})$ by $R_{s_{i_1},\ldots,s_{i_k}}(\mathbf{y}_{i_1},\ldots,\mathbf{y}_{i_k})$; s_i' is a Θ-skeleton of depth at most r such that $s_i' = t_i(s_1,\ldots,s_m)$; and $\bar{\mathbf{y}}_i = (\mathbf{y}_{j_1},\ldots,\mathbf{y}_{j_k})$ for $\mathbf{x}_i = x_{j_1},\ldots,x_{j_k}$.
- $\hat{\Sigma}_t^2$ consists of all tgds of the form

$$\phi_{s_1,\ldots,s_m}(\mathbf{y},\ldots,\mathbf{y}_m) \rightarrow EQ_{s_i,s_j}(\mathbf{y}_i,\mathbf{y}_j)$$

for $\phi(\mathbf{x}) \rightarrow x_i = x_j \in \Sigma_t$ or $\phi(\mathbf{x}) \rightarrow x_j = x_i \in \Sigma_t$, with $\mathbf{x} = x_1,\ldots,x_m$, where $s_1,\ldots,s_m$ are Θ-skeletons of depth at most r, $s_i \neq \bullet$, and each $\mathbf{y}_k$ is a sequence of $arity(s_k)$ fresh variables. Note that, if s_i and s_j both have depth at least 1, then $\hat{\Sigma}_t^2$ contains also the above tgd with i,j interchanged.
- $\hat{\Sigma}_t^3$ consists of all tgds of the form

$$\phi_{s_1,\ldots,s_m}(\mathbf{y},\ldots,\mathbf{y}_m) \rightarrow y_i = y_j$$

for $\phi(\mathbf{x}) \rightarrow x_i = x_j \in \Sigma_t$ with $\mathbf{x} = x_1,\ldots,x_m$, where $s_1,\ldots,s_m$ are Θ-skeletons of depth at most r, $s_i = s_j = \bullet$, and each $\mathbf{y}_k$ is a sequence of $arity(s_k)$ fresh variables.
- $\hat{\Sigma}_t^4$ consists of all tgds of the form

$$R_{s_1,\ldots,s_n}(\mathbf{y}_1,\ldots,\mathbf{y}_n) \wedge EQ_{s_k,s'}(\mathbf{y}_k,\mathbf{z}) \rightarrow$$
$$R_{s_1,\ldots,s_{k-1},s',s_{k+1},\ldots,s_n}(\mathbf{y}_1,\ldots,\mathbf{y}_{k-1},\mathbf{z},\mathbf{y}_{k+1},\ldots,\mathbf{y}_n)$$

where $R \in T$; $s_1,\ldots,s_n,s'$ are Θ-skeletons of depth at most r, with $n = |R|$, $depth(s_k) > 0$; each $\mathbf{y}_i$ is a sequence of $arity(s_i)$ fresh variables, and $\mathbf{z}$ is a sequence of $arity(s')$ fresh variables.

In addition, for each conjunctive query $q(\mathbf{x}) = \exists \mathbf{y} \psi(\mathbf{x}, \mathbf{y})$ over $\mathbf{T}$ with $\mathbf{x} = x_1, \ldots, x_n$ and $\mathbf{y} = y_1, \ldots, y_m$, we denote by $\hat{q}(\mathbf{x})$ the union of conjunctive queries over $\hat{\mathbf{T}}$ of the form $\exists \mathbf{z}_1 \ldots \mathbf{z}_m \psi_{s_1,\ldots,s_n,s'_1,\ldots,s'_m}(x_1, \ldots, x_n, \mathbf{z}_1, \ldots, \mathbf{z}_m)$, where $s_1 = \ldots = s_n = \bullet$; $s'_1, \ldots, s'_m$ are Θ-skeletons of depth at most r; and each $\mathbf{z}_i$ is a sequence of fresh variables of length $arity(s'_i)$.

Theorem 7. *If $\mathcal{M} = (\mathbf{S}, \mathbf{T}, \Sigma_{st}, \Sigma_t)$ is a GLAV+(WA-GLAV, EGD) schema mapping and q a conjunctive query over $\mathbf{T}$, then XR-certain$(q, I, \mathcal{M}) =$ XR-certain$(\hat{q}, I, \hat{\mathcal{M}})$, where $\hat{\mathcal{M}}$ and $\hat{q}$ are the skeleton rewriting of $\mathcal{M}$ and q.*

The proof of Theorem 7 uses a variant of the *chase* procedure (cf. [11]), which we call *exhaustive chase*, and which bears similarity to the oblivious Skolem chase introduced in [20]. Distinguishing features of the exhaustive chase procedure are that (i) skolem terms are used to represent labeled nulls, (ii) when an egd is applied and an equality between two null values, or between a null value and a constant, is derived, appropriate substitutions are applied and the resulting facts are added to the chase instance *without removing any previously derived facts*; (iii) when an egd is applied and an equality between two constants is derived, the chase continues, and *violations are reported only at the end of the process*.

It is worth noting that, when a source instance I has a solution w.r.t. a schema mapping $\mathcal{M}$ and the exhaustive chase terminates successfully on input I, the output J may not be a solution for I. However, it can be shown that J is homomorphically equivalent to a solution, and, in fact, to a universal solution. Moreover, the exhaustive chase terminates whenever the standard chase (as considered in [11]) terminates. In particular, the result of the exhaustive chase can be used to compute the certain answers of a conjunctive query over the target.

The target instance that is the result of chasing a source instance I w.r.t. the skeleton rewriting $\hat{\mathcal{M}}$ of a schema mapping $\mathcal{M}$ can be related, in a precise way, to the result of the exhaustive chase of I w.r.t. $\mathcal{M}$. Specifically, an instance I has a solution w.r.t. a schema mapping $\mathcal{M}$ if and only if it has a solution w.r.t. $\hat{\mathcal{M}}$, and furthermore there is a bijection between the respective chase results.

The proof of Theorem 7 uses the above facts about the exhaustive chase and its relationship to the standard chase, as well as its close relationship to skeleton rewritings, in order to show that XR-certain$(q, I, \mathcal{M}) =$ XR-certain$(\hat{q}, I, \hat{\mathcal{M}})$.

7   Concluding Remarks

In this paper, we introduced the framework of exchange-repairs and explored the XR-certain answers as an alternative non-trivial semantics of queries in the context of data exchange. Exchange-repair semantics differ from other proposals for handling inconsistencies in data exchange in that, conceptually, the inconsistencies are repaired at the source rather than the target. This allows the shared origins of target facts to be reflected in the answers to target queries.

This framework brings together data exchange, database repairs, and disjunctive logic programming, thus enhancing the interaction between three different areas of research. Moreover, the results reported here pave the way for using DLP solvers, such as DLV, for query answering under the exchange-repair semantics.

Acknowledgements. The research of all authors was partially supported by NSF Grant IIS-1217869. Kolaitis' research was also supported by the project "Handling Uncertainty in Data Intensive Applications" under the program THALES.

References

[1] Afrati, F.N., Kolaitis, P.G.: Repair checking in inconsistent databases: algorithms and complexity. In: Fagin, R. (ed.) ICDT. ACM International Conference Proceeding Series, vol. 361, pp. 31–41. ACM (2009)

[2] Alviano, M., Faber, W., Leone, N., Perri, S., Pfeifer, G., Terracina, G.: The disjunctive datalog system dlv. In: Datalog, pp. 282–301 (2010)

[3] Arenas, M., Barceló, P., Libkin, L., Murlak, F.: Relational and XML Data Exchange. Synthesis Lectures on Data Management. Morgan & Claypool Publishers (2010)

[4] Arenas, M., Bertossi, L.E., Chomicki, J.: Consistent query answers in inconsistent databases. In: Vianu, V., Papadimitriou, C.H. (eds.) PODS, pp. 68–79. ACM Press (1999)

[5] Bertossi, L.E.: Database Repairing and Consistent Query Answering. Synthesis Lectures on Data Management. Morgan & Claypool Publishers (2011)

[6] Calì, A., Lembo, D., Rosati, R.: On the decidability and complexity of query answering over inconsistent and incomplete databases. In: Neven, F., Beeri, C., Milo, T. (eds.) PODS, pp. 260–271. ACM (2003)

[7] Calì, A., Lembo, D., Rosati, R.: Query rewriting and answering under constraints in data integration systems. In: Gottlob, G., Walsh, T. (eds.) IJCAI, pp. 16–21. Morgan Kaufmann (2003)

[8] ten Cate, B., Fontaine, G., Kolaitis, P.G.: On the data complexity of consistent query answering. In: Deutsch, A. (ed.) ICDT, pp. 22–33. ACM (2012)

[9] Chomicki, J., Marcinkowski, J.: Minimal-change integrity maintenance using tuple deletions. Inf. Comput. 197(1-2), 90–121 (2005)

[10] Duschka, O.M., Genesereth, M.R.: Answering recursive queries using views. In: Mendelzon, A.O., Özsoyoglu, Z.M. (eds.) PODS, pp. 109–116. ACM Press (1997)

[11] Fagin, R., Kolaitis, P.G., Miller, R.J., Popa, L.: Data exchange: semantics and query answering. Theor. Comput. Sci. 336(1), 89–124 (2005)

[12] Fuxman, A., Miller, R.J.: First-order query rewriting for inconsistent databases. J. Comput. Syst. Sci. 73(4), 610–635 (2007)

[13] Gelfond, M., Lifschitz, V.: The stable model semantics for logic programming. In: Kowalski, R.A., Bowen, K.A. (eds.) ICLP/SLP, pp. 1070–1080. MIT Press (1988)

[14] Grahne, G., Onet, A.: Data correspondence, exchange and repair. In: Segoufin, L. (ed.) ICDT. pp. 219–230. ACM International Conference Proceeding Series. ACM (2010)

[15] Janhunen, T., Oikarinen, E.: Capturing parallel circumscription with disjunctive logic programs. In: Alferes, J.J., Leite, J. (eds.) JELIA 2004. LNCS (LNAI), vol. 3229, pp. 134–146. Springer, Heidelberg (2004)

[16] Kolaitis, P.G., Panttaja, J., Tan, W.C.: The complexity of data exchange. In: Vansummeren, S. (ed.) PODS, pp. 30–39. ACM (2006)

[17] Lenzerini, M.: Data integration: A theoretical perspective. In: Popa, L., Abiteboul, S., Kolaitis, P.G. (eds.) PODS, pp. 233–246. ACM (2002)

[18] Leone, N., Pfeifer, G., Faber, W., Eiter, T., Gottlob, G., Perri, S., Scarcello, F.: The dlv system for knowledge representation and reasoning. ACM Trans. Comput. Log. 7(3), 499–562 (2006)

[19] Marileo, M.C., Bertossi, L.E.: The consistency extractor system: Answer set programs for consistent query answering in databases. Data Knowl. Eng. 69(6), 545–572 (2010)

[20] Marnette, B.: Generalized schema-mappings: from termination to tractability. In: Paredaens, J., Su, J. (eds.) PODS, pp. 13–22. ACM (2009)

Rules and Ontology Based Data Access

Guohui Xiao[1], Martin Rezk[1], Mariano Rodríguez-Muro[2], and Diego Calvanese[1]

[1] Faculty of Computer Science, Free University of Bozen-Bolzano, Italy
[2] IBM Watson Research Center, USA

Abstract. In OBDA an ontology defines a high level global vocabulary for user queries, and such vocabulary is mapped to (typically relational) databases. Extending this paradigm with rules, e.g., expressed in SWRL or RIF, boosts the expressivity of the model and the reasoning ability to take into account features such as recursion and n-ary predicates. We consider evaluation of SPARQL queries under rules with linear recursion, which in principle is carried out by a 2-phase translation to SQL: (1) The SPARQL query, together with the RIF/SWRL rules, and the mappings is translated to a Datalog program, possibly with linear recursion; (2) The Datalog program is converted to SQL by using recursive common table expressions. Since a naive implementation of this translation generates inefficient SQL code, we propose several optimisations to make the approach scalable. We implement and evaluate the techniques presented here in the *Ontop* system. To the best of our knowledge, this results in the first system supporting all of the following W3C standards: the OWL 2 QL ontology language, R2RML mappings, SWRL rules with linear recursion, and SPARQL queries. The preliminary but encouraging experimental results on the NPD benchmark show that our approach is scalable, provided optimisations are applied.

1 Introduction

In Ontology Based Data Access (OBDA) [5], the objective is to access data trough a conceptual layer. Usually, this conceptual layer is expressed in the form of an OWL or RDFS ontology, and the data is stored in relational databases. The terms in the conceptual layer are mapped to the data layer using so-called globas-as-view (GAV) mappings, associating to each element of the conceptual layer a (possibly complex) query over the data sources. GAV mappings have been described as Datalog rules in the literature [17] and formalized in the R2RML W3C standard [8]. Independently of the mapping language, these rules entail a *virtual* RDF graph that uses the ontology vocabulary. This virtual graph can then be queried using an RDF query language such as SPARQL.

There are several approaches for query answering in the context of OBDA, and a number of techniques have been proposed [17,16,13,21,9,3]. One of such techniques, and the focus of this paper, is query answering by *query rewriting*. That is, answer the queries posed by the user (e.g., SPARQL queries) by translating them into queries over the database (e.g., SQL). This kind of technique has several desirable features; notably, since all data remains in the original source there is no redundancy, the system immediately reflects any changes in the data, well-known optimizations for relational databases can be used, etc. It has been shown that through this technique one can obtain performance comparable or sometimes superior to other approaches when the ontology

R. Kontchakov and M.-L. Mugnier (Eds.): RR 2014, LNCS 8741, pp. 157–172, 2014.

language is restricted to OWL 2 QL [22]. While the OWL 2 QL specification (which subsumes RDFS in expressive power) offers a good balance between expressivity and performance, there are many scenarios where this expressive power is not enough.

As a motivating example and to illustrate the main concepts in this paper, suppose we have a (virtual) RDF graph over a database with information about direct flights between locations and their respective cost. Suppose we have a `flight` relation in the database, and we want to find out all the possible (direct and non-direct) routes between two locations such that the total cost is less than 100 Euros. This problem is a particular instance of the well-known reachability problem, where we need to be able to compute the transitive closure over the `flight` relation respecting the constraint on the flight cost. While SPARQL 1.1 provides *path expressions* that can be used to express the transitive closure of a property, it may be cumbersome and prone to errors, especially in the presence of path constraints such as the cost in our example.

Computational complexity results show that unless we limit the form of the allowed rules, on-the-fly query answering by rewriting into SQL Select-Project-Join (SPJ) queries is not possible [6,2]. However, as target language for query rewriting, typically only a fragment of the expressive power of SQL99 has been considered, namely unions of SPJ SQL queries. We propose here to go beyond this expressive power, and we advocate the use of SQL99's Common Table Expressions (CTEs) to obtain a form of linear recursion in the rewriting target language. In this way, we can deal with recursive rules at the level of the ontology, and can reuse existing query rewriting optimisations developed for OBDA to provide efficient query rewriting into SQL99. The languages that we target are those that are used more extensively in the context of OBDA for Semantic Web application, i.e., RIF and SWRL as rule language, SPARQL 1.0 as query language, and R2RML as relational databases to RDF mapping language.

The contributions of this paper can be summarized as follows: *(i)* We provide translations from SWRL, R2RML, and SPARQL into relational algebra extended with a fixed-point operator that can be expressed in SQL99's Common Table Expressions (CTEs); *(ii)* We show how to extend existing OBDA optimisation techniques that have been proven effective in the OWL 2 QL setting to this new context. In particular, we show that so called *T-mappings* for recursive programs exist and how to construct them. *(iii)* We provide an implementation of such technique in the open source OBDA system *Ontop*, making it the first system of its kind to support all the following W3C recommendations: OWL 2 QL, R2RML, SPARQL, and SWRL; *(iv)* We provide a preliminary evaluation of the techniques using an extension of the NPD benchmark (a recently developed OWL 2 QL benchmark) with rules, and show that the proposed solution competes and sometimes outperforms existing triple stores.

2 Preliminaries

2.1 RDF

The Resource Description Framework (RDF) is a standard model for data interchange on the Web [15]. The language of RDF contains the following pairwise disjoint and countably infinite sets of symbols: $\mathbf{I}$ for *IRIs*, $\mathbf{L}$ for *RDF literals*, and $\mathbf{B}$ for *blank nodes*. *RDF terms* are elements of the set $\mathbf{T} = \mathbf{I} \cup \mathbf{B} \cup \mathbf{L}$. An *RDF knowledge base* (also called

RDF graph) is a collection of triples of the form (s, p, o), where $s \in \mathbf{I}$, $p \in \mathbf{I} \cup \mathbf{B}$, and $o \in \mathbf{T}$. A triple (s, p, o) intuitively expresses that s and o are related by p; when p is the special role rdf:type, the triple $(s, \text{rdf:type}, o)$ means that s is an instance of o.

It is sometimes convenient to define conversions between RDF graphs and sets of (Datalog) facts. Thus, given an RDF graph G, the corresponding set of Datalog facts is:

$$\mathcal{A}(G) = \{o(s) \mid (s, \text{rdf:type}, o) \in G\} \cup \{p(s, o) \mid (s, p, o) \in G, p \neq \text{rdf:type}\}$$

And given a set A of facts, the corresponding RDF graph is:

$$\mathcal{G}(A) = \{(s, \text{rdf:type}, o) \mid o(s) \in A\} \cup \{(s, p, o) \mid p(s, o) \in A\}$$

Note that $\mathcal{G}(A)$ discards the facts that are not unary or binary.

2.2 SPARQL

SPARQL is the standard RDF query language. For formal purposes we will use the algebraic syntax of SPARQL similar to the one in [18] and defined in the standard[1]. However, to ease the understanding, we will often use graph patterns (the usual SPARQL syntax) in the examples. The SPARQL language that we consider shares with RDF the set of symbols: constants, blank nodes, IRIs, and literals. In addition, it adds a countably infinite set $\mathbf{V}$ of variables. The *SPARQL algebra* is constituted by the following graph pattern operators (written using prefix notation): *BGP* (basic graph pattern), *Join*, *LeftJoin*, *Filter*, and *Union*. A *basic graph pattern* is a statement of the form: $BGP(s, p, o)$. In the standard, a BGP can contain several triples, but since we include here the join operator, it suffices to view BGPs as the result of *Join* of its constituent triple patterns. Observe that the only difference between blank nodes and variables in BGPs, is that the former do not occur in solutions. So, to ease the presentation, we assume that BGPs contain no blank nodes. Algebra operators can be nested freely. Each of these operators returns the result of the sub-query it describes.

Definition 1 (SPARQL Query). *A SPARQL query is a pair (V, P), where V is a set of variables, and P is a SPARQL algebra expression in which all variables of V occur.*

We will often omit V when it is understood from the context . A *substitution*, θ, is a *partial* function $\theta : \mathbf{V} \mapsto \mathbf{T}$. The domain of θ, denoted by $dom(\theta)$, is the subset of $\mathbf{V}$ where θ is defined. Here we write substitutions using postfix notation. When a query (V, P) is evaluated, the result is a set of substitutions whose domain is contained in V. For space reasons, we omit the semantics of SPARQL, and refer to [11] for the specification of how to compute the answer of a query Q over an RDF graph G, which we denote as $[\![Q]\!]_G$.

Example 1 (Flights, continued). Consider the flight example in the introduction. The low cost flights from Bolzano can be retrieved by the query:

---

[1] http://www.w3.org/TR/rdf-sparql-query/#sparqlAlgebra

```
Select ?x  Where {
  ?x :tripPlanFrom :Bolzano .  ?x :tripPlanTo ?y .
  ?x :tripPlanPrice ?z .  Filter(?z < 100)
}
```

The corresponding SPARQL algebra expression is as follows:

```
Filter (?z < 100)(
  Join(BGP(?x :tripPlanFrom :Bolzano .)
  Join(BGP(?x :tripPlanTo ?y .) BGP(?x :tripPlanPrice ?z .))))
```

2.3 Rules: RIF and SWRL

We describe now two important rule languages, SWRL and RIF, and the semantics of their combination with RDF graphs.

The Semantic Web Rule Language (SWRL) is a widely used Semantic Web language combining a DL ontology component with rules[2]. Notice that the SWRL language allows only for the use of unary and binary predicates. SWRL is implemented in many systems, such as, Pellet, Stardog, and HermiT.

The Rule Interchange Format (RIF) is a W3C recommendation [12] defining a language for expressing rules. The standard RIF dialects are Core, BLD, and PRD. RIF-Core provides "safe" positive Datalog with built-ins; RIF-BLD (Basic Logic Dialect) is positive Horn logic, with equality and built-ins; RIF-PRD (Production Rules Dialect) adds a notion of forward-chaining rules, where a rule fires and then performs some action. In this paper we focus on RIF-Core [4], which is equivalent to Datalog without negation, but supports an F-Logic style frame-like syntax: $s[p_1 \rightarrow o_1, p_2 \rightarrow o_2, \ldots, p_n \rightarrow o_n]$ is a shorthand for the conjunction of atoms $\bigwedge_{p_i = \mathtt{rdf:type}} o_i(s) \wedge \bigwedge_{p_i \neq \mathtt{rdf:type}} p_i(s, o_i)$. Observe that the RIF language allows for additional n-ary Datalog predicates besides the unary concept names and binary role names from the RDF vocabulary. In this paper, we make the restriction that variables cannot be used in the place of the predicates. For instance, neither $s[?p \rightarrow o]$ nor $s[\mathtt{rdf:type} \rightarrow ?o]$ are allowed.

For the sake of simplicity, in the following we will use Datalog notation, where (SWRL or RIF) rules are simply written as

$$l_0 :- l_1, \ldots, l_m$$

where each l_i is an atom. Therefore we refer to a set of rules as Datalog rules. Recall that a Datalog program Π that does not contain negation has a unique minimal model, which can be computed via a repeated exhaustive application of the rules in it in a bottom-up fashion [14]. We denote such model, $\mathsf{MM}(\Pi)$.

An *RDF-rule combination* is a pair (G, Π), where G is an RDF graph and Π is a set of Datalog rules.

Definition 2. *The RDF graph induced by an RDF-rule combination* (G, Π) *is defined as* $\mathcal{G}(\mathsf{MM}(\mathcal{A}(G) \cup \Pi))$.

---

[2] http://www.w3.org/Submission/SWRL/

Example 2 (Example 1, continued). The following rules model the predicates `plan`, representing the transitive closure of flights, including the total price of the trip, and `tripPlanFrom/To/Price`, which project `plan` into triples:

$$\texttt{plan}(from, to, price, plan\_url) \quad :\text{-} \quad \texttt{flightFrom}(fid, from), \texttt{flightTo}(fid, to),$$
$$\texttt{flightPrice}(fid, price),$$
$$plan\_url = \texttt{CONCAT}(\texttt{"http://flight/"}, fid)$$

$$\texttt{plan}(from, to, price, plan\_url) \quad :\text{-} \quad \texttt{plan}(from, to_1, price_1, plan\_url_1),$$
$$\texttt{flightFrom}(fid, to_1), \texttt{flightTo}(fid, to),$$
$$\texttt{flightPrice}(fid, price_2),$$
$$price = price_1 + price_2,$$
$$plan\_url = \texttt{CONCACT}(plan\_url_1, \texttt{"/"}, fid)$$

$$\texttt{tripPlanFrom}(plan\_url, from) \quad :\text{-} \quad \texttt{plan}(from, to, price, plan\_url)$$
$$\texttt{tripPlanTo}(plan\_url, to) \quad :\text{-} \quad \texttt{plan}(from, to, price, plan\_url)$$
$$\texttt{tripPlanPrice}(plan\_url, price) \quad :\text{-} \quad \texttt{plan}(from, to, price, plan\_url)$$

Observe that rules not only boost the modelling capabilities by adding recursion, but also allow for a more elegant and succinct representation of the domain using n-ary predicates.

2.4 SPARQL and Rules: Entailment Regime

The RIF entailment regime specifies how RIF entailment can be used to redefine the evaluation of basic graph patterns. The evaluation of complex clauses is computed by combining already computed solutions in the usual way. Therefore, in this section we can restrict the attention to queries that consist of a single BGP.

The semantics provided in [10] is defined in terms of pairs of RIF and RDF interpretations. These models are then used to define satisfiability and entailment in the usual way. Combined entailment extends both entailment in RIF and entailment in RDF. To ease the presentation, we will present a simplified version of such semantics based on Datalog models seen as RDF graphs (c.f. Section 2.1).

Definition 3. *Let Q be a BGP, G an RDF graph, and Π a set of rules. The evaluation of Q over G and Π, denoted $[\![Q]\!]_{G,\Pi}$, is defined as the evaluation of Q over the induced RDF graph of (G, Π), that is*

$$[\![Q]\!]_{G,\Pi} = [\![Q]\!]_{\mathcal{G}(MM(\mathcal{A}(G)\cup\Pi))}.$$

Example 3 (Example 2, continued). Suppose we have the following triples in our RDF graph:

```
AF22 :flightFrom :Bolzano .    AF23 :flightTo :Dublin .
AF22 :flightTo    :Milano .    AF22 :flightPrice 45 .
AF23 :flightFrom :Milano .     AF23 :flightPrice 45 .
```

It is easy to see that these triples, together with the rules in Example 2 "extend" the original RDF with the following triples:

```
http://flight/AF22/AF23 :tripPlanFrom :Bolzano .
http://flight/AF22/AF23 :tripPlanTo    :Dublin .
http://flight/AF22/AF23 :tripPlanPrice 90 .
```

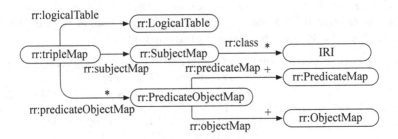

Fig. 1. A well formed R2RML mapping node

This implies that we get the following two substitutions by evaluating the query in Example 1: $\{x \mapsto \texttt{http://flight/AF22}\}, \{x \mapsto \texttt{http://flight/AF22/AF23}\}$.

2.5   R2RML: Mapping Databases to RDF

R2RML is a W3C standard [8] defining a language for mapping relational databases into RDF data. Such mappings expose the relational data as RDF triples, using a structure and vocabulary chosen by the mapping author.

An R2RML mapping is expressed as an RDF graph (in Turtle syntax), where a well-formed mapping consists of one or more trees called *triple maps* with a structure as shown in Figure 1. Each tree has a root node, called *triple map node*, which is linked to exactly one *logical table* node, one *subject map* node and one or more *predicate object map* nodes. Intuitively, each triple map states how to construct a set of triples (subject, predicate, object) using the information contained in the logical table (specified as an SQL query).

The R2RML syntax is rather verbose, therefore, due to the space limitations, in this paper we represent triple maps using standard Datalog rules of the form:

$$predicate(subject, object) \text{ :- } body \qquad\qquad concept(subject) \text{ :- } body$$

where *body* is a conjunction of atoms that refers to the database relations, possibly making use of auxiliary relations representing the SQL query of the mapping, when the semantics of such query cannot be captured in Datalog. For the formal translation from R2RML to Datalog, we refer to [23].

Example 4 (Example 2, continued). We present the R2RML rules mapping a relational database to the relations `flightFrom` and `flightPrice`. Recall that the relations `tripPlanFrom`, `tripPlanTo`, etc. are defined by rules. Suppose we have a table `flight` in the database, with attributes: *id, departure, arrival, segment,* and *cost*. Then the mappings are as follows:

$$\texttt{flightFrom}(id, departure) \text{ :- } \texttt{flight}(id, departure, arrival, cost)$$
$$\texttt{flightPrice}(id, cost) \qquad\text{ :- } \texttt{flight}(id, departure, arrival, cost)$$

Next we define the RDF graph induced by a set of mapping rules and a database.

Definition 4 (Virtual RDF Graph via R2RML Mapping). *Let $\mathcal{M}$ be a set of R2RML mappings (represented in Datalog), and I a relational database instance. Then the virtual RDF graph $\mathcal{M}(I)$ is defined as the RDF graph corresponding to the minimal model of $\mathcal{M} \cup I$, i.e., $\mathcal{M}(I) = \mathcal{G}(MM(\mathcal{M} \cup I))$.*

3 Answering SPARQL over Rules and Virtual RDF

In this section we describe how we translate SPARQL queries over a rule-enriched vocabulary into SQL. The translation consists of two steps: *(i)* translation of the SPARQL query and RIF rules into a recursive Datalog program, and *(ii)* generation of an SQL query (with CTEs) from the Datalog program.

3.1 SPARQL to Recursive Datalog

The translation we present here extends the one described in [18,19,23], where the authors define a translation function τ from SPARQL queries to *non-recursive* Datalog programs with stratified negation. Due to space limitations, we do not provide the details of the translation τ, but illustrate it with an example, and refer to [18,19] for its correctness. Note, in this paper we only consider BGPs corresponding to atoms in SWRL or RIF rules; in other words, triple patterns like $(t_1, ?x, t_2)$ or $(t, \texttt{rdf:type}, ?x)$ are disallowed (cf. the restrictions on RIF in Section 2.3).

Example 5. Consider the query (V, P) in Example 1, for which we report below the algebra expression in which we have labeled each sub-expression P_i of P.

`Filter (?z < 100)(`	# P_1
$\quad$ `Join(`	# P_2
$\quad\quad$ `BGP(?x :tripPlanFrom :Bolzano .)`	# P_3
$\quad\quad$ `Join(`	# P_4
$\quad\quad\quad$ `BGP(?x :tripPlanTo ?y .)`	# P_5
$\quad\quad\quad$ `BGP(?x :tripPrice   ?z .))))`	# P_6

The Datalog translation contains one predicate ans_i representing each algebra sub-expression P_i. The Datalog program $\tau(V, P)$ for this query is as follows:

$$
\begin{array}{ll}
\mathrm{ans}_1(x) & :\text{- } \mathrm{ans}_2(x,y,z), \texttt{Filter}(z > 100) \\
\mathrm{ans}_2(x,y) & :\text{- } \mathrm{ans}_3(x), \mathrm{ans}_4(x,y,z) \\
\mathrm{ans}_3(x) & :\text{- } \texttt{tripPlanFrom}(x,\texttt{:Bolzano}) \\
\mathrm{ans}_4(x,y,z) & :\text{- } \mathrm{ans}_5(x,y), \mathrm{ans}_6(x,z) \\
\mathrm{ans}_5(x,y) & :\text{- } \texttt{tripPlanTo}(x,y) \\
\mathrm{ans}_6(x,z) & :\text{- } \texttt{tripPrice}(x,z)
\end{array}
$$

The overall translation of a SPARQL query and a set of rules to recursive Datalog is defined as follows.

Definition 5. *Let $Q = (V, P)$ be a SPARQL query and Π a set of rules. We define the translation of Q and Π to Datalog as the Datalog program $\mu(Q, \Pi) = \Pi \cup \tau(V, P)$.*

Observe that while $\tau(V, P)$ is not recursive, Π and therefore $\mu(Q, \Pi)$ might be so.

Proposition 1. *Let (G, Π) be an RDF-rule combination, Q a SPARQL query, and θ a solution mapping. Then*

$$\theta \in [\![Q]\!]_{G,\Pi} \quad \textit{if and only if} \quad \mu(Q, \Pi) \cup \mathcal{A}(G) \models \mathrm{ans}_Q(\theta)$$

where ans_Q is the predicate in $\mu(Q, \Pi)$ corresponding to Q.

Proof. Let $Q = (V, P)$. By definition, $\tau(V, P)$ is a stratified Datalog program and we assume that it has a stratification $(S_0, \ldots, S_n)$. As Π is a positive program, clearly $(\Pi \cup \mathcal{A}(G), S_0, \ldots, S_n)$ is a stratification of $\mu(Q, \Pi) \cup \mathcal{A}(G)$. Then the following statements are equivalent:

- $\theta \in [\![Q]\!]_{G,\Pi} = [\![Q]\!]_{\mathcal{G}(\mathrm{MM}(\mathcal{A}(G) \cup \Pi))}$
- *(By the correctness of the translation $\tau(V, P)$ [19])*
  $\mathrm{ans}_Q(\theta) \in \mathrm{MM}(\tau(V, P) \cup \mathrm{MM}(\mathcal{A}(G) \cup \Pi))$
- *(By the definition of GL-reduction)*
  $\mathrm{ans}_Q(\theta) \in \mathrm{MM}(\tau(V, P)^{\mathrm{MM}(\mathcal{A}(G) \cup \Pi)})$
- *(By the definition of model of a stratified Datalog program)*
  $\mathrm{ans}_Q(\theta) \in \mathrm{MM}(\tau(V, P) \cup \mathcal{A}(G) \cup \Pi) = \mathrm{MM}(\mu(Q, \Pi) \cup \mathcal{A}(G))$
- $\mu(Q, \Pi) \cup \mathcal{A}(G) \models \mathrm{ans}_Q(\theta)$ □

Considering that R2RML mappings are represented in Datalog, we immediately obtain the following result.

Corollary 1. *Let $Q = (V, P)$ be a SPARQL query, Π a set of rules, $\mathcal{M}$ a set of R2RML mappings, I a database instance, and θ a solution mapping. Then*

$$\theta \in [\![Q]\!]_{\mathcal{M}(I),\Pi} \quad \textit{if and only if} \quad \mu(Q, \Pi) \cup \mathcal{M} \cup I \models \mathrm{ans}_Q(\theta)$$

3.2 Recursive Datalog to Recursive SQL

In this section, we show how to translate to SQL the recursive Datalog program obtained in the previous step. Note that translation from Datalog to SQL presented in [23] does not consider recursive rules. Here we extend such translation to handle recursion by making use of SQL common table expressions (CTEs). We assume that the extensional database (EDB) predicates (i.e., those not appearing in the heads of Datalog rules) are stored in the database.

Fixpoint Operator, Linear Recursive Query and CTE. We recall first how relational algebra can be extended with a fixpoint operator [1]. Consider an equation of the form $R = f(R)$, where $f(R)$ is a relational algebra expression over R. A least fixpoint of the equation above, denoted $\mathrm{LFP}(R = f(R))$, is a relation S such that

- $S = f(S)$
- if R is any relation such that $R = f(R)$, then $S \subseteq R$.

In order to ensure the existence of a fixpoint (and hence of the least fixpoint), f must be monotone.

Consider now the equation $R = f(R_1, \ldots, R_n, R)$, making use of a relational algebra expression over R and other relations. Such equation is a *linear* recursive expression, if R occurs exactly once in the expression f. Then, we can split f into a non-recursive part f_1 not mentioning R, and a recursive part f_2, i.e., $f(R_1, \ldots, R_n, R) = f_1(R_1, \ldots, R_n) \cup f_2(R_1, \ldots, R_n, R)$. In this case, $\mathsf{LFP}(R = f(R_1, \ldots, R_n, R))$ can be expressed using a *common table expression* (CTE) of SQL99:

```
WITH RECURSIVE R AS {
   [block for base case f₁]
UNION
   [block for recursive case f₂]
}
```

We remark that CTEs are already implemented in most of the commercial databases, e.g, Postgres, MS SQL Server, Oracle, DB2, H2, HSQL.

Example 6. Suppose we have the database relation `Flight` (f, for short), with attributes *id*, *source*, *destination*, and *cost* (i, s, d, c, for short) and we want to compute all the possible routes such that the total cost is less that 100. To express this relation `plan` we can use the following equation with least fixpoint operator:

$$\texttt{plan} = \mathsf{LFP}(f^* = \pi_{var_1}(f) \cup \pi_{var_2}(\rho_{count}(\sigma_{fil}(f \bowtie f^*))))$$

where

$$var_1 = f.s, f.d, f.c \qquad count = (f.c + f^*.c)/c$$
$$var_2 = f.s, f^*.d, c \qquad fil = f.c + f^*.c < 100, \; f.s = f^*.d$$

It can be expressed as the following CTE:

```
WITH RECURSIVE plan AS (
   SELECT   f.s,  f.d,  f.c  FROM f
UNION
   SELECT   plan.s,  f.d,  f.c + plan.c  AS c
   FROM f,  plan
   WHERE f.c + plan.c < 100 AND f.s = plan.d
)
```

Now we proceed to explain how to translate a recursive program into a fixpoint relation algebra expression.

The *dependency graph* for a Datalog program is a directed graph representing the relation between the predicate names in a program. The set of nodes are the relation symbols in the program. There is an arc from a node a to a node b if and only if a appears in the body of a rule in which b appears in the head. A program is *recursive* if there is a cycle in the dependency graph.

We say that a Datalog program Π is SQL99 compatible if *(i)* there are no cycles in the dependency graph of Π apart from self-loops; and *(ii)* the recursive predicates are restricted to linear recursive ones. For the sake of simplicity, and ease the presentation, we assume that non-recursive rules have the form

$$p(\boldsymbol{x}) :\text{-} atom_1(\boldsymbol{x}_1), \ldots, atom_n(\boldsymbol{x}_n), cond(\boldsymbol{z}) \tag{1}$$

where each $atom_i$ is a relational atom, and $cond(\boldsymbol{z})$ is a conjunction of atoms over built-in predicates. Moreover, for each recursive predicate p, there is a pair of rules defining p of the form

$$\begin{aligned}
p(\boldsymbol{x}) &:\text{-} atom_0(\boldsymbol{x}_0), p(\boldsymbol{y}), cond_1(\boldsymbol{z}_1) \\
p(\boldsymbol{x}) &:\text{-} atom_1(\boldsymbol{x}_1), \ldots, atom_n(\boldsymbol{x}_n), cond_2(\boldsymbol{z}_2)
\end{aligned} \tag{2}$$

In addition, we assume that equalities between variables in the rules are made explicit in atoms of the form $x = y$. Thus, there are no repeated variables in non-equality atoms.

The intuition behind the next definition is that for each predicate p of arity n in the program, we create a relational algebra expression $\mathsf{RA}(p)$.

Definition 6. *Let p be a predicate and Π a set of Datalog rules.*

- *If p is an extensional predicate, then*

$$\mathsf{RA}(p(\boldsymbol{x})) = p$$

- *If p is a non-recursive intensional predicate, let Π_p be the set of rules in Π defining p. For such a rule r, which is of the form (1), let*

$$\mathsf{RA}(r) = \sigma_{cond}(\mathsf{RA}(atom_1(\boldsymbol{x}_1)) \bowtie \cdots \bowtie \mathsf{RA}(atom_n(\boldsymbol{x}_n)))$$

where cond is the condition corresponding to the conjunction of atoms $cond(\boldsymbol{z})$ in (1). Then

$$\mathsf{RA}(p(\boldsymbol{x})) = \bigcup_{r \in \Pi_p} \mathsf{RA}(r)$$

- *If p is a recursive intensional predicate defined by a pair of rules of the form (2), then*

$$\begin{aligned}
\mathsf{RA}(p(\boldsymbol{x})) = \mathsf{LFP}(p = &\sigma_{cond_1}(\mathsf{RA}(atom_0(\boldsymbol{x}_0)) \bowtie p) \cup \\
&\sigma_{cond_2}(\mathsf{RA}(atom_1(\boldsymbol{x}_1)) \bowtie \cdots \bowtie \mathsf{RA}(atom_n(\boldsymbol{x}_n))))
\end{aligned}$$

where again $cond_1$ and $cond_2$ are the conditions corresponding to the conjunctions of atoms $cond_1(\boldsymbol{z}_1)$ and $cond_2(\boldsymbol{z}_2)$ in the two rules defining p.

The next proposition shows that if the rule component of an RDF rule combination is SQL99 compatible, then the Datalog transformation of the combination is also SQL99 compatible.

Proposition 2. *Let (G, Π) be an RDF-rule combination, $Q = (V, P)$ a SPARQL query, $\mathcal{M}$ a set of R2RML mapping. If Π is SQL99 compatible, then $\mu(Q, \Pi) \cup \mathcal{M}$ is also SQL99 compatible.*

Proof (Sketch). This can be easily verified by checking the layered structure of $\mu(Q, \Pi) \cup \mathcal{M} = \tau(V, P) \cup \Pi \cup \mathcal{M}$. Observe that (1) $\tau(V, P)$ and $\mathcal{M}$ are non-recursive Datalog programs, and (2) there is no arc from the head predicates of Π (resp. $\tau(V, P)$) to the body predicates of $\mathcal{M}$ (resp. Π) in the dependency graph of $\mu(Q, \Pi) \cup \mathcal{M}$. Therefore no additional cycles in the dependency graph will be introduced except the ones already in Π (if any).

4   Embedding Entailments in Mappings Rules

A first naive implementation of the technique described above was unable to generate SQL queries, due to the blowup caused by the processing of mappings, i.e., by *unfolding* the predicates defined in the heads of mapping rules with the corresponding queries in the bodies. Thus, it was necessary to optimize the rules and the mappings before unfolding the query. In this section, we present two optimizations based on the key observation that we can optimize the rules together with R2RML mappings independently of the SPARQL queries.

1. For the recursive rules, we can pre-compute the relational algebra expression (i.e., the recursive common table expressions (CTEs) in SQL).
2. For the non-recursive rules, we introduce a method to embed entailments into the mapping rules.

4.1   Recursion Elimination

The presence of recursion in the Datalog representation of the SPARQL query gives rise to a number of issues e.g., efficiently unfolding the program using SLD resolution, and managing the different types of variables. For this reason, before generating SQL for the whole set of rules, *(i)* we pre-compute CTEs for recursive predicates, making use of the expressions in relational algebra extended with fixpoints provided in Definition 6, *(ii)* we eliminate the recursive rules and replace the recursive predicates by fresh predicates; these fresh predicates are defined by cached CTEs.

4.2   T-Mappings for SWRL Rules

We introduce now an extension of the notion of *T-mappings* [20] to cope with SWRL rules. T-mappings have been introduced in the context of OBDA systems in which queries posed over an OWL 2 QL ontology that is mapped to a relational database, are rewritten in terms of the database relations only. They allow one to embed in the mapping assertions entailments over the data that are caused by the ontology axioms, and thus to obtain a more efficient Datalog program. In our setting, T-mappings extend the set of mapping to embed entailments caused by (recursive) rules into the mapping rules. Formally:

Definition 7 (SWRL T-Mappings). *Let $\mathcal{M}$ be a set of mappings, I a database instance, and Π a set of SWRL rules. A T-mapping for Π w.r.t. $\mathcal{M}$ is a set $\mathcal{M}_\Pi$ of mappings such that: (i) every triple entailed by $\mathcal{M}(I)$ is also entailed by $\mathcal{M}_\Pi(I)$; and (ii) every fact entailed by $\Pi \cup \mathcal{A}(\mathcal{M}(I))$ is also entailed by $\mathcal{M}_\Pi(I)$.*

A T-mapping for SWRL rules can be constructed iteratively, unlike OWL 2 QL-based T-mappings, using existing mappings to generate new mappings that take into account the implications of the rules[3]. In Algorithm 1, we describe the construction process which is similar to the classical semi-naive evaluation for Datalog programs.

---

[3] Recall that SWRL allows only for unary and binary predicates.

Algorithm 1. T-Mapping($\Pi,\mathcal{M}$)

Input: a set Π of (SWRL) rules; a set $\mathcal{M}$ of R2RML mappings
Output: T-Mapping $\mathcal{M}_\Pi$ of $\mathcal{M}$ w.r.t. Π
$\Delta\mathcal{M} \leftarrow \mathcal{M};$ $\mathcal{M}_\Pi \leftarrow \emptyset;$
while $\Delta\mathcal{M} \neq \emptyset$ **do**
    $\mathcal{M}_\Pi \leftarrow \mathcal{M}_\Pi \cup \Delta\mathcal{M};$ $\Delta\mathcal{M}' \leftarrow \emptyset;$
    foreach *mapping* $m \in \Delta\mathcal{M}$ **do**
        foreach *rule* $r \in \Pi$ **do**
            if m *and* r *resolves* **then**
                $\Delta\mathcal{M}' \leftarrow \Delta\mathcal{M}' \cup res(m,r);$ ▷ `res(m,r) is a set of`
                `mappings`
    $\Delta\mathcal{M} \leftarrow \Delta\mathcal{M}';$
return $\mathcal{M}_\Pi$

Theorem 1. *Let $\mathcal{M}$ be a set of mappings and Π a set of SWRL rules. Then there exists always a T-mapping for Π w.r.t. $\mathcal{M}$.*

Proof (sketch). Let I be a database instance. The Datalog program $\mathcal{M} \cup \Pi \cup I$ does not contain negation, therefore, it has a unique minimal model, which can be computed via a repeated exhaustive application of the rules in it in a bottom-up fashion.

Since the rules in $\mathcal{M}$ act as links between the atoms in Π and the ground facts in I, it is clear that if a predicate A in Π does not depend on an intensional predicate in $\mathcal{M}$, every rule containing A can be removed without affecting the minimal model of $\mathcal{M} \cup \Pi \cup I$. Thus, we can safely assume that every predicate in Π depends on an intensional predicate in $\mathcal{M}$. Thus, if every predicate in a rule is defined (directly or indirectly) by mappings in $\mathcal{M}$, we can always replace the predicate by its definition, leaving in this way rules whose body uses only database relations.   □

5   Implementation

The techniques presented here are implemented in the *Ontop*[4] system. *Ontop* is an open-source project released under the Apache License, developed at the Free University of Bozen-Bolzano and part of the core of the EU project Optique[5]. *Ontop* is available as a plugin for Protege 4, as a SPARQL end-point, and as OWLAPI and Sesame libraries. To the best of our knowledge, *Ontop* is the first system supporting all the following W3C recommendations: OWL 2 QL, R2RML, SPARQL, and SWRL[6]. Support for RIF and integration of SWRL and OWL 2 QL ontologies will be implemented in the near future.

In Figure 2, we depict the new architecture that modifies and extends our previous OBDA approach, by replacing OWL 2 QL with SWRL. First, during loading time, we

---

[4] http://ontop.inf.unibz.it
[5] http://www.optique-project.eu/
[6] SWRL is a W3C submission, but not a W3C recommendation yet.

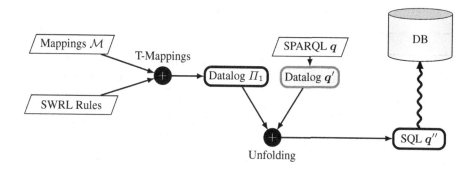

Fig. 2. SWRL and Query processing in the *Ontop* system

translate the SWRL rules and the R2RML mappings into a Datalog program. This set of rules is then optimised as described in Section 4. This process is query independent and is performed only once when *Ontop* starts. Then the system translates the SPARQL query provided by the user into another Datalog program. None of these Datalog programs is meant to be executed. They are only a formal and succinct representation of the rules, the mappings, and the query, in a single language. Given these two Datalog programs, we unfold the query with respect to the rules and the mappings using SLD resolution. Once the unfolding is ready, we obtain a program whose vocabulary is contained in the vocabulary of the datasource, and therefore can be translated to SQL. The technique is able to deal with all aspects of the translation, including URI and RDF Literal construction, RDF typing, and SQL optimisation. However, the current implementation supports only a restricted form of queries involving recursion: SPARQL queries with recursion must consist of a single triple involving the recursive predicate. This preliminary implementation is meant to test performance and scalability.

6 Evaluation

To evaluate the performance and scalability of *Ontop* with SWRL ontologies, we adapted the NPD benchmark. The NPD benchmark [7] is based on the *Norwegian Petroleum Directorate*[7] *Fact Pages*, which contains information regarding the petroleum activities on the Norwegian continental shelf. We used PostgreSQL as the underlying relational database system. The hardware consisted of an HP Proliant server with 24 Intel Xeon X5690 CPUs (144 cores @3.47GHz), 106GB of RAM and a 1TB 15K RPM HD. The OS is Ubuntu 12.04 LTS.

The original benchmark comes with an OWL ontology[8]. In order to test our techniques, we translated a fragment of this ontology into SWRL rules by *(i)* converting the OWL axioms into rules whenever possible; and *(ii)* manually adding linear recursive rules. The resulting SWRL ontology contains 343 concepts, 142 object properties, 238

---

[7] http://www.npd.no/en/
[8] http://sws.ifi.uio.no/project/npd-v2/

Table 1. Evaluation of *Ontop* on NPD benchmark

		Load	q_1	q_2	q_3	q_4	q_5	q_6	q_7	q_8	q_9	q_{10}	q_{11}	q_{12}	r_1
NPD	*Ontop*	16.6	0.1	0.09	0.03	0.2	0.02	1.7	0.1	0.07	5.6	0.1	1.4	2.8	0.25
	Stardog	-	2.06	0.65	0.29	1.26	0.20	0.34	1.54	0.70	0.06	0.07	0.11	0.15	-
NPD	*Ontop*	17.1	0.12	0.13	0.10	0.25	0.02	3.0	0.2	0.2	5.7	0.3	6.7	8.3	27.8
(×2)	Stardog	-	5.60	1.23	0.85	1.89	0.39	2.29	2.41	1.47	0.34	0.36	1.78	1.52	-
NPD	*Ontop*	16.7	0.2	0.3	0.17	0.67	0.05	18.08	0.74	0.35	6.91	0.55	162.3	455.4	237.6
(×10)	Stardog	-	8.89	1.43	1.17	2.04	0.51	4.12	5.84	5.30	0.42	0.72	3.03	3.86	–

data properties, 1428 non-recursive SWRL rules, and 1 recursive rule. The R2RML file includes 1190 mappings. The NPD query set contains 12 queries obtained by interviewing users of the NPD data.

We compared *Ontop* with the only other system (to the best of our knowledge) offering SWRL reasoning over on-disk RDF/OWL storage : *Stardog 2.1.3*. Stardog[9] is a commercial RDF database developed by Clark&Parsia that supports SPARQL 1.1 queries and OWL 2/SWRL for reasoning. Since Stardog is a triple store, we needed to materialize the virtual RDF graph exposed by the mappings and the database using *Ontop*. In order to test the scalability w.r.t. the growth of the database, we used the data generator described in [7] and produced several databases, the largest being approximately 10 times bigger than the original NPD database. The materialization of NPD (x2) produced 8,485,491 RDF triples and the materialization of NPD (x10) produced 60,803,757 RDF triples. The loading of the last set of triples took around one hour.

The results of the evaluation (in seconds) are shown in Table 1. For queries q_1 to q_{12}, we only used the non-recursive rules and compared the performance with Stardog. For the second group (r_1), we included recursive rules, which can only be handled by *Ontop*.

Discussion. The experiments show that the performance obtained with *Ontop* is comparable with that of Stardog and in most queries *Ontop* is faster. There are 4 queries where *Ontop* performs poorly compared to Stardog. Due to space limitations, we will analyze only one of these 4; however, the reason is the same in each case. Consider query 12, which is a complex query that produces an SQL query with 48 unions. The explosion in the size of the query is produced by interaction of long hierarchies below the concepts used in the query and multiple mappings for each of these concepts. For instance npdv:Wellbore has 24 subclasses, and npdv:name has 27 mappings defining it. Usually just a join between these two should generate a union of $24 \times 27 = 648$ SQL queries. *Ontop* manages to optimize this down to 48 unions but more work needs to be done to get better performance. This problem is not a consequence of the presence of

---

[9] http://stardog.com/

rules, but is in the very nature of the OBDA approach, and is one of the main issues to be studied in the future. For Stardog, the situation is slightly easier as it works on the RDF triples directly and does not need to consider mappings.

7 Conclusion

In this paper we have studied the problem of SPARQL query answering in OBDA in the presence of rule-based ontologies. We tackle the problem by rewriting the SPARQL queries into recursive SQLs. To this end we provided a translation from SWRL rules into relational algebra extended with fixed-point operators that can be expressed in SQL99's Common Table Expressions (CTEs). We extended the existing T-mapping optimisation technique in OBDA, proved that for every non-recursive SWRL program there is a T-mapping, and showed how to construct it. The techniques presented in this paper were implemented in the system *Ontop*. We evaluated its scalability and compared the performance with the commercial triple store Stardog. Result shows that most of the SQL queries produced by *Ontop* are of high quality, allowing fast query answering even in the presence of big data sets and complex queries.

Acknowledgement. This paper is supported by the EU under the large-scale integrating project (IP) Optique (Scalable End-user Access to Big Data), grant agreement n. FP7-318338. We thank Héctor Pérez-Urbina for his support in evaluating Stardog.

References

1. Aho, A.V., Ullman, J.D.: The universality of data retrieval languages. In: Proc. of the 6th ACM SIGPLAN-SIGACT Symp. on Principles of Programming Languages (POPL 1979), pp. 110–120 (1979)
2. Artale, A., Calvanese, D., Kontchakov, R., Zakharyaschev, M.: The *DL-Lite* family and relations. J. of Artificial Intelligence Research 36, 1–69 (2009)
3. Bienvenu, M., Ortiz, M., Simkus, M., Xiao, G.: Tractable queries for lightweight description logics. In: Proc. of the 23rd Int. Joint Conf. on Artificial Intelligence (IJCAI 2013). IJCAI/AAAI (2013)
4. Boley, H., Kifer, M.: A guide to the basic logic dialect for rule interchange on the Web. IEEE Trans. on Knowledge and Data Engineering 22(11), 1593–1608 (2010)
5. Calvanese, D., De Giacomo, G., Lembo, D., Lenzerini, M., Poggi, A., Rodriguez-Muro, M., Rosati, R.: Ontologies and databases: The *DL-Lite* approach. In: Tessaris, S., Franconi, E., Eiter, T., Gutierrez, C., Handschuh, S., Rousset, M.-C., Schmidt, R.A. (eds.) Reasoning Web. LNCS, vol. 5689, pp. 255–356. Springer, Heidelberg (2009)
6. Calvanese, D., De Giacomo, G., Lembo, D., Lenzerini, M., Rosati, R.: Tractable reasoning and efficient query answering in description logics: The *DL-Lite* family. J. of Automated Reasoning 39(3), 385–429 (2007)
7. Calvanese, D., Lanti, D., Rezk, M., Slusnys, M., Xiao, G.: Data generation for OBDA systems benchmarks. In: Proc. of The 3rd OWL Reasoner Evaluation Workshop (ORE 2014). CEUR-WS.org (2014)

8. Das, S., Sundara, S., Cyganiak, R.: R2RML: RDB to RDF mapping language. W3C Recommendation, World Wide Web Consortium (September 2012),
http://www.w3.org/TR/r2rml/

9. Eiter, T., Ortiz, M., Simkus, M., Tran, T.K., Xiao, G.: Query rewriting for Horn-SHIQ plus rules. In: Proc. of the 26th AAAI Conf. on Artificial Intelligence (AAAI 2012). AAAI Press (2012)

10. Glimm, B., Ogbuji, C.: SPARQL 1.1 Entailment Regimes. W3C Recommendation, World Wide Web Consortium (March 2013),
http://www.w3.org/TR/sparql11-entailment/

11. Harris, S., Seaborne, A.: SPARQL 1.1 Query Language. W3C Recommendation, World Wide Web Consortium (March 2013), http://www.w3.org/TR/sparql11-query

12. Kifer, M., Boley, H.: RIF Overview (Second Edition). W3C working group note 5 February 2013, World Wide Web Consortium (2013),
http://www.w3.org/TR/2013/NOTE-rif-overview-20130205/

13. Kontchakov, R., Lutz, C., Toman, D., Wolter, F., Zakharyaschev, M.: The combined approach to ontology-based data access. In: Proc. of the 22nd Int. Joint Conf. on Artificial Intelligence (IJCAI 2011), pp. 2656–2661 (2011)

14. Lloyd, J.W.: Foundations of Logic Programming, 2nd Extended edn. Springer, Heidelberg (1987)

15. Manola, F., Mille, E.: RDF primer. W3C Recommendation, World Wide Web Consortium (February 2004), http://www.w3.org/TR/rdf-primer-20040210/

16. Pérez-Urbina, H., Motik, B., Horrocks, I.: Tractable query answering and rewriting under description logic constraints. J. of Applied Logic 8(2), 186–209 (2010)

17. Poggi, A., Lembo, D., Calvanese, D., De Giacomo, G., Lenzerini, M., Rosati, R.: Linking data to ontologies. J. on Data Semantics X, 133–173 (2008)

18. Polleres, A.: From SPARQL to rules (and back). In: Proc. of the 16th Int. World Wide Web Conf. (WWW 2007), pp. 787–796 (2007)

19. Polleres, A., Wallner, J.P.: On the relation between SPARQL 1.1 and Answer Set Programming. J. of Applied Non-Classical Logics 23(1-2), 159–212 (2013)

20. Rodríguez-Muro, M., Calvanese, D.: Dependencies: Making ontology based data access work in practice. In: Proc. of the 5th Alberto Mendelzon Int. Workshop on Foundations of Data Management (AMW 2011). CEUR Electronic Workshop Proceedings, vol. 749 (2011),
http://ceur-ws.org/

21. Rodriguez-Muro, M., Calvanese, D.: High performance query answering over DL-Lite ontologies. In: Proc. of the 13th Int. Conf. on the Principles of Knowledge Representation and Reasoning (KR 2012), pp. 308–318 (2012)

22. Rodríguez Muro, M., Kontchakov, R., Zakharyaschev, M.: Ontology-based data access: Ontop of databases. In: Alani, H., et al. (eds.) ISWC 2013, Part I. LNCS, vol. 8218, pp. 558–573. Springer, Heidelberg (2013)

23. Rodriguez-Muro, M., Rezk, M.: Efficient SPARQL-to-SQL with R2RML mappings. Tech. rep., Free University of Bozen-Bolzano (January 2014),
http://www.inf.unibz.it/~mrezk/pdf/sparql-sql.pdf

Semantic Search for Earth Observartion Products using Ontology Services [*]

Maria Karpathiotaki[1], Kallirroi Dogani[1], Manolis Koubarakis[1],
Bernard Valentin[2], Paolo Mazzetti[3], Mattia Santoro[3], and Sabina Di Franco[3]

[1] National and Kapodistrian University of Athens, Greece
{mkarpat,kallirroi,koubarak}@di.uoa.gr
[2] Space Applications Services, Belgium
bernard.valentin@spaceapplications.com
[3] Institute of Atmospheric Pollution Research - National Research Council, Italy
paolo.mazzetti@cnr.it, {m.santoro,difranco}@iia.cnr.it

Abstract. Access to Earth Observation products remains difficult for
end users in most domains. Although various search engines have been
developed, they neither satisfy the needs of scientific communities for
advanced search of EO products, nor do they use standardized vocabularies reusable from other organizations. To address this, we present
the Prod-Trees platform, a semantically-enabled search engine for EO
products enhanced with EO-netCDF, a new standard for accessing Earth
Observation products.

1 Introduction and Motivation

The demand for aerial and satellite imagery, and products derived from them
has been increasing over the years, in parallel with technological advances that
allow producing a bigger variety of data with an increasing quality and accuracy.
As a consequence of these advances, and the multiplication of deployed sensors,
the amount of Earth Observation (EO) data collected and stored has exploded.

However, access to EO products remains difficult for end users in most scientific domains. Various search engines for EO products, generally accessible
through Web portals, have been developed. For example, see the interfaces offered by the European Space Agency portal for accessing data of Copernicus, the
new satellite programme of the European Union[1] or the EOWEB portal of the
German Aerospace Center (DLR)[2]. Typically these search engines allow searching for EO products by selecting some high level categories (e.g., the mission
from which the product was generated, the satellite instrument that was used
etc.) and specifying basic geographical and temporal filtering criteria. Although
this might suit the needs of very advanced users who know exactly what dataset

---

[*] This work was supported by the Prod-Trees project funded by ESA ESRIN.
[1] http://gmesdata.esa.int/web/gsc/home
[2] https://centaurus.caf.dlr.de:8443/eoweb-ng/template/default/welcome/
entryPage.vm

R. Kontchakov and M.-L. Mugnier (Eds.): RR 2014, LNCS 8741, pp. 173–178, 2014.

they are looking for, other scientific communities or the general public require more application-oriented means to find EO products. Other related work in this area is the RESTo framework[3] (REstful Semantic search Toolkit for geOspatial), a web interface that allows EO search through OpenSearch APIs using natural language queries.

In this paper, we present a semantically-enabled search engine for EO products currently under development by the project Prod-Trees funded by the European Space Agency. The system uses semantic technologies to allow users to search for EO products in an application-oriented way using free-text keywords (as in search engines like Google), their own domain terms or both, in conjuction with the well-known interfaces already available for expert users. A specific innovation of the presented system is the use of a new standard called EO-netCDF, currently under development in Prod-Trees and expected to be submitted to OGC, for accessing EO products annotated with netCDF. netCDF is a well-known standard consisting of set of self-describing, machine-independent data formats and software libraries that support the creation, access, and sharing of array-oriented scientific data.[4]

The Prod-Trees system has been developed using state of the art semantic technologies developed by the partners of the project: the company Space Applications Services, the National and Kapodistrian University of Athens and the research institute CNR.

2   The Prod-Trees Platform

The Prod-Trees platform is a semantically-enabled EO products search engine. It allows end-users to search for EO products using filtering criteria provided by the EO-netCDF specification and the EO vocabulary designed and implemented in the Prod-Trees project[5]. Figure 1 depicts the architecture of the platform, which partially re-uses components from the RARE platform[6].

The web interface of the Prod-Trees platform allows the users to submit free-text queries, navigate to the ontology browser, select applications terms defined in the supported ontologies and finally, search for EO product by specifing EO-netCDF parameters and controlled (bounding box, time, range) search criteria. When the user has filled the search form, the Query Analyzer is responsible for displaying a number of different interpretations for the inserted free-text. After the user has selected the semantics she wants to be used for the search, the backend service is called, generates one or more queries and sends them to GI-cat through its EO-netCDF Profiler. GI-cat searches for the matching EO products and returns back the metadata. Depending on the nature of each product (JPG, XML, HDF, etc.), this may be either visualized on-line or downloaded on the local system. The following paragraphs describe in more detail the components of the Prod-Trees architecture and their interaction.

---

[3] http://mapshup.info/resto/

[4] http://www.unidata.ucar.edu/software/netcdf/

[5] http://deepenandlearn.esa.int/tiki-index.php?page=Prod-Trees+Project

[6] http://deepenandlearn.esa.int/tiki-index.php?page=RARE%20Project

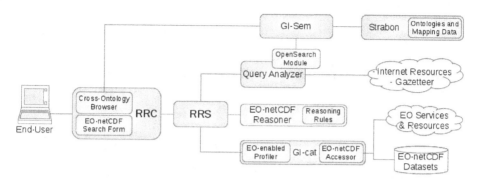

Fig. 1. The Prod-Trees Platform architecture

The ***Rapid Response[7] Client (RRC)*** provides the user interface to the Prod-Trees platform and communicates with several backend services. It displays a search form, where a user can give as input EO-specific search criteria or free text and can navigate to the supported ontologies through the ***Cross-Ontology Browser.*** This component is a browser for ontologies expressed in SKOS that allows the users to exploit the knowledge contained in the supported ontologies. It provides relevant information for each concept and highlights the connections between different (but related) concepts belonging to the same or other ontologies. Its role is to support the user in the query creation phase, as a disambiguation and discovery tool. The browser is accessed via the RRC search page.

GI-Sem [4] is a middleware which is in charge of interconnecting heterogeneous and distributed components. Its main role in the Prod-Trees platform is to create a connection between the Cross-Ontology Browser and the supported ontologies. GI-Sem performs remote queries to Strabon and returns the results to the Cross-Ontology Browser. It can also be omitted from the system by using a version of the Cross Ontology Browser that calls Strabon directly.

Strabon [3] is a well-known spatiotemporal RDF store. It holds the supported ontologies and the cross-ontology mappings appropriately encoded in RDF. The supported SKOS ontologies are the GSCDA, GEOSS, GEMET and NASA GCMD. The mappings between these ontologies were created using an algorithm developed in the scope of Prod-Trees [2].

All the interactions with the backend modules go through the ***Rapid Response Server (RRS)***. In case a query string entered by the user needs to be disambiguated, the RRS invokes the ***Query Analyzer (QA)***. The QA processes the query string, identifying the words that may be mapped to application terms, location names (toponyms), time constraints, or other types of named entities. In order to carry out this task, the QA interacts with GI-Sem (using an

---

[7] The name "Rapid Response" comes from project RARE where the main application of the developed system was rapid response for various emergencies (e.g., humanitarian or environmental). Similarly, for the Rapid Response Server mentioned below.

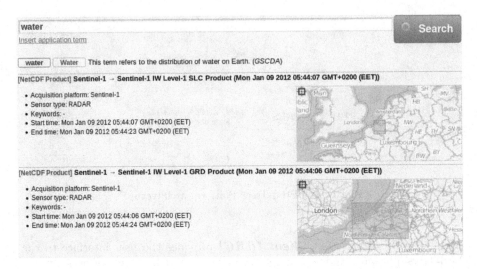

Fig. 2. Search results for the keyword "water"

OpenSearch[8] interface), Internet Resources such as gazetteers, as well as external databases such as Wordnet.

After the disambiguation process, if the user has selected an ontology concept, the RRS interacts with the ***EO-netCDF Reasoner*** to obtain the filter criteria for the search. The reasoner uses reasoning rules to map an ontology concept to EO-netCDF search criteria. These rules have been built manually with the consultation of experts in the context of the project Prod-Trees and the previous project RARE. RSS uses the returned results to build an appropriate query that is sent to GI-cat.

GI-cat [1] is an implementation of a catalogue service, which can be used to access various distributed sources of Earth Observation products. In Prod-Trees, it has been extended to support products compliant with the EO-netCDF convention. Thus, it provides an EO-netCDF enabled discovery and access engine, so that products annotated with EO-netCDF are searchable and accessible to the users.

3   Demonstration Overview

We will now present three core scenarios that show how a user can perform a semantic search for EO products using the Prod-Trees platform.

In the first scenario the user inserts a free-text query, for example "water". The system replies by presenting a number of different interpretations for the inserted text, which are provided by the Query Analyzer during the disambiguation phase. This way it is clear for the user what are the semantics of the text on

---

[8] http://www.opensearch.org/Home

which the search will be based. The default interpretation for "water" maps this text to the concept "water" of GSCDA ontology. In case the user is not satisfied with this interpretantion, she can select another one from a proposed list, for example "water use", "water temperature", "ocean level" and more. Another option is to use the inserted text without any specific interpretation. In this case, a simple text-based search will be performed. The EO-netCDF reasoner is used to map the concept "water" of GSCDA to EO-netCDF parameters with specific values. This is done using appropriate mapping rules which allow us to connect concepts of an ontology (in this case water of GSCDA) to EO-netCDF parameters with specific values (in this case combinations of satellite sensor type, resolution, polarization etc.). As a result, GI-cat returns only the EO products that include EO-netCDF parameters with these values. Figure 2 displays the first two results of the keyword search for "water".

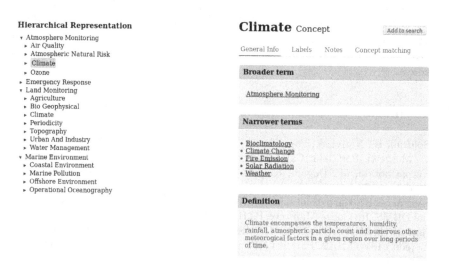

Fig. 3. The Cross-Ontology Browser displaying the GSCDA ontology

Instead of the text queries, the user can also use the ontology browser to select terms he wants to search for. Figure 3 displays the interface of the browser. The selected concept is copied back to the initial text area. Assuming the user has selected the concept "agriculture" of GEOSS ontology, she can add then more keywords (toponyms, date etc.) to the text area in order to restrict the search, for example "agriculture Bahamas 2010". Keywords with toponyms are also disambiguated using the Geonames gazetteer. Afterwards, the workflow is similar to the one described above.

Finally, the third scenario will show how to search using EO-related search criteria. This option might be more appropriate for expert users. In particular, the user can search using specific metadata attributes such as sensor type, bounding box, time, etc. and by specifying one or more EO-netCDF parameters.

The search will be based on these attributes and will return only EO products that satisfy them. For example, selecting the parameter "Sensor Type" an optional value would be "optical" or "radar". As the EO-netCDF parameter is provided directly by the user, the EO-netCDF reasoner is bypassed and only the GI-cat component is invoked to return the relevant resources.

A video demonstrating the above functionality is available at http://bit.ly/ProdTreesPlatform.

4   Conclusions

In this paper we presented the Prod-Trees platform, a semantically-enabled search engine for EO products. Given the huge growth of EO data collected daily, Prod-Trees addresses the data glut problem of this domain. In addition, we go one step further by proposing a new standard called EO-netCDF for accessing EO products annotated with netCDF. As a result, Prod-Trees brings a standardized solution that substantially improves the ability of EO experts to explore, understand and, finally, exploit the vast amount of data that is available nowadays.

References

1. Boldrini, E., Nativi, S., Papeschi, F., Santoro, M., Bigagli, L., Vitale, F., Angelini, V., Mazzetti, P.: GI-cat Catalog Service ver. 6.0 Specification. Draft Specification
2. Karpathiotaki, M., Dogani, K., Koubarakis, M.: Ontology Mapping on netCDF. Prod-Trees Technical Note
3. Kyzirakos, K., Karpathiotakis, M., Koubarakis, M.: Strabon: A Semantic Geospatial DBMS. In: Cudré-Mauroux, P., Heflin, J., Sirin, E., Tudorache, T., Euzenat, J., Hauswirth, M., Parreira, J.X., Hendler, J., Schreiber, G., Bernstein, A., Blomqvist, E. (eds.) ISWC 2012, Part I. LNCS, vol. 7649, pp. 295–311. Springer, Heidelberg (2012)
4. Santoro, M., Mazzetti, P., Nativi, S., Fugazza, C., Granell, C., Diaz, L.: Methodologies for Augmented Discovery of Geospatial Resources. In: Discovery of Geospatial Resources: Methodologies, Technologies and Emergent Applications, ch. 9

Airport Context Analytics

Eli Katsiri[1,2], George Papastefanatos[2], Manolis Terrovitis[2], and Timos Sellis[3]

[1] Department of Electrical and Computer Engineering,
Democritus University of Thrace, Xanthi, 67100, Greece
[2] Institute for the Management of Information Systems,
Research and Innovation Centre in Information, Communication and Knowledge
Technologies - "Athena",
Artemidos 6 & Epidaurou, Marousi, 15125, Athens, Greece
{eli,gpapas,mter}@imis.athena-innovation.gr,
[3] School of Computer Science and Information Technology,
RMIT University, Australia
timos@rmit.edu.au

Abstract. Airports today can constitute a perfect environment for developing novel digital marketplaces offering location-specific and semantically rich context-aware services, such as personalized marketing campaigns, last minute, discounted airline tickets while helping users access the airport and speed through the airport process.

Underpinning the above vision is the ability to target service content to users' current context, e.g., their location, intent, environment, in real time. The contribution of this work is that it uses a *pervasive computing* system with three key ingredients: (a) a data model, comprising user and service content entities, (b) a user context model and (c) rules for simple *pattern matching* on service content and user context *event streams*. This modus operandi is encapsulated inside a SOA architecture, the *Common Airport Portal - CAP* and it is illustrated through the description of a real application, Offers and Coupons Services that was deployed recently at Athens International Airport (AIA) (http://airpoint.gr).

Keywords: airport information systems, context-awareness, real-time analytics, personalisation, rule-based reasoning, system implementation.

1 Introduction

Airports nowadays form an integral part of urban spaces, and have become more than just a place where you fly from. They provide the technology substrate for commercial stakeholders (e.g., airport companies, shipping companies, retail shops, etc.) to offer added-value services to potential consumers of the airport community (visitors, passengers and airport employees) and for airport stakeholders (e.g., airport authorities) to improve airport infrastructure, redesign existing services and take strategic decisions about the future.

An airport can collect and provide valuable information in the form of event streams (feeds), to location-constrained stakeholders about passenger flows, user

R. Kontchakov and M.-L. Mugnier (Eds.): RR 2014, LNCS 8741, pp. 179–184, 2014.

travel status, alerts, flight data and time-schedules; this data is instrumental in developing spatial, temporal and context-aware personalized services, seamlessly operating over a variety of distribution channels such as web, sms, email, IPTV, info-kiosks. Such added-value services include but are not restricted to, *personalised route calculation* from a user's current position to their departure gate, *personalised marketing campains* involving offers and discount coupons, *last minute air-tickets* and *personalised alerts*. The above services are personalised and context-aware, i.e., relevant to user role, device, location, departure gate etc. Personalised marketing campaigns provide offers and *dynamic recommendations* based on what other users with similar profile are currently purchasing. These are relevant to the above parameters while abiding by constraints that are inherent in the airport domain: for example, *passengers flying to middle Eastern destinations should not be targeted for offers on alcoholic beverages,* while, *airport visitors escorting passengers should not be targeted for Duty Free discounts.*

Three successful paradigms exist in the literature that can be leveraged in order to achieve these goals: *Data Analytics* [1], *context-awareness and personalisation* [2] and *Service Oriented Architecture (SOA)* [3]: Big data helps businesses understand the data and extract patterns in order to become smarter. Context-awareness advocates that applications should be aware of user context in order to best serve them. Service Oriented Architecture (SOA), provides structured collections of discrete software modules, known as services that collectively provide the complete functionality of an application, with the ability of being reusable and composable into complex applications.

Combining the above paradigms leads to the definition of **airport context analytics**: User generated data such as position and device usage can be collected, analyzed and correlated with other airport sources in order to make service content provisioning aware of user context. The contribution of this work is that it uses the *Rete algorithm* [20] as the analytics engine. Rete is a very fast algorithm for matching data tuples ("facts") against productions ("rules"). By reducing certain types of redundancy through the use of node sharing and by storing partial matches when performing joins between fact types, Rete avoids the complete re-evaluation of all facts each time changes are made to working memory, thus increasing performance and scalability.

The rest of this paper is structured as follows: Section 2 presents a context model for the airport domain. Section 3 discusses the ACA service its implementation using web services. Section 4 discusses literature and concludes.

2   Context Model

Due to restrictions very specific to the airport community culture, e.g. the sensitive nature of passenger flight data, context that is not available can be *inferred*. For this reason the context model contains both *static* (user profile and preferences) and *dynamic* predicates. Dynamic predicates can be either low-level, directly available from the sources, or high-level, indirectly derived from low-level attributes, by mining or inference. Examples of low-level dynamic predicates are *location* (user and device locations) and *user activity* (user clicks, user requests, coupon redeems). Examples of high-level predicates are *user colocation*. Dynamic predicates that are *infrequently changing*, (frequent flyer) are stored in generic placeholders in the database, as opposed to *frequently changing* ones (location) that is stored in memory. Next, a formal definition using Hoare logic is given.

Data Model comprises *User* and *Generic Service Content*, representing generic content that is eligible for processing and provisioning within the airport environment (e.g., Offers, Coupons, LM tickets, etc.). An *Offer* is a type of advertisement of discounted products on sale while a *Discount Coupon* needs to be *redeemed* at the time of purchase. *LM Tickets* are specialized Offers for airline tickets departing within the next 48 hours.

Static context entities include *Role(rid, description)* and *Target_Group*. An AIA User is an employee of the airport authority, Airport User is an employee of the associated companies and a Company Admin is a privileged user that can authorize service content updates. *Target_Group* represents a combination of roles. Also included are: *Product_Category_Prefs*, *Company_Prefs*, *Distribution_Channel_Prefs*.

Dynamic, frequently changing, context entities include
Location(zone_id, description, < x, y, z >), where $< x, y, z >$ are positions encoded in the AIA GIS coordinate system and *zone_id* are locations at airport zone granularity. *Intent(uid, description)* refers to airport uses: an AIA User (Role) may be traveling as a regular passenger (Intent) and therefore be eligible for Duty Free offers. *Trip_Status* models the nature of the trip, e.g., {*business, economy, traveling_with_family*}. *Trip_Phase* results from associating airport zones with the airport processes: {*before_check_in, after_security, at_lounge, at_departure_gate*}. *User_Activity* can be inferred from logged information and user location (e.g.,*has_requested_offer*).

Dynamic, infrequently changing, context entities represent features that are mined from historic context instances and include *Frequent_Traveler* (*uid, description*), *Frequent_Shopper* (*uid, description*), *Technology_Savviness* (*uid, description, level*). The latter is determined by *device type* (conventional or smart phone), *frequency of service use* and *method of coupon redemption* (printed or electronic coupons.) Such knowledge can be used, for example, for *creating discounts campaigns targeted only to frequent shoppers*.

Data Model and contextualization. User and Service content, when associated with context entities, becomes *contextualized*. For example, the contextualized *Generic_Service_Content* predicate is specified as follows:

$$Generic\_Service\_Content(gscid, args, tgid, rid, dcid, oid)$$

where {*tgid, rid, pcid, compid, dcid, oid*} represent the context predicates:
{*Target_Group, Role, Product_Category_Prefs, Distribution_Channel_Prefs, Opt_ins*}.
Matching contextualized service content to eligible users is implemented by the following simple rule:

$$User(uid, profile, context\_id[])$$
$$\wedge Offer(gsid, oid, prod\_cat)$$

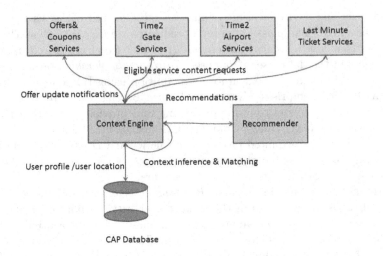

Fig. 1. ACA Service Component Diagram

$$\wedge Generic\_Service\_Content(gsid, context\_id[\,])$$
$$\wedge User.context\_id[\,] = offer.context\_id[\,]$$

where *context_id*[] represent matching context predicates between users and service content (e.g., offers for AIA Users).

3   Airport Context Analytics Service (ACA)

The ACA Service modus operandi is the following:

1. **Offer Matching:** On reception of a request from any of the services shown in Figure 1 (top), containing (a) a description of the offer or coupon or related content to be dispatched, (b) the unique id of the current user and (c) a set of optional eligibility parameters, respond by returning a list of requested content that is eligible for that user, based on the supplied parameters. This is the primary operation.

2. **Recommendation Matching:** On reception of a request from one of the CAP components shown in Figure 3, containing (a) an ordered list of recommendations to be dispatched, (b) the unique id of the current user and (c) a set of optional eligibility parameters, respond by returning a subset of the ordered list of recommendations that is eligible for that user, based on the supplied parameters.

3. **Offer Notification:** On insertion, deletion, update, cancellation of an Offer, Coupon, LMT Offer or similar content entity, respond by repeating the matching process and generating a new set of eligible content. This mechanism is based on publish/subscribe.

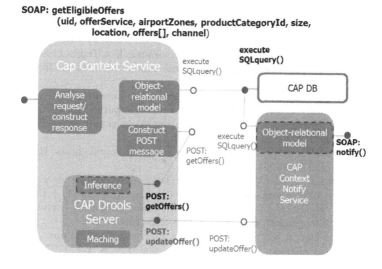

Fig. 2. ACA Implementation

Implementation (Figure 2) includes three basic web services: *CAP Drools Server, CAP Context Service, CAP Context Notify Service.* CAP Drools Server integrates *Drools* [13], a business rule management system based on the *Rete algorithm.* CAP Context Service receives the service requests, extracts service parameters and queries the CAP database for the most up-to-date data. Next it constructs contextualized facts, encodes them in HTTP POST messages and inserts them in the Drools Server, triggering a matching cycle. The CAP Context Notify Service implements publish/subscribe. First, it registers with the CAP Event Server, subscribing to databases changes (inserts, updates, deletes, cancellations, rejects) to the tables Users, Offers and Coupons. Next, it listens asynchronously for any such events, in which case it invokes the CAP Drools Server, making all relevant updates to the affected facts and re-triggering a matching cycle. In this way, when queried by the service layer, it is up-to-date.

4 Related work

Several definitions for context have been proposed in the literature [4–8], including the author's previous work [21, 22]. Certain mobile applications [15–18] use context for selecting the best communication channel per device and application. [11] proposes a platform for executing web services that adapt to application QoS, under changing conditions. [19] discusses an airport knowledge-base system designed with the CommonKADS methodology. None of these works are directly applicable here. CAP (http://airpoint.gr/en) goes far beyond previous efforts bringing an integrated solution, a new paradigm for service management in indoors spaces.

References

1. Jagadish, H.V.: Big Data: It's not just the analytics,
   http://wp.sigmod.org/?p=430
2. Weiser, M.: Ubiquitous Computing. IEE Computer, Hot Topics (1993)
3. Arsanjani, A.: Service-oriented Modeling and Architecture. How to identify, specify and realize services for your SOA,
   http://www.ibm.com/developerworks/library/ws-soa-design1
4. Schilit, B., Adams, N., Want, R.: Context-aware computing applications. In: First International Workshop on Mobile Computing Systems and Applications, pp. 85–90 (1994)
5. Schilit, B.: Disseminating active map information to mobile hosts. IEEE Network 8(5), 22–23 (1994)
6. Dey, K., Abowd, G.D., Wood, A.: CyberDesk: A framework for providing Self-Integrating context-aware services. Knowledge-base Systems 11, 3–13 (1999)
7. Brown, P.J.: The Stick-e Document: A framework for creating context-aware applications. Electronic Publishing 96, 259–272 (1996)
8. Abowd, G.D., Dey, A.K.: Towards a better understanding of context and context-awareness. In: Gellersen, H.-W. (ed.) HUC 1999. LNCS, vol. 1707, pp. 304–307. Springer, Heidelberg (1999)
9. Mahout: Scalable machine learning and Data Mining,
   http://mahout.apache.org/
10. Welcome to Apache Hadoop, http://hadoop.apache.org/
11. PAWS: Processes with adaptive web services, http://www.paws.elet.polimi.it/
12. Truong, H.L., Dustdar, H.: A survey on context-aware Web Service Systems. International Journal of Web Information Systems of Context and Context-Awareness 5(1), 5–31 (2009)
13. Drools: The business logic integration platform, http://www.jboss.org/drools
14. Sanchez-Pi, N., Carbo, J., Molina, J.M.: Building a Knowledge Base System for an Airport Case of Use. In: Corchado, J.M., Rodríguez, S., Llinas, J., Molina, J.M. (eds.) DCAI 2008. AISC, vol. 50, pp. 739–747. Springer, Heidelberg (2009)
15. Capra, L., Emmerich, W., Mascolo, C.: Carisma: Context-aware reflective middleware system for mobile applications. IEEE Transactions on Software Engineering 29, 929–944 (2003)
16. Chakraborty, D., Lei, H.: Pervasive enablement of business processes. In: Second IEEE Annual Conference on Pervasive Computing and Communications (PERCOM 2004), Orlando, Florida, USA (2004)
17. Team, T., Mais, M.: Multichannel adaptive information systems. In: International Conference on Web Information Systems Engineering, Rome, Italy (2003)
18. Capiello, C., Comuzzi, M., Mussi, E., Pernici, B.: Context Management for Adaptive Information Systems
19. Schreiber, G., Akkermans, H., Anjewierden, A., de Hoog, R., Shadbolt, N., Van de Velde, W., Wielinga, B.: Knowledge Engineering and Management, The CommonKADS Methodology. MIT Press, Cambridge (1999)
20. Forgy, C.: Rete: a fast algorithm for the many pattern/many object pattern match problem. In: Expert Systems, pp. 324–341. IEEE Computer Society Press (1990)
21. Katsiri, E.: Knowledge-base representation and reasoning for the autonomic management of pervasive healthcare. In: Eighth Join Conference on Knowledge-Based Engineering (JCKBSE 2008), August 25-28. University of Piraeus, Greece (2008)
22. Katsiri, E., Bacon, J., Mycroft, A.: Linking sensor data to context-aware applications using abstract events. International Journal of Pervasive Computing and Communications 3(4), 347–377 (2007)

Navigating among Educational Resources in the Web of Linked Data

Dimitrios A. Koutsomitropoulos, Georgia D. Solomou, and Aikaterini K. Kalou

High Performance Information Systems Laboratory (HPCLab),
Computer Engineering and Informatics Dpt., School of Engineering,
University of Patras, Building B, 26500 Patras-Rio, Greece
{kotsomit,solomou,kaloukat}@hpclab.ceid.upatras.gr

Abstract. Linked Data seem to play a seminal role in the establishment of the Semantic Web as the next-generation Web. This is even more important for digital object collections and educational institutions that aim not only to promote and disseminate their content but also to aid its discoverability and contextualization. Having already 'semantified' a popular digital repository system, DSpace, in this paper we show how repository metadata can be exposed as Linked Data, thus enhancing their machine understandability and contributing to the LOD cloud. Our effort comes complete with an updated UI that allows for reasoning-based search and navigation between linked resources within and outside the scope of the digital repository.

1   Introduction

Linked and Open Data (LOD) [2] appear to be the "silver-bullet" in the forming Semantic Web ecosystem, that promise to breathe new life to the latter's benefits for real-world web applications. This is often combined with lightweight semantics [3] so that known scalability problems and reasoning inefficiencies could be sidestepped and still to retain some essence of the knowledge discovery capabilities of ontologies. However, tried-and-true systems like digital repositories for educational and other institutions need a little more incentive to embark on such a migration and to get tempted to adopt this new paradigm.

In this paper we present our work for publishing Linked Data and navigating among resources of a popular digital repository system, DSpace. Semantics play a crucial role and this is exhibited by an OWL 2 inference-based knowledge acquisition mechanism that lies at the core of this implementation, aka *Semantic Search for DSpace* [6]. Challenges for imposing semantic searching over otherwise semantically-oblivious systems are well-known and have been discussed earlier (e.g. [5]). Further, Linked Data provision requires a careful replication design for existing resource descriptions; a data linking and resolution mechanism; and a content negotiation strategy to serve information both to end-users and machines.

R. Kontchakov and M.-L. Mugnier (Eds.): RR 2014, LNCS 8741, pp. 185–190, 2014.

Next, in Section 2 we describe the process of publishing and minting Linked Data out of DSpace resources; Section 3 presents the semantic querying interface and the Linked Data facility; and Section 4 summarizes our conclusions and future work.

Semantic Search is hosted and maintained as a Google Code project[1] and is listed as an official DSpace add-on[2]. A working demo of this implementation is also available[3].

2   Linking Data for DSpace

2.1   Publication and Linking of Entities

Linked Data principles are in essence a few simple rules that foster the idea of an interlinked 'Web of Data'. In our context this means that resource URIs need to be dereferenceable, to provide meaningful information for users and services alike and to give references (or links) to other related entities whenever possible.

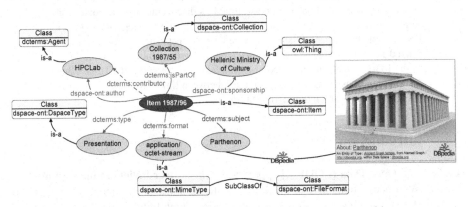

Fig. 1. Example repository item and its relationships to other entities

In DSpace the main unit of information is the 'item', i.e. a publication or learning object that is described with a set of metadata based on Dublin Core (DC). During the mapping to OWL however, we identify additional implicit entities and assign resolvable URIs to them too (see below, section 2.2). Further, these entities are linked together or refer to other external datasets like DBpedia (see section 3.2). Fig.1 illustrates a sample instance of the resulting DSpace ontology and the way it gets interlinked with other entities and/or datasets. Using the Jersey framework[4], the reference implementation of the Java API for RESTful services, both HTML as well as

---

[1] http://code.google.com/p/dspace-semantic-search/
[2] https://wiki.duraspace.org/display/DSPACE/Extensions+and+
Addons+Work#Extensionsand AddonsWork-SemanticSearchforDSpacev2.0
[3] http://apollo.hpclab.ceid.upatras.gr:8000/dspace-ld
[4] https://jersey.java.net/

RDF/XML representations are accommodated and the following URI pattern has been established:

— `http://{repositoryURL}/semantic-search/resource/{entity-id}`
— `http://{repositoryURL}/semantic-search/page/{entity-id}`(HTML)
— `http://{repositoryURL}/semantic-search/data/{entity-id}`(RDF)

Moreover, these URIs are dereferenceable by using Jersey's content negotiation capability and performing HTTP *303 See Other* redirects.

2.2   Exposing Educational Metadata into OWL

The first step towards providing Linked Data is to unleash resource information and metadata that are hidden within databases. To maximize the semantic value of exported metadata as well as to maintain interoperability, a careful and elaborate process has to be conducted. This process includes an exhaustive mapping of the inherent DSpace metadata application profile to the DC metadata terms (DCTerms), part of which is shown in Table 1. This mapping provides for the following:

— Other than the default, additional elements are exported.
— Map everything under the DCTerms namespace, rather than mixing DC with DCTerms, which is generally not advisable [1].
— Provide additional Learning Object mappings to DCTerms (in case LOM metadata exist).
— Assign types to non-literal values.

Mapped metadata can then be exposed and harvested through the supported OAI-PMH interface [7].

Table 1. Snippet of the performed mapping of DSpace internal metadata to DCTerms

DSpace internal metadata representation	Provided mapping	Notes
`dc.contibutor.author`	`dspace-ont:author` `dcterms:contributor`	*'author' is not compatible with QDC. However it is a subproperty of 'contributor'*
`dc.subject`	`dcterms:subject`	*We use the DCTerms namespace*

`dc.identifier.uri`	`dcterms:identifier type=`
	`http://www.w3.org/2001/XML`
	`Schema#anyURI`
		Literals get typed when possible

Bitstream metadata	`dcterms:format` `dcterms:extent`	*Not exposed by default*

`lom.intendedenduser`
`role`
	`dcterms:audience`
	`type="lom:IntendedEndUser`
	`Role"`
		LOM specific mapping

Having exported as much available DSpace metadata as possible, the next step is to translate them into full-fledged RDF/OWL triples. During this semantic translation, certain implicit entities (like items, collections, authors, sponsors) are reified and become individual nodes themselves instead of mere literals, thus resulting into an OWL 2 DSpace ontology [5]. Most of these entities are assigned resolvable identifiers, so that it would be easy for them to get dereferenced within the individual *Navigation Pane* (see below). In addition, this is when URLs to the DBpedia Lookup service[5] are injected, in order to enrich reified entities such as authors, contributors, sponsors and item types.

3   Semantic Search and Navigation

In what follows, we summarize the semantic search interface's main features (3.1) and then we detail the newly implemented Linked Data facility (3.2). A more thorough account of the architecture, methodology and design principles behind Semantic Search v2 can be found in [6].

3.1   An Interface for Semantic Querying Educational Resources

The main idea used by our semantic search interface, lies behind the deconstruction of a semantic query into smaller building parts (query *atoms*) that are assigned to different fields of a dynamic UI. Query crumbs that are provided through these fields are then assembled by the underlying mechanism to create valid Manchester Syntax expressions [4] (see fig. 2). Each such expression is an atomic or anonymous class in the Web Ontology Language (OWL) [8] and its (both inferred and asserted) members are the answers to the query. Search results are presented as a browsable list of linkable entities/resources. Effective querying of the knowledge base is accomplished by interconnecting to an appropriate inference engine, capable of reasoning with OWL 2 ontologies.

Fig. 2. The auto-complete and query construction mechanisms of the semantic search interface

---

[5] http://wiki.dbpedia.org/Lookup

New features in the upcoming version 2.4 include: *a)* syntax highlighting for Manchester Syntax, implemented through the CodeMirror Javascript component, *b)* a history subsystem that keeps track of the last ten input queries, *c)* a RESTful Linked Data provider that exposes resources' metadata as resolvable entities, and *d)* a DBpedia URL injection facility.

When an entity is selected, the corresponding individual's navigation pane that gathers the resource's semantic metadata is produced on the fly. This ontological information is formed as Linked Data and is presented in either an HTML or an RDF format, depending on whether the request was made by a person or a service, respectively. It is important to notice that this holds not only for DSpace items themselves, but also for every other (implicit) entity in the repository model that gets reified during the semantic translation phase.

3.2 Navigation and Data Linking

The main objective of the navigation pane is not only to give a detailed reference to the resources' ontological information (semantic metadata) but also to allow users to further explore and navigate among interlinked information in the LOD cloud. To achieve this, information is structured in the form of resolvable URIs, as much as possible.

All non-literal metadata values are denoted as URIs, which can be dereferenced on the web. In particular, each class redirects back to the semantic search page with the specific class already predefined and its members appearing on the result list. In the case of object properties, the corresponding values are resolvable entities that lead to the particular entity's navigation pane. And even for data properties, where mere text values mostly apply, a datatype of `xsd:AnyURI` is rendered as a resolvable link. This is useful for example to maintain context with the original DSpace item view or with external references, such as the DBpedia Lookup service.

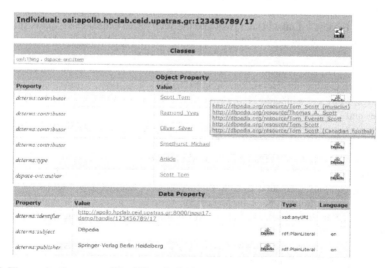

Fig. 3. The navigation pane - The DBpedia lookup service is triggered for author "Tom Scott"

More specifically, a DBpedia icon is automatically placed next to specific reified entities (contributor, author, type, sponsor) and next to the object property values (within an item's navigation pane) that correspond to these entities. When the icon is clicked, the DBpedia Lookup service is triggered for this entity, leading to a keyword-based search against DBpedia. This label matching process inevitably includes a certain extent of ambiguity. In order to resolve this, a dynamic tooltip is presented to the user, including up to five matching DBpedia resource URIs (see fig. 3). Moreover, the `dcterms:subject` and `dcterms:publisher` literal values are also linked to DBpedia in the same way.

All information gathered in the navigation pane can also be obtained in a machine-readable RDF format. An RDF icon next to the entity points to our REST Linked Data service and requests the RDF representation of the entity's data.

4   Conclusions and Future Work

We soon intend to release this feature-set in the upcoming version of our Semantic Search plugin for DSpace. The combination of a reasoning-based knowledge acquisition mechanism with a Linked Data service can help educational institutions to provide new discovery capabilities for their content and to be part of the greater LOD cloud effortlessly. DBpedia is naturally a nodal point of the latter, but interconnecting with other data sources would also be useful, like for example DBLP. What is more, data from these sources can be brought back into our model, so that we could reason with them and reveal a whole new set of correlations between repository assets and the outside world.

References

1. DCMI Usage Board: DCMI Metadata Terms. DCMI Recommendation (2008), http://dublincore.org/documents/2008/01/14/dcmi-terms/
2. Heath, T., Bizer, C.: Linked Data: Evolving the Web into a Global Data Space. In: Synthesis Lectures on the Semantic Web: Theory and Technology, 1st edn., vol. 1(1), pp. 1–136 (2011)
3. Hendler, J.: Why the Semantic Web will Never Work. In: Keynote at the 8th Extended Semantic Web Conference, Heraklion, Greece (2011)
4. Horridge, M., Patel-Schneider, P.S.: Manchester Syntax for OWL 1.1. In: 4th OWL Experiences and Directions Workshop, Gaithersburg, Maryland (2008)
5. Koutsomitropoulos, D., Solomou, G., Alexopoulos, A., Papatheodorou, T.: Semantic Metadata Interoperability and Inference-Based Querying in Digital Repositories. Journal of Information Technology Research 2(4), 36–52 (2009)
6. Koutsomitropoulos, D., Solomou, G., Papatheodorou, T.: Semantic query answering in digital repositories: Semantic Search v2 for DSpace. International Journal of Metadata, Semantics and Ontologies 8(1), 46–55 (2013)
7. Lagoze, C., Van de Sompel, H., Nelson, M., Warner, S.: The Open Archive Initiative Protocol for Metadata Harvesting (2002), http://www.openarchives.org/OAI/2.0/openarchivesprotocol.htm
8. Motik, B., Parsia, B., Patel-Schneider, P.F. (eds.): OWL 2 Web Ontology Language XML Serialization, 2nd edn. W3C Recommendation (2012)

Investigating Information Diffusion in a Multi-Social-Network Scenario via Answer Set Programming

Giuseppe Marra[1], Antonino Nocera[1], Francesco Ricca[2], Giorgio Terracina[2], and Domenico Ursino[1]

[1] DIIES, University Mediterranea of Reggio Calabria, Via Graziella, Località Feo di Vito, 89122 Reggio Calabria, Italy
[2] Dipartimento di Matematica, University of Calabria, Via Pietro Bucci, 89136 Rende (CS), Italy

Abstract. Information Diffusion is a classical problem in Social Network Analysis, where it has been deeply investigated for single social networks. In this paper, we begin to study it in a multi-social-network scenario, where many social networks coexist and are strictly connected to each other, thanks to those users who join more social networks. In this activity, Answer Set Programming provided us with a powerful and flexible tool for an easy set-up and implementation of our investigations.

1 Introduction

Information Diffusion has been largely investigated in Social Network Analysis [5,7,8,10,11]. However, all the investigations about this problem performed in the past analyze single social networks, whereas the current scenario is multi-social-network [2,4]. Here, many social networks coexist and are strictly connected to each other, thanks to those users who join more social networks, acting as bridges among them. But, what happens to the Information Diffusion problem when passing to this new scenario? New aspects must be taken into account and new considerations are in order. In this paper[1] we investigate the problem of Information Diffusion in a multi-social-network scenario (MSNS, for short). For this purpose, first we propose a graph-based model for an MSNS. This model takes into account the existence of more social networks, as well as the presence of bridges and topics of interest for MSNS users. Then, we provide a formal definition of the Information Diffusion problem in an MSNS. In order to implement our approach, and perform an analysis on real world data, we applied Answer Set Programming [1,6] (ASP). ASP is an ideal framework for the rapid development and implementation of programs solving complex problems [9] given its declarativity, expressive power, and availability of efficient ASP systems [12]. After describing the ASP specification solving the Information Diffusion problem, we also present the results of an experimental campaign conducted on an

---

[1] A preliminary version was submitted to SEBD 2014, which has informal proceedings.

R. Kontchakov and M.-L. Mugnier (Eds.): RR 2014, LNCS 8741, pp. 191–196, 2014.

MSNS based on four social networks, namely LiveJournal, Flickr, Twitter and YouTube. As far as the system for running our logic program is concerned, we used the ASP system DLV [12]. Our experimental campaign allowed us to draw an identikit of the best nodes for spreading information in an MSNS.

2   Our ASP-Based Approach to Information Diffusion in an MSNS

An MSNS Ψ, consisting of n social networks $\{S_1, S_2, \ldots, S_n\}$, can be modeled by a pair $\langle G, T \rangle$. Here, T is a list $\{t_1, t_2, \ldots, t_p\}$ of topics of interest for the users of Ψ. It is preliminarily obtained by performing the union/reconciliation of the topics related to the social networks of Ψ. G is a graph and can be represented as $G = \langle V, E \rangle$. V is the set of nodes. A node $v_i \in V$ represents a user account in a social network of Ψ. $E = E_f \cup E_m$ is a set of edges. E_f is the set of friendship edges; E_m is the set of me edges. An edge $e_j \in E$ is a triplet $\langle v_s, v_t, L_j \rangle$. v_s and v_t are the source and the target nodes of e_j, whereas L_j is a list of p pairs $\langle t_{j_k}, w_{j_k} \rangle$, where t_{j_k} is a topic and w_{j_k} is a real number between 0 and 1 representing the corresponding weight. This weight depends on both t_{j_k} and the ability of the user associated with v_t to propagate, to the user associated with v_s, information related to t_{j_k}.

Thus, an MSNS models a context where several social networks coexist and are strictly connected to each other, thanks to those users who join more social networks. Indeed, when a user joins more social networks, her multiple accounts allow these networks to be connected. We call *bridge user* each user joining more social networks, *bridge (node)* each account of such a user and me *edge* each edge connecting two bridges.

The Information Diffusion problem in an MSNS takes as input: *(i)* An MSNS Ψ, consisting of n social networks $\{S_1, \ldots, S_n\}$. *(ii)* A list D of n elements. The generic element D_h of D consists of a tuple $\langle S_h, p_h, c_h \rangle$. Here, p_h denotes the priority of S_h and is an indicator of the relevance of this social network in Ψ. It is an integer from 1 to n, where 1 (resp., n) is the maximum (resp., minimum) priority. c_h is the minimum desired coverage for S_h, i.e., the minimum number of nodes of S_h which must be reached by the information to spread throughout Ψ. *(iii)* A list τ of q elements. The generic element $\tau[k]$ of τ is a pair $\langle t_k, \omega_k \rangle$. Here, t_k corresponds to the kth element of the set of topics T of Ψ. ω_k is a real number, belonging to the interval $[0, 1]$ and indicating the weight of t_k in the information to spread throughout Ψ. The Information Diffusion problem in Ψ requires to find the minimum set of the nodes of Ψ allowing the maximization of the coverage of the social networks of Ψ, taking into account the minimum allowed network coverage, the network priorities (as expressed in D), and the topics characterizing the information to spread (as expressed in τ).

Our ASP-based solution of this problem is based on the following guidelines. First, a support graph $G' = \langle N', E' \rangle$ is constructed starting from G. Specifically, there is a node $n' \in N'$ for each node $n_i \in N$, and an edge $e'_j = \langle v_s, v_t, w_j \rangle \in E'$ for each edge $e_j = \langle v_s, v_t, L_j \rangle \in E$. w_j is obtained as: $w_j = \sum_{k=1}^{p} w_{j_k} \omega_k$. In other words, w_j measures the relevance of e'_j in the current Information Diffusion activity. It depends on both the importance of each topic of T in the information to spread throughout Ψ and the ability, of the user associated with v_t, to propagate, to the user associated with v_s, the topics of T. The Information Diffusion model adopted in our approach is the well known Linear Threshold one [8]. In this model, a node is considered *active* if the sum of the weights of the friendship edges from it to the already active nodes is higher than a certain threshold. In an MSNS, the Linear Threshold model must be extended in such a way as to consider me edges. Given a me edge from a bridge b_s to a bridge b_t, our extension requires that the information to spread propagates from b_s to b_t if the weight associated with the me edge is higher than a certain threshold (different from the one concerning friendship edges). Starting nodes are randomly selected. However, since in an MSNS the number of bridges is extremely low [3] w.r.t. the number of non-bridges, we first state the percentage of bridges and non-bridges which must be present in the set of starting nodes, and then randomly select the two kinds of node accordingly.

The problem we are considering is extremely complex. Since this paper represents a first attempt of investigating the possibility of using ASP in this context, in the following we perform some simplifications. Specifically, we assume that all the topics of Ψ have the same weight in the information to spread; this implies the removal of the topic dependency of the problem which, thus, becomes only structural. The important consequence of this choice is that all the friendship edges in G' have the same weight (we assign a weight equal to 1 to them). As for me edges, we assume that all of them always propagate the information to spread. Also in this case, the consequence is that we assign a weight equal to 1 to all me edges. A final simplification regards the node activation policy. In fact, we assume that a node is activated when at least two edges, outgoing from it, are pointing to already activated nodes. This corresponds to set the threshold to 2 in the Linear Threshold Information Diffusion model mentioned above.

It is important to stress that the adoption of ASP allowed an easy and fast set-up of the approach implementation, while attaining acceptable performances. The logic program designed to solve our problem is as follows (see [1] for a nice introduction to ASP):

```
1. in(X) v out(X) <- starting_node(X).
2. active(X) <- in(X).
3. active(X) <- active(X1),edge(X,X1,me).
4. active(X) <- edge(X,X1,friendship),edge(X,X2,friendship), X1!=X2,
        active(X1),active(X2).
5. hasActiveNodes(Sn)<-node(N,Sn),active(N).
6. <- D(Sh,Ph, Ch), Ch!=0, not hasActiveNodes(Sh).
7. <- D(Sh,Ph,Ch), #count{N:active(N),node(N,Sh)}<Ch.
8. <~ in(X). [1:2]
9. <~ node(N,Sh), not active(N). [1:1]
```

Here, the input is given as a set of facts of the form edge(X,X1,K) modeling edges from X to X1, where K specifies the edge kind (me or friendship); node(N,Sn) denotes the set of nodes in the social network S_n; starting_node(X) is the set of starting nodes. Finally, D(Sh,Ph,Ch) identifies the desiderata. Rule 1. guesses the nodes that must be selected in the best solution. Rules 2. to 4.

Table 1. Effectiveness of our approach

Number of Social Networks	2	3	4
Percentage of activated nodes	70%	81%	81%
Number of necessary starting nodes	2	3	3

compute active nodes, based on the guess; in particular, a node is active if either it is a selected one, or it reaches an active node through a me edge, or, finally, it reaches two active nodes through friendship edges. Constraint rules 5. to 7. impose admissibility conditions, as specified in D. Weak constraint rules 8. and 9. implement the optimality requirements for consistent solutions, so that the minimum sets of nodes providing consistent solutions are identified first (8.), and, among them, the ones minimizing non-active nodes are selected (9.).

3   Experimental Campaign

To test our Information Diffusion approach we performed an experimental campaign on an MSNS consisting of four social networks, namely LiveJournal, Flickr, Twitter and YouTube. Our MSNS has 93177 nodes and 146957 edges. All the corresponding data can be downloaded from the following address: www.ursino.unirc.it/RR2014.html. The password the Referee must specify is "85749236". We performed a large number of runs of our ASP program. In these runs we considered different configurations of the starting nodes. They differed for the number of nodes, the percentage of bridges (this, very important, parameter ranged from 0 to 100 with a step of 10), and the number of the social networks to cover (this number ranged from 2 to 4). We constructed more sets of starting nodes for the same configuration in such a way as to reduce the influence of possible outliers. The whole number of runs we have performed was 576. These runs allowed us to carry out several investigations, the most significant of whom are reported below (the other ones cannot be shown due to space limitations).

In a first test we computed the percentage of the nodes of our MSNS activable by our approach that were really activated by it, as well as the number of nodes necessary for this activation, against the number of social networks to cover. The corresponding results are shown in Table 1. This table evidences that our approach is really effective. Interestingly, 3 starting nodes are sufficient to cover 4 social networks. This could not have been obtained without the presence of bridges. In fact, without this kind of node, at least 8 nodes would have been necessary to cover 4 social networks.

In a second test we computed the variation of the average percentage of bridges present in the optimal solution of runs against the variation of the average percentage of bridges present in the sets of starting nodes. Obtained results are

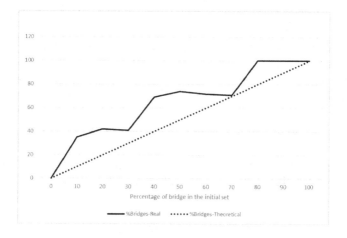

Fig. 1. Average percentage of bridges in the optimal solutions

Table 2. Composition of optimal solutions

(i)	(ii)	(iii)	(iv)	(v)	(vi)	(vii)	(viii)	(ix)	(x)
75%	24%	86%	22%	75%	86%	0 %	11%	14%	87%

shown in Figure 1. Observe that the percentage of bridges in the optimal solutions is generally higher, or much higher, than the percentage of bridges in the sets of starting nodes. This information is precious for drawing an identikit of the most influential nodes for Information Diffusion in an MSNS.

In a third test we computed the following statistics about the composition of optimal solutions: (i) average percentage of bridges; (ii) average percentage of the direct neighbors of bridges; (iii) average percentage of power users; (iv) average percentage of the direct neighbors of power users; (v) average percentage of nodes being both bridges and power users; (vi) average percentage of nodes being bridges or power users; (vii) average percentage of nodes being bridges but not power users; (viii) average percentage of nodes being power users but not bridges; (ix) average percentage of nodes being neither bridges nor power users; (x) average Jaccard coefficient[2] of bridges and power users. The corresponding results are reported in Table 2. From the analysis of this table we can observe that 99% of the nodes in the optimal solutions are either bridges or direct neighbors of bridges. Analogously, 98% of the nodes in the optimal solutions are either power users or direct neighbors of power users. Furthermore, all the bridges involved in the optimal solutions are power users, and almost all the power users involved in the optimal solutions are bridges. Finally, only a little fraction of the nodes present in the optimal solutions are neither bridges nor power users.

---

[2] We recall that the Jaccard Coefficient $J(A, B)$ between two sets A and B is defined as $J(A, B) = \frac{A \cap B}{A \cup B}$.

4   Conclusion

In this paper we have investigated Information Diffusion problem in a Multi-Social-Network Scenario. We have applied ASP and analyzed the properties of real MSNSs using real-world data. In the future, we plan to remove the simplifications applied to the approach introduced in this paper, design other predictive models for Information Diffusion, and, finally, apply ASP for extending Social Network Analysis investigations from single social networks to MSNSs.

Acknowledgments. This work was partially supported by Aubay Italia S.p.A., by the project BA2Kno (Business Analytics to Know) funded by the Italian Ministry of Education, University and Research, and by Istituto Nazionale di Alta Matematica "F. Severi" - Gruppo Nazionale per il Calcolo Scientifico.

References

1. Baral, C.: Knowledge Representation, Reasoning and Declarative Problem Solving. Cambridge University Press (2003)
2. Berlingerio, M., Coscia, M., Giannotti, F., Monreale, A., Pedreschi, D.: The pursuit of hubbiness: Analysis of hubs in large multidimensional networks. J. Comput. Science 2(3), 223–237 (2011)
3. Buccafurri, F., Foti, V.D., Lax, G., Nocera, A., Ursino, D.: Bridge analysis in a social internetworking scenario. Inf. Sci. 224, 1–18 (2013)
4. Buccafurri, F., Lax, G., Nocera, A., Ursino, D.: Moving from social networks to social internetworking scenarios: The crawling perspective. Inf. Sci. 256, 126–137 (2014)
5. Domingos, P., Richardson, M.: Mining the network value of customers. In: Proceedings of the Seventh ACM SIGKDD KDD, San Francisco, CA, USA, August 26-29, pp. 57–66. ACM (2001)
6. Gelfond, M., Lifschitz, V.: Classical negation in logic programs and disjunctive databases. New Generation Comput. 9(3/4), 365–386 (1991)
7. Goldenberg, J., Libai, B., Muller, E.: Talk of the network: A complex systems look at the underlying process of word-of-mouth. Marketing Letters 12(3), 211–223 (2001)
8. Granovetter, M.: Threshold models of collective behavior. New Generation Comput. 83(6), 1127–1138 (1978)
9. Grasso, G., Leone, N., Manna, M., Ricca, F.: Asp at work: Spin-off and applications of the dlv system. In: Balduccini, M., Son, T.C. (eds.) Logic Programming, Knowledge Representation, and Nonmonotonic Reasoning. LNCS, vol. 6565, pp. 432–451. Springer, Heidelberg (2011)
10. Guille, A., Hacid, H., Favre, C., Zighed, D.A.: Information diffusion in online social networks: a survey. SIGMOD Record 42(2), 17–28 (2013)
11. Kempe, D., Kleinberg, J.M., Tardos, É.: Maximizing the spread of influence through a social network. In: Proceedings of the Ninth ACM SIGKDD, Washington, DC, USA, August 24-27, pp. 137–146. ACM (2003)
12. Leone, N., Pfeifer, G., Faber, W., Eiter, T., Gottlob, G., Perri, S., Scarcello, F.: The dlv system for knowledge representation and reasoning. ACM Trans. Comput. Log. 7(3), 499–562 (2006)

Web Stream Reasoning Using Probabilistic Answer Set Programming*

Matthias Nickles[1,2] and Alessandra Mileo[1]

[1] INSIGHT Centre for Data Analytics
National University of Ireland, Galway
{matthias.nickles,alessandra.mileo}@deri.org
[2] Department of Information Technology
National University of Ireland, Galway

Abstract. We propose a framework for reasoning about dynamic Web data, based on probabilistic Answer Set Programming (ASP). Our approach, which is prototypically implemented, allows for the annotation of first-order formulas as well as ASP rules and facts with probabilities, and for learning of such weights from examples (parameter estimation). Knowledge as well as examples can be provided incrementally in the form of RDF data streams. Optionally, stream data can be configured to decay over time. With its hybrid combination of various contemporary AI techniques, our framework aims at prevalent challenges in relation to data streams and Linked Data, such as inconsistencies, noisy data, and probabilistic processing rules.

Keywords: Web Reasoning, Uncertainty Stream Reasoning, Answer Set Programming, RDF, Probabilistic Inductive Logic Programming, Machine Learning.

1 Introduction and Related Work

Many real-world applications on the Web involve data streams (e.g., messaging events, web searches, or sensor data), but while stream processing and data stream mining are already established research areas, *stream reasoning* [28] is still a very young research field. Challenges in this regard are not only incremental reasoning in the presence of rapidly changing dynamic information, but also provisions for inconsistencies, incoherence and noise in stream data, and stream reasoning using probabilistic background knowledge (e.g., probabilistic rules). Probabilistic logic programing, and the ability to learn facts and rules from possibly incomplete or noisy data, can provide an attractive approach to stream reasoning, since it combines the deduction capability and declarative nature of logic programming with probabilistic inference abilities traditionally modeled using Bayesian or Markov networks. In particular nonmonotonic probabilistic (inductive) logic programming [18,1,25,4,16] is promising in this regard, as it already provides for concepts useful for dealing with dynamic knowledge by means of, e.g., default reasoning. In this paper, we present a novel approach to probabilistic inductive logic programming based on Answer Set Programming (ASP) [9]. In contrast to existing approaches, it provides a unified framework for probabilistic inference as well

---

* This research is sponsored by Science Foundation Ireland (SFI) grant No. SFI/12/RC/2289

R. Kontchakov and M.-L. Mugnier (Eds.): RR 2014, LNCS 8741, pp. 197–205, 2014.

as for parameter estimation (hypotheses weight learning) from examples, using incrementally provided stream data as input. In the design of our framework and in use case examples provided, we focus on streams on the Web of Data, given the opportunities it offers in terms of linking data for sharing and re-use. Although we cannot report yet on scalability and performance results, we are already able to exhibit how the prototypical implementation of our approach deals with streaming RDF and uncertain sensor data.

Our framework consists of a probabilistic reasoning core based on ASP (for inference and induction) and a client for incrementally feeding RDF stream data (prefiltered by CEQLS [12]) into the core reasoning module. The stream reasoning architecture of our prototype software is described in Sect. 3, whereas our probabilistic logic programming language and reasoning core is presented in the next section (a shortened account of a more detailed description in [16]).

Works related to the reasoning core include [26,17,5,7,10,11,22,20,24]. While none of these approaches supports nonmonotonic logic programming, [18,1,25,4] are based on nonmonotonic logics. Like P-log [1], our approach computes probability distributions over answer sets. However, P-log (as well as [25]) does not allow for annotating arbitrary formulas (including FOL formulas) with probabilities. [4] allows to associate probabilities (only) with abducibles and to learn both rules and probabilistic weights from given data (in form of literals).

While a large number of stream processing and stream data mining approaches exist (such as [12]), including some which allow for probabilistic inference, e.g., using Bayesian Networks or fuzzy logic (e.g., [23,29]), only a few stream frameworks for the Web provide true (logical) reasoning capabilities (if we do not count simple SPARQL entailment) - see [14] for a recent survey. Some approaches allow for non-monotonic reasoning [8,15,6], but none accounts for probabilistic uncertainty. [2] allows for (OWL-based) deductive reasoning and machine learning (prediction of assertions, learned from RDF streams), but not nonmonotonic reasoning or logic programming. To our best knowledge, no other approach combines nonmonotonic logic programming with probabilistic inference and inductive learning in a unified framework such as ours.

2   Probabilistic Inductive Answer Set Programming

Before we turn to stream reasoning, we briefly introduce our language for probabilistic non-monotonic logic programming, called Probabilistic Answer Set Programming (*PrASP*). A more detailed introduction can be found in [19,16].

PrASP is a Nilsson-style probabilistic logic language, with the main enhancement in relation to normal ASP and other probabilistic approaches to ASP being the possibility to annotate any formulas (formulas in first-order syntax as well as AnsProlog rules and facts) with probabilities. Given a probabilistic logic program, it can infer unconditional as well as conditional probabilities of any formulas (if the program is probabilistically consistent), as well as learn the probabilities of formulas from examples.

A PrASP program is a non-empty finite set $\{([p_i]f_i)\}$ of annotated formulas. The $[p_i]$ are the *weights* of the formulas and directly represent point-valued probabilities. If the weight is omitted, weight 1 is assumed. Such weighted formulas can intuitively seen as constraints which specify which possible worlds (in the form of answer sets) are indeed possible, and with which probability. The formulas f_i are either in FOL syntax (supported by means of a transformation into ASP syntax described in [13]) or plain AnsProlog syntax, e.g., [0.5] win :- coin(heads). In addition to that, conditional probabilities are supported, using syntax [p|c] f (this asserts $Pr(f|c) = p$).

A PrASP program induces a probability distribution over possible worlds. The possible worlds correspond to the answer sets of a certain plain answer set program (the so-called *spanning program*) which is automatically generated from the original PrASP program and which reflects the (unweighted) nondeterminism induced by the PrASP program. To make this tractable, any formulas which do not contribute to the probability of the query (the formula whose probability or truth value we want to infer from the given PrASP program) are omitted from the spanning program, by means of a dependency analysis. After generating the spanning program, the probability distribution over possible worlds (the answer sets of the spanning program) is basically computed by solving a system of linear equalities involving probabilities (the formula weights). However, because this system typically has multiple or even infinitely many solutions and direct computation can be very costly, the PrASP inferencer applies several optimization steps besides dependency analysis, including optional sampling and the maximum entropy principle [27].

Given a probabilistic distribution over possible worlds, probabilistic inference (computing probabilities of the forms $Pr(\phi)$ and $Pr(\phi|c)$) becomes a model counting task where each model has a weight: we can compute the probability of any query formula ϕ by summing up the probabilities (weights) of those possible worlds (answer sets) where ϕ is true. For a more detailed description of the semantics and inference process, please refer to [16,19]. Examples in the context of stream reasoning are provided in Sect. 4.

As a very simple example for a PrASP program (which does not require any of the inference optimization steps mentioned above), consider the following dice game:

```
face(1..6).
[[:]] result(F) :- face(F).
1{ result(F): face(F) }1.
win :- result(6).
[0.8|win] happy.
:- happy, not win.
```

At this, [[:]] result(F) :- face(F) is syntactic sugar for defining a uniform distribution over the six possible dice throwing results (each face comes up with the same probability). The spanning program (which contains a little syntactic noise introduced by the current generation algorithm) is

```
face(1..6).
result(6) :- {not result(6)}0,true.
result(5) :- {not result(5)}0,true.
result(4) :- {not result(4)}0,true.
result(3) :- {not result(3)}0,true.
result(2) :- {not result(2)}0,true.
result(1) :- {not result(1)}0,true.
1{result(F):face(F)}1 :- true.
win :- result(6).
happy :- {not happy}0,true.
:- happy, not win.
```

...which defines the following list of possible worlds (predicate face/1 omitted): $\{result(1)\}, \{result(2)\}, \{result(3)\}, \{result(4)\}, \{result(5)\}, \{result(6), win\},$ $\{result(6), win, happy\}$.

Over these possible worlds, a probability distribution is computed (using the formula weights in the PrASP program) which is then used to calculate query probabilities. Concretely, the probabilities of the possible worlds above are approximately $0.167, 0.167, 0.167, 0.167, 0.167, 0.033, 0.133$. With this distribution, query happy results in approximately 0.133 (by summation of the probabilities of those possible worlds in which happy holds), and query $Pr(happy|result(6))$ results in 0.8, which is simply computed as $\frac{Pr(happy \wedge result(6))}{result(6)}$.

PrASP can also be used for parameter estimation tasks (learning of formula weights from examples). Generally, the task of parameter learning in probabilistic inductive logic programming is to find probabilistic parameters (weights) of logical formulas (hypotheses) which maximize the likelihood given some data (learning examples) [21]. In our case, the hypothesis H (a set of formulas without weights) is provided by an expert, optionally together with some PrASP program as background knowledge B. The goal is then to discover weights w of the formulas H such that $Pr(E|H_w \cup B)$ is maximized given example formulas $E = e_1, e_2, \ldots$. Formally, we compute

$$argmax_w(Pr(E|H_w \cup B)) = argmax_w(\prod_{e_i \in E} Pr(e_i|H_w \cup B)) \qquad (1)$$

(Making the usual i.i.d. assumption regarding the individual examples in E. H_w denotes the hypothesis weighted with weight vector w.) PrASP learns such weights using a variant of the Barzilai and Borwein method [3], a gradient approach with possibly superlinear convergence. For details, see [19,16].

3   Probabilistic Reasoning about Data Streams on the Web

PrASP can be provided with data (beliefs, learning examples) in the same way as a conventional inductive logic programming tool, but additionally it can reason about streamed RDF data. The overall PrASP stream reasoning architecture is shown in Fig. 1. With stream data as input, PrASP acts like an uncertainty reasoning server which is step-by-step supplied with a sequence (of undetermined, theoretically even infinite, length) of new beliefs (certain or uncertain facts and rules) and learning examples (typically unweighted literals). Streamed beliefs are incrementally added to an initially loaded PrASP program, whereas streamed example formulas add to the set of learning examples (see Sect. 2). Optionally, each streamed belief or example can have a specified lifetime (number of streaming steps after which the belief or example is forgotten), in order to allow for time-decay and sliding window reasoning (but it is also possible to retract formulas explicitly). After each step, PrASP returns updated query and/or learning results. While this resembles systems for reactive answer set programming like oClingo [8], no reactive ASP solver is currently used (as these are incompatible with probabilistic reasoning) - instead, the current PrASP prototype implementation uses standard ASP solvers and recognizes repeated window prefixes (windows being the floating subsequences of streams visible to the reasoner) using a caching mechanism, in order to make stream reasoning more efficient.

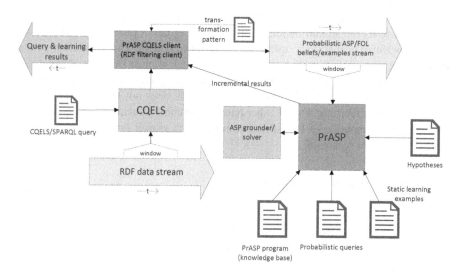

Fig. 1. PrASP probabilistic stream reasoning framework

RDF streams are preprocessed using CQELS [12], an extended SPARQL query tool for stream processing. Query results delivered by CQELS are translated into PrASP syntax using a simple pattern language which we insert directly into queries written using CQELS' SPARQL dialect. An example for such an extended stream query (translation pattern in bold typeface) is

```
PREFIX lv: <http://deri.org/floorplan/>
SELECT ?person1 ?loc1
```

PRASP E5 atPos(?person1,?loc1,TIMESTAMP).
  [OMIT http://deri.org/dblp/persons/]

```
FROM NAMED <http://deri.org/floorplan/> WHERE {
    GRAPH <http://deri.org/floorplan/>
    {?loc1 lv:connected ?loc2}
    STREAM <http://deri.org/streams/rfid> [NOW]
    {?person1 lv:detectedAt ?loc1} }
```

The translation pattern (the bold lines above) specifies how the query results received from CQELS are translated into a form which can be processed by PrASP (essentially ASP facts or rules, optionally with weights). The query refers to streams of RFID location sensor data - e.g., a single ground fact within the stream send to PrASP could be atPos(paul_erdoes, d010, 1), telling that person paul_erdoes is at location (room) d010 at time 1. E5 in the translation pattern specifies that the formula should be provided as a *learning example* with restricted time of validity (floating window size), namely only for the duration of five time steps. After each reception of a new learning example, PrASP updates the weights of given hypotheses - please see the next section for how this looks concretely.

Alternatively to dealing with streams of machine learning examples as above, it is also possible to let the stream incrementally supply certain or probabilistic *beliefs* and let PrASP perform probabilistic inference using these beliefs. The translation pattern within the query is analogous, but uses Bn [weight] in place of En as above.

Due to the use of CQELS as a filter engine, the RDF data flow is similar to that in the StreamRule framework [15], however StreamRule does not allow for any uncertainty reasoning or learning. On the other hand, uncertainty stream reasoning using our prototype implementation of PrASP is currently much slower than any form of non-probabilistic stream reasoning.

4   Examples and Conclusions

Our implementation of PrASP is still preliminary, but we can already illustrate how our system deals with incrementally provided finite fragments of a data stream and reasons about uncertain knowledge, typical of open and uncontrolled environment such as the Web of Data and the Internet of Things (IoT).

The   first   example   considers   streams   of   RFID   location sensor   data   in   form   of   RDF   triples   of   the   form http://deri.org/dblp/persons/Paul_Erdoes detectedAt D011   where D011 is the number of the room in which the person was detected.[1]

Background knowledge (in form of a simple PrASP program in FOL syntax) contains only one rule, the commonsense rule

```
L1 == L2 <- atPos(P,L1,T) & atPos(P,L2,T) & person(P)
            & location(L1) & location(L2) & time(T).
```

which states that a person cannot be at two different locations at the same time.

We use detected positions returned from the stream query processing engine[2] as examples (or observations) for learning the location of a person expressed as a hypothesis. This is a key capability under the assumption that sensor data might be noisy or inconsistent. This learning process starts from the hypothesis and updates its weight based on the observed examples.

In the sample execution trace below, PrASP is launched with hypothesis atPos(paul_erdoes, d010, 1) and it updates the weight of this hypothesis after each incoming example (the lines starting with E5):

```
E5 atPos(paul_erdoes,d010,1).
Server: [0.9999999999999999] atPos(paul_erdoes, d010, 1).
E5 atPos(paul_erdoes,d011,1).
Server: [0.4999999999999996] atPos(paul_erdoes, d010, 1).
E5 atPos(cecil_kochler,e017,1).
Server: [0.4999999999999996] atPos(paul_erdoes, d010, 1).
...
```

---

[1] URI prefix http://deri.org/floorplan/ as per example query in Sect. 3 has been omitted for readability.

[2] Using a translation pattern like that in Sect. 3, but now we assume that multiple position dates with the same time stamp can occur in the stream, e.g., from different sensors.

The next example (also about localization but using different background knowledge and data) shows how probabilistic reasoning can make use of semantically richer streams of weighted (that is, uncertain) beliefs, such as weighted sensor data. However, since data streams with RDF triples do not provide any triple weights, we now assume stream data directly in PrASP syntax[3].

We consider the problem of estimating the position of a moving target (such as a car) on a static map (such as a road grid) by using uncertain input data about the position of the car at different timestamps and the sensed speed, which are added incrementally. Sensor data is assumed to be uncertain because of factors such as noise or fusion of multiple sensors. The expressiveness of the framework enables non-monotonic probabilistic inference such as considering variation of speed w.r.t. a default speed, or the possibility of having invalid locations (such as inaccessible areas).

The reasoning capabilities illustrated in this example are closely related to stream reasoning since we can deal with aspects such as assertion and retraction of beliefs, and time-decay model as well as sliding window. For simplicity, we consider a 3×3 location grid and a time range 1..5 within a time window.

The program below encodes the localization problem through a generate and test specification that is typical of Answer Set Programming. The generation part (line 5) specifies that a target can be in only one location at a time (provided that this location is not invalid). The test part is modeled by two parts: one of them specifies what locations are invalid (line 6) so that default negation in the generation part can eliminate them from the model; the other one defines an integrity constraint (line 7) stating that a location cannot be plausible if it is not coherent with the speed. What it means to be speed-coherent is inductively defined in lines 8-10. Uncertain a priori information about the target being at position loc(1,1) of the grid at time 1 is stated in line 11.

```
1   time(1..5).
2   row(1..3).
3   col(1..3).
4   location(loc(X,Y)) :- row(X), col(Y).
5   1{atPos(L,T) : location(L) : not invalid(L)}1 :- time(T).
6   invalid(L) :- 1{wall(L), locked(L) }, location(L).
7   :- atPos(L,T), not speed_coherent(L,T), time(T).
8   speed_coherent(L,1) :- atPos(L,1), location(L).
9   speed_coherent(L,T) :- speed(X,T-1), atPos(L1,T-1),
         distanceT(L,L1,X,T).
10  distanceT(loc(X,Y),loc(U,V),#abs(X-U) + #abs(Y-V),T)
         :- atPos(loc(X,Y),T), atPos(loc(U,V),T-1).
11  [0.8] atPos(loc(1,1),1).
12  sensedspeed(1,1).
13  speed(S,T) :- sensedspeed(S,T).
14  speed(X,T) :- speed(X,T-1), not n_speed(X,T), time(T).
15  n_speed(X,T) :- sensedspeed(Z,T), speed(X,T-1), Z!=X.
```

In order to explore inference capabilities of our system, we ask for the weights of certain facts (queries), given a number of incrementally supplied uncertain belief updates. Firstly, if we ask our system to infer probabilities for the target being at position

---

[3] These weights could also be attached automatically when translating triples to PrASP syntax.

loc(2,2) at times 1, 3 and 5 respectively, assuming default speed is equal to 1, we obtain

```
[0.03870967741935094] atPos(loc(2,2),1).
[0.4483870967741945] atPos(loc(2,2),3).
[0.2989247311827963] atPos(loc(2,2),5).
```

We now want to observe how the variations of speed (provided as belief updates), may affect our inferred beliefs about possible target positions. Adding as new belief that a sensed speed equal to 3 was observed at time 2 with weight 0.7
(B [0.7] sensedspeed(3,2)), results in new (lower) probabilities for the target locations at time steps 3 and 5:

```
[0.21601941747572903] atPos(loc(2,2),3).
[0.14401294498381956] atPos(loc(2,2),5).
```

Providing another new belief B1 [0.7] sensedspeed(1,4) indicates speed is equal to one at time four with weight 0.7, while the number after the B indicates that this belief decays after one time step. This update followed by a similar belief but with different weight (B1 [0.5] sensedspeed(1,4)) changes the results of the weighted beliefs for the positions at time 5 first to ca. [0.26502748141] atPos(loc(2,2),5) and afterwards to [0.23302618816682907] atPos(loc(2,2),5).

We have introduced a framework for uncertainty reasoning about Web stream data, including both deductive and inductive reasoning (parameter estimation), based on ASP. Our framework is motivated by our belief that semantically rich reasoning about dynamic data on the Web requires support for non-monotonic reasoning approaches such as belief revision and default reasoning, as well as the ability to deal with uncertainty and data inconsistencies, making Web stream reasoning an ideal application field for probabilistic ASP. Future work will mainly focus on the improvement of our current early-stage prototype implementation, in order to make it usable for uncertainty reasoning about very large, real-world data streams.

References

1. Baral, C., Gelfond, M., Rushton, N.: Probabilistic reasoning with answer sets. Theory Pract. Log. Program. 9(1), 57–144 (2009)
2. Barbieri, D.F., Braga, D., Ceri, S., Valle, E.D., Huang, Y., Tresp, V., Rettinger, A., Wermser, H.: Deductive and inductive stream reasoning for semantic social media analytics. IEEE Intelligent Systems, 32–41 (2010)
3. Barzilai, J., Borwein, J.M.: Two point step size gradient methods. IMA J. Numer. Anal. (1988)
4. Corapi, D., Sykes, D., Inoue, K., Russo, A.: Probabilistic rule learning in nonmonotonic domains. In: Leite, J., Torroni, P., Ågotnes, T., Boella, G., van der Torre, L. (eds.) CLIMA XII 2011. LNCS, vol. 6814, pp. 243–258. Springer, Heidelberg (2011)
5. Cussens, J.: Parameter estimation in stochastic logic programs. In: Machine Learning (2000)
6. Do, T.M., Loke, S.W., Liu, F.: Answer set programming for stream reasoning. In: Butz, C., Lingras, P. (eds.) Canadian AI 2011. LNCS, vol. 6657, pp. 104–109. Springer, Heidelberg (2011)

7. Getoor, L.: Learning probabilistic relational models. In: Choueiry, B.Y., Walsh, T. (eds.) SARA 2000. LNCS (LNAI), vol. 1864, pp. 322–1309. Springer, Heidelberg (2000)

8. Gebser, M., Grote, T., Kaminski, R., Obermeier, P., Sabuncu, O., Schaub, T.: Answer set programming for stream reasoning. CoRR abs/1301.1392 (2013)

9. Gelfond, M., Lifschitz, V.: The stable model semantics for logic programming. In: Proc. of the 5th Int'l Conference on Logic Programming, vol. 161 (1988)

10. Kersting, K., Raedt, L.D.: Bayesian logic programs. In: Proceedings of the 10th International Conference on Inductive Logic Programming (2000)

11. Laskey, K.B., Costa, P.C.: Of klingons and starships: Bayesian logic for the 23rd century. In: Proceedings of the Twenty-First Conference on Uncertainty in Artificial Intelligence (2005)

12. Le-Phuoc, D., Dao-Tran, M., Xavier Parreira, J., Hauswirth, M.: A native and adaptive approach for unified processing of linked streams and linked data. In: Aroyo, L., Welty, C., Alani, H., Taylor, J., Bernstein, A., Kagal, L., Noy, N., Blomqvist, E. (eds.) ISWC 2011, Part I. LNCS, vol. 7031, pp. 370–388. Springer, Heidelberg (2011)

13. Lee, J., Palla, R.: System F2LP – computing answer sets of first-order formulas. In: Erdem, E., Lin, F., Schaub, T. (eds.) LPNMR 2009. LNCS, vol. 5753, pp. 515–521. Springer, Heidelberg (2009)

14. Margara, A., Urbani, J., van Harmelen, F., Bal, H.: Streaming the web: Reasoning over dynamic data. In: Web Semantics: Science, Services and Agents on the World Wide Web (2014)

15. Mileo, A., Abdelrahman, A., Policarpio, S., Hauswirth, M.: StreamRule: A nonmonotonic stream reasoning system for the semantic web. In: Faber, W., Lembo, D. (eds.) RR 2013. LNCS, vol. 7994, pp. 247–252. Springer, Heidelberg (2013)

16. Mileo, A., Nickles, M.: Probabilistic inductive answer set programming by model sampling and counting. In: 1st Int'l Workshop on Learning and Nonmonotonic Reasoning (2013)

17. Muggleton, S.: Learning stochastic logic programs. Electron. Trans. Artif. Intell. 4(B), 141–153 (2000)

18. Ng, R.T., Subrahmanian, V.S.: Stable semantics for probabilistic deductive databases. Inf. Comput. 110(1), 42–83 (1994)

19. Nickles, M., Mileo, A.: Probabilistic inductive logic programming based on answer set programming. In: Procs. 15th International Workshop on Non-Monotonic Reasoning (2014)

20. Poole, D.: The independent choice logic for modelling multiple agents under uncertainty. Artificial Intelligence 94, 7–56 (1997)

21. De Raedt, L., Kersting, K.: Probabilistic inductive logic programming. In: De Raedt, L., Frasconi, P., Kersting, K., Muggleton, S.H. (eds.) Probabilistic Inductive Logic Programming. LNCS (LNAI), vol. 4911, pp. 1–27. Springer, Heidelberg (2008)

22. Raedt, L.D., Kimmig, A., Toivonen, H.: Problog: A probabilistic prolog and its application in link discovery. In: IJCAI, pp. 2462–2467 (2007)

23. Ré, C., Letchner, J., Balazinksa, M., Suciu, D.: Event queries on correlated probabilistic streams. In: Procs. 2008 SIGMOD International Conference on Management of Data (2008)

24. Richardson, M., Domingos, P.: Markov logic networks. Mach. Learning 62(1-2) (2006)

25. Saad, E., Pontelli, E.: Hybrid probabilistic logic programming with non-monotoic negation. In: Twenty First International Conference on Logic Programming (2005)

26. Sato, T., Kameya, Y.: Prism: a language for symbolic-statistical modeling. In: Proceedings of the 15th International Joint Conference on Artificial Intelligence (IJCAI 1997) (1997)

27. Thimm, M., Kern-Isberner, G.: On probabilistic inference in relational conditional logics. Logic Journal of the IGPL 20(5), 872–908 (2012)

28. Valle, E.D., Ceri, S., van Harmelen, F., Fensel, D.: It's a streaming world! reasoning upon rapidly changing information. IEEE Intelligent Systems 24(6), 83–89 (2009)

29. Wasserkrug, S., Gal, A., Etzion, O., Turchin, Y.: Complex event processing over uncertain data. In: Procs. 2nd International Conference on Distributed Event-based Systems (2008)

Efficient Federated Debugging of Lightweight Ontologies⋆

Andreas Nolle[1], Christian Meilicke[2], Heiner Stuckenschmidt[2],
and German Nemirovski[1]

[1] Business Informatics, Department of Business and Computer Science
Albstadt-Sigmaringen University, Germany
{nolle,nemirovskij}@hs-albsig.de
[2] Knowledge Representation and Knowledge Management Group,
Computer Science Institute, University of Mannheim, Germany
{chistian,heiner}@informatik.uni-mannheim.de

Abstract. In the last years ontologies have been applied increasingly as
a conceptual view facilitating the federation of numerous data sources
using different access methods and data schemes. Approaches such as
ontology-based data integration (OBDI) are aimed at this purpose. According to these approaches, queries formulated in an ontology describing
the knowledge domain as a whole are translated into queries formulated
in vocabularies of integrated data sources. In such integrative environments the increasing number of heterogeneous data sources increases the
risk of inconsistencies. These inconsistencies become a serious obstacle
for leveraging the full potential of approaches like OBDI since inconsistencies can be hardly identified by existing reasoning algorithms, which
mostly have been developed for processing of locally available knowledge
bases. In this paper we present an alternative approach for efficient federated debugging. Our solution relies on the generation of so called clash
queries that are evaluated over all integrated data sources. We further explain how these queries can be used for pinpointing those assertions that
cause inconsistencies and discuss finally some experimental evaluation
results of our implementation.

Keywords: Inconsistency Detection, Clash Queries, *DL-Lite$_A$*, Federated Querying, Ontology-based Data Integration (OBDI), Query Rewriting, Backward-Chaining.

1 Introduction

Dealing with distributed and heterogeneous data sources has become an important research topic since the amount of available data grows continuously in

---

⋆ We refer the interested reader to our extended version of this paper
(http://www.researchgate.net/publication/263051384_Efficient_Federated_
Debugging_of_Lightweight_Ontologies) comprising some additional parts like
an introduction to inconsistencies in description logics, conjunctive queries and
DL-Lite, a complete clash type definition, a complete list of derived clash queries,
an example for clash query federation, the proof of Proposition 1, and a section
about related work.

R. Kontchakov and M.-L. Mugnier (Eds.): RR 2014, LNCS 8741, pp. 206–215, 2014.

companies and in the public sector. To handle the resulting challenges of data integration the approach of *ontology-based data access* (OBDA) has been proposed. In OBDA an ontology serves as conceptual view that comprises and possibly extends the semantics of each integrated data source. Mappings between this conceptual view and the different data schemes that describe the diverse data sources are used to transform original queries referring to the ontology into queries referring to the related vocabulary of each data source. Thus, on formulating queries clients do not have to be aware of each specific data schema. In the traditional OBDA approach, the data sources itself are assumed to be relational databases that are accessible via SQL. However, the approach of OBDA can also be adapted to all kinds of data sources and is in this setting also known under the designation *ontology-based data integration* (OBDI) [1,9,11].

Given in this context a set of distributed, heterogeneous *DL-Lite$_A$* knowledge bases. Even though each data source is self-consistent, the integrative knowledge base over all (or some of) these distributed sources may contain inconsistencies. We will illustrate this by the following example. Given a central ontology $\mathcal{T}$ and two distributed data sources DS$_A$ and DS$_B$. For the sake of simplicity we assume that DS$_A$ and DS$_B$ use the same ontology $\mathcal{T}$. Note that our example can easily be extended to the case where DS$_A$ and DS$_B$ use different terminologies that are linked by equivalence or subsumption axioms in the central ontology $\mathcal{T}$.

Example 1. Our terminology $\mathcal{T}$ contains the following axioms that describe persons and their blood relationships:

$$Woman \sqsubseteq Person \qquad \exists hasRelative \sqsubseteq Person$$
$$Man \sqsubseteq Person \qquad \exists hasRelative^- \sqsubseteq Person$$
$$Woman \sqsubseteq \neg Man \qquad hasAncestor \sqsubseteq hasRelative$$
$$(\textsf{funct hasBirthday}) \qquad hasDescendant \sqsubseteq hasRelative$$
$$\rho(\textsf{hasBirthday}) \sqsubseteq \texttt{xsd:dateTime} \qquad hasDescendant \sqsubseteq hasAncestor^-$$
$$Person \sqsubseteq \delta(\textsf{hasBirthday}) \qquad hasAncestor \sqsubseteq \neg hasAncestor^-$$
$$(\textsf{funct } hasDNA) \qquad hasDescendant \sqsubseteq \neg hasDescendant^-$$
$$(\textsf{funct } hasDNA^-) \qquad gaveBirthTo \sqsubseteq hasDescendant$$
$$\exists hasDNA^- \sqsubseteq DNA \qquad Person \sqsubseteq \exists hasAncestor$$
$$Person \sqsubseteq \exists hasDNA \qquad \exists gaveBirthTo \sqsubseteq Woman$$

The two data sources mentioned above contain the following assertions:

DS$_A$	DS$_B$
Man(**Homer**)	*gaveBirthTo*(**Homer, Lisa**)
Man(**Bart**)	*gaveBirthTo*(**Marge, Lisa**)
Woman(**Lisa**)	*hasRelative*(**Maggie, Lisa**)
Woman(**Marge**)	...

Since in DS_B **Homer** is defined as someone who *gaveBirthTo* somebody, according to $\mathcal{T}$ **Homer** is implicitly defined to be a *Women*. However, at the same time we have *Man*(**Homer**) $\in DS_A$. Due to $Woman \sqsubseteq \neg Man \in \mathcal{T}$ we have obviously a contradiction between DS_A and DS_B.

To detect such contradictions the data of each integrated data source needs to be taken into account. Using traditional approaches, like tableau-based reasoning algorithms, requires to have all the data at a place before the algorithm can be applied. This makes such approaches hardly applicable in the context of huge amounts of distributed data. To facilitate identification of contradictions in distributed data sources, we propose an alternative approach. We identify all possible types of inconsistencies and formulate appropriate queries in terms of the central ontology. Furthermore, we reformulate these queries in order to take into account all logical consequences for each of the concepts, roles and attributes addressed in these queries and evaluate the rewritten queries, more precisely its query atoms at each integrated data source. To enable high efficiency of reasoning tasks and query answering we exploit a specific family of Description Logics, called *DL-Lite*, which has been especially developed for this aim.

The rest of the paper is organized as follows. In Section 2.1 we describe the task of inconsistency detection in *DL-Lite$_\mathcal{A}$* knowledge bases. Sections 2.2 and 2.3 shows our approach for clash query generation and federation of clash queries, correspondingly. In Section 2.4 we further describe an algorithm for inconsistency detection and generation of its explanations. Before concluding this paper in Section 4, we discuss some experimental evaluation results in Section 3.

2   Inconsistency Detection

In this section we first explain different clash types that may occur in a *DL-Lite$_\mathcal{A}$*-based knowledge base. Basing on that, we define a translation function that is used in our approach to generate queries for inconsistency detection. Before describing our algorithm for inconsistency detection and generation of its explanations we elucidate the previously defined clash queries for distributed environments.

2.1   Inconsistency Detection in *DL-Lite$_\mathcal{A}$* Knowledge Bases

The consistency of a knowledge base can be determined by searching for obvious contradictions (also known as *clashes*) in the ABox. According to the work of Lembo et al. [7], in a *DL-Lite$_\mathcal{A}$* knowledge base clashes can be caused by only six different reasons, where ABox assertions contradicting TBox assertions, more precisely negative inclusions, value-domains, role or attribute functionalities.

Detection of such clashes requires that not only explicit but also implicit knowledge has to be taken into consideration. Since in our approach the focus is on distributed environments like in OBDI, implicit knowledge can be not only derived from the ontologies of each data source but also from the conceptual and centralized view. To obtain this complete knowledge especially by querying there exist two different ways to compute the certain answers.

One way is the materialization of ABoxes, where a materialized ABox is an original ABox extended by all assertions that can be additionally implied by the TBox(es) defined locally but also centrally. Queries will then be evaluated against the materialized ABox. This method, which is known as *forward-chaining* (or also bottom-up), requires the duplication of information. Like in data warehousing the redundant data have to be kept up-to-date on each modification and requires therewith additional resources. For that reasons such an approach is intractable in our application scenario.

Instead, we apply the method of *backward-chaining* (also known as top-down) where the ABoxes can be kept in the original state. The original query is reformulated (rewritten) with respect to the TBox to the effect that all knowledge relevant for the computation of the certain answers to that query is compiled into a set of rewritten queries (unions of conjunctive queries). Roughly speaking, if a query atom addresses individuals of a specific concept, the rewritten queries will contain atoms addressing all possible concepts, roles and attributes that also provide individuals of the originally requested concept. Especially for *DL-Lite$_{\mathcal{A}}$* such rewritings can be done in PTIME and query answering in AC^0 each in size of the TBox and ABox, respectively [5, 9].

To utilize this feature of computing certain answers to a query by its rewriting our approach of inconsistency detection comprises the generation of so called *clash queries*.

2.2 Clash Query Generation

According to the clash definitions given by Lembo et al. [7] and based on the work of Calvanese et al. [1] we are able to define a translation function τ that generates queries for inconsistency detection from negative inclusions, functionality assertions or value-domain inclusions in $\mathcal{T}$, denoted by $\mathcal{T}_n$. If any of such clash queries delivers non-empty result sets, we can conclude that the delivered individuals cause inconsistencies with respect to elements in $\mathcal{T}_n$ addressed by the atoms of the generated query. For *DL-Lite$_{\mathcal{A}}$*, the translation function τ and all required kinds of generated clash queries are listed below in Datalog notation:

(i) $\tau(B_1 \sqsubseteq \neg B_2) = q(x) \leftarrow b_1, b_2$, where $b_i = A_i(x)$ if $B_i = A_i$, $b_i = P_i(x, \_)$ if $B_i = \exists P_i$, $b_i = P_i(\_, x)$ if $B_i = \exists P_i^-$, and $b_i = U_i(x, \_)$ if $B_i = \delta(U_i)$

(ii) $\tau(R_1 \sqsubseteq \neg R_2) = q(x, y) \leftarrow r_1, r_2$, where $r_i = P_i(x, y)$ if $R_i = P_i$, $r_i = P_i(y, x)$ if $R_i = P_i^-$, and $r_i = U_i(x, y)$ if $R_i = U_i$

(iii) $\tau(\rho(U) \sqsubseteq T_i) = q(x, y) \leftarrow U(x, y), T_i \neq datatype(y)$

(iv) $\tau((\text{funct } R)) = q(x, y, z) \leftarrow r_1, r_2, y \neq z$, where $r_1 = P(x, y)$ if $R = P$, $r_1 = P(y, x)$ if $R = P^-$, $r_1 = U(x, y)$ if $R = U$, $r_2 = P(x, z)$ if $R = P$, $r_2 = P(z, x)$ if $R = P^-$, and $r_2 = U(x, z)$ if $R = U$

For our running example defined in Section 1 the following clash query can be derived: $q(x) \leftarrow Man(x), Woman(x)$. Now we have to apply the rewriting techniques introduced above, which results in the following Datalog program:

$$q(x) \leftarrow Woman(x), Man(x)$$
$$q(x) \leftarrow gaveBirthTo(x, \_), Man(x)$$

By identification of query parts that return some results due to an inconsistency in the knowledge base, it is possible to pinpoint those ABox assertions that are responsible for the inconsistency. In case of our example above, the last query part will return **Homer** and for that reason we know that $\{gaveBirthTo(\mathbf{Homer}, \_),$ $Man(\mathbf{Homer})\}$ is an explanation for the inconsistency. Since the translation function may produces query atoms containing unbound variables (denoted by $\_$), the derived ABox assertions are not complete. This is the case especially for existential restrictions in clash query type (i). For this special case a subsequent query to select the unbound values can be formulated. Another option is to expand this specific type of generated clash queries by new distinguished variables for each unbound variable.

2.3   Clash Query Federation

Considering the distributive environment, the generated clash queries will be evaluated through a simple federation algorithm for our first experiments. The processing of a query starts with its rewriting according to the method of backward-chaining, taking the semantics of the central ontology into account. In the more general case mappings to external ontologies are also taken into account within this step. This results in unions of conjunctive queries that are equivalent to the original query. Since a query, i.e., its atoms may address several data sources each query atom is sent to all sources and its results are federated.

2.4   Generating Explanations

Algorithm 1 is based on similar Consistent algorithms proposed by Calvanese et al. [1,2] and summarizes our approach for computing all federated inconsistency explanations for a $DL\text{-}Lite_A$-based knowledge base $\mathcal{K} = \langle \mathcal{T}, \mathcal{A} \rangle$. We first iterate over $\mathcal{T}_n$, which is the set of all negative inclusions, functionality assertions and value-domain inclusions in $\mathcal{T}$. In Algorithm 1 $\mathcal{T}_n$ is set by the function Determine$\mathcal{T}_n$. For each element α in $\mathcal{T}_n$ we apply the translation function τ to generate the corresponding clash query. Since the semantics of $DL\text{-}Lite_A$ does not contain specializations of elements in functionality assertions, we only have to rewrite clash queries for negative inclusions and value-domain inclusions according to the method of backward-chaining. Implementations of such rewriting algorithms are for example PerfectRef given by Calvanese et al. [1] or TreeWitness constituted by Kontchakov et al. [6], which is more efficient than PerfectRef. Both algorithms are part of the −ontop− framework[1] that is used within our implementation. For the experimental evaluation in Section 3 we used the TreeWitness.

---
[1] http://ontop.inf.unibz.it

Algorithm 1. InconsistencyDetection($\mathcal{K}$)

Input: $DL\text{-}Lite_{\mathcal{A}}$ knowledge base $\mathcal{K} = \langle \mathcal{T}, \mathcal{A} \rangle$
Output: all inconsistency explanations $\mathcal{C}$
begin
$\quad$ $\mathcal{C} \leftarrow \emptyset$;
$\quad$ $\mathcal{T}_n \leftarrow$ Determine$\mathcal{T}_n(\mathcal{T})$;
$\quad$ **foreach** $\alpha \in \mathcal{T}_n$ **do**
$\quad\quad$ $q \leftarrow \tau(\alpha)$;
$\quad\quad$ $q^S \leftarrow \emptyset$;
$\quad\quad$ **if** α *is a negative inclusion or a value-domain inclusion* **then**
$\quad\quad\quad$ $q^S \leftarrow$ Rewrite$(q, \mathcal{T})$;
$\quad\quad$ **else**
$\quad\quad\quad$ $q^S \leftarrow \{q\}$;
$\quad\quad$ **foreach** $\varphi \in q^S$ **do**
$\quad\quad\quad$ $R^S \leftarrow$ Answ$(\varphi, \mathcal{A})$;
$\quad\quad\quad$ **if** $R^S \neq \emptyset$ **then**
$\quad\quad\quad\quad$ $\mathcal{C}' \leftarrow$ TransformIntoAssertions(R^S, φ);
$\quad\quad\quad\quad$ $\mathcal{C} \leftarrow \mathcal{C} \cup \mathcal{C}'$;
$\quad$ **return** $\mathcal{C}$;
end

This rewriting step results in unions of conjunctive queries (q^S). We execute each conjunctive query φ of q^S separately, as explained in the previous section. If the result of φ is not empty, we use the query result R^S and the query itself to transform all result tupels into a set of clashing ABox assertions $\mathcal{C}'$, that represents a set of all inconsistency explanations related to that query result (TransformIntoAssertions in Algorithm 1). We omit the TBox elements in these explanations since we assume that $\mathcal{T}$ is commonly accepted and kept constantly. We collect all explanations $\mathcal{C}'$ of each conjunctive query φ in a overall set $\mathcal{C}$ of all inconsistency explanations.

Since $\mathcal{T}_n$, the set of all negative inclusions, functionality assertions and value-domain inclusions in $\mathcal{T}$ is finite and the termination of Rewrite$(q, \mathcal{T})$ (such as PerfectRef or TreeWitness) is assumed to be already established, the termination of this algorithm is given.

Proposition 1. *Let* $\mathcal{K} = \langle \mathcal{T}, \mathcal{A} \rangle$ *be a DL-Lite$_{\mathcal{A}}$ knowledge base, where $\mathcal{A}$ is the union of distributed data sources. Then* InconsistencyDetection$(\mathcal{K})$ *generates the set of all inconsistency explanations for* $\mathcal{K}$.

Repairing the detected inconsistencies, i.e., deciding which assertions should be eliminated, is beyond the scope of this paper. Several approaches have been already proposed to solve this problem [3]. Depending on the specifics of the setting one might, for example, be interested to remove a minimum number of assertions causing inconsistencies by computing a smallest minimal hitting set over all explanations. In most of these approaches it is required to have access to the set of all inconsistency explanations, which are generated by our approach.

3   Experimental Evaluation

In order to evaluate the performance of our approach we compare our implementation of algorithm InconsistencyDetection, called ClashSniffer, to the reasoning system Pellet [10]. Pellet offers a specific service for computing inconsistency explanation and can thus be directly compared to our approach. Moreover, we are conducting experiments with the Black-Box algorithm for computing explanations, which is implemented as component in the OWL API, using HermiT [4] as underlying reasoner. We have artificially generated some RDF datasets comprising 500, 5000, 10000, 50000 and 100000 ABox assertions according to the TBox definition of our running example. The collection of datasets is available at http://www.researchgate.net/publication/263051841_ClashSniffer_Evaluation_Datasets. Each dataset contains some assertions that will cause inconsistencies and are generated randomly with a rate about 2% of the complete number of assertions. All possible clash types referred in Section 2.1 may occur within these datasets. Since the OWL 2 QL profile[2] is based on DL-Lite, we use it as specification language of our defined TBox.

Pellet and HermiT can only be applied to the non-distributed version of the dataset. For that reason we run our algorithm both in a local setting using a central repository that contains the complete dataset (ClashSniffer$_L$), and in a distributed environment (ClashSniffer$_F$). In the distributed environment the ABox assertions are randomly distributed over four data sources, represented by instances of Virtuoso (Open-Source Edition)[3]. In this setting we sent each query atom to all data sources and federated the results.

The results of our experimental evaluation are depicted in Table 1. It illustrates that first of all, the runtimes for the local and the distributed settings of our algorithm differ significantly. This is caused by the latency in the network and also by the fashion how the federated queries are executed. Since in our implementation we use ARQ, a query engine for Apache Jena, for each tupel that is returned as an answer for a query atom of a federated query, a new subquery for the next query atom that is related to the first one will be generated for all assigned data sources by default. Its also interesting to see that the runtimes for the distributed settings increase linear with respect to the problem size. This is not the case for the local setting, where the size of the ABox has only a minor impact on the overall runtime.

A surprising result is related to the performance of Pellet and HermiT. For the smallest dataset Pellet requires significantly more time to compute all explanations than our algorithm, in both the local and the distributed setting. Pellet takes more than five hours to compute explanations for the data set comprising

---

[2] http://www.w3.org/TR/owl2-profiles
[3] http://virtuoso.openlinksw.com/dataspace/doc/dav/wiki/Main/

Table 1. Experimental Evaluation Results

ABox Size		500	5,000	10,000	50,000	100,000
Pellet	Execution (ms)	1,713,901	>18,000,000	–	–	–
	♯Explanations	20 (13)	–	–	–	–
HermiT	Execution (ms)	[Error]	–	–	–	–
	♯Explanations	–	–	–	–	–
Clash Sniffer$_L$	Execution (ms)	359	563	718	1,313	2,171
	♯Explanations	13	113	191	1,050	2,121
Clash Sniffer$_F$	Execution (ms)	29,360	234,878	464,973	2,227,466	4,541,140
	♯Explanations	13	113	191	1,050	2,121

5000 assertions. We stopped the experiment after five hours. Contrary to Pellet, HermiT ends up on 500 assertions with an `OutOfMemoryError` despite of an assigned memory of 4GB. We are currently missing an appropriate explanation for this behaviour.

Comparing the generated inconsistency explanations of Pellet to the ones that are produced by our algorithm, it can be observed that both approaches detected the same explanations. However, Pellet generates (particularly concerning inconsistencies on attributes) explanation sets that are not minimal, i.e., some of the generated sets are supersets of (minimal) explanations. Since supersets of the same explanation are computed in some cases, Pellet produces a higher number of explanations. We have manually analysed the explanations generated by Pellet for the test case with 500 assertions. After mapping each superset of an explanation on the minimal explanation that was contained in the superset, the results for Pellet and our approach were the same. Restricting the number of explanations to small numbers (e.g., 5 or 10) the Black-Box approach is also capable of generating an output. However, again, the generated sets are often proper supersets of an explanation and the algorithm cannot generate such sets if we increase the number of requested explanations. Moreover, our algorithm generates in both, the local and the distributed setting, the same explanations, which is in line with our theoretical considerations.

4   Conclusions and Future Work

In this paper we have described an approach of efficient inconsistency detection in distributed knowledge bases based on $DL\text{-}Lite_A$. The described approach relies on the generation of clash queries that are evaluated over all integrated data sources. We have further depicted an algorithm that detects existing inconsistencies and generates explanations to them. We have also shown experimental evaluation results for our algorithm.

Since in this paper we cover only one part of ontology debugging, namely the identification of inconsistencies and its explanations, in our future work we will address the generation of extended explanations, e.g., comprising the corresponding data sources, and the task of proposing some repair plans. Furthermore, we will evaluate our approach by using "real" instead of artificially generated data sets. In addition we plan to compare our approach of inconsistency detection also against some *DL-Lite* tailored solutions, such as the OBDA management system Mastro[4]. Due to the fact that we have used just a simple federation algorithm we will integrate the proposed approach into the federation engine ELITE [8]. ELITE was developed with the purposes of efficient and complete processing of federated queries in distributed environments. Especially the use of the R-Tree-based index of ELITE guarantee that only those query parts are evaluated that probably deliver some results. By this means the task of inconsistency detection in distributed environments can be solved more efficient.

Acknowledgments. This research has been partially supported by the SE-MANCO project (http://www.semanco-project.eu), which is being carried out with the support of the Seventh Framework Programme "ICT for Energy Systems" 2011–2014, under the grant agreement no. 287534.

References

1. Calvanese, D., De Giacomo, G., Lembo, D., Lenzerini, M., Rosati, R.: Tractable reasoning and efficient query answering in description logics: The DL-Lite family. Journal of Automated Reasoning 39(3), 385–429 (2007)
2. Calvanese, D., De Giacomo, G., Lembo, D., Lenzerini, M., Rosati, R.: Data complexity of query answering in description logics. Artificial Intelligence 195, 335–360 (2013)
3. Haase, P., Qi, G.: An analysis of approaches to resolving inconsistencies in DL-based ontologies. In: Proceedings of the International Workshop on Ontology Dynamics (IWOD 2007), pp. 97–109 (2007)
4. Horrocks, I., Motik, B., Wang, Z.: The HermiT OWL Reasoner. In: Proceedings of the 1st International Workshop on OWL Reasoner Evaluation (ORE 2012), Manchester, UK (2012)
5. Kiryakov, A., Damova, M.: Storing the Semantic Web: Repositories. Handbook of Semantic Web Technologies 1, 231–297 (2011)
6. Kontchakov, R., Lutz, C., Toman, D., Wolter, F., Zakharyaschev, M.: The combined approach to ontology-based data access. In: Proceedings of the Twenty-Second International Joint Conference on Artificial Intelligence, vol. 3, pp. 2656–2661. AAAI Press (2011)
7. Lembo, D., Lenzerini, M., Rosati, R., Ruzzi, M., Savo, D.F.: Query rewriting for inconsistent DL-Lite ontologies. In: Rudolph, S., Gutierrez, C. (eds.) RR 2011. LNCS, vol. 6902, pp. 155–169. Springer, Heidelberg (2011)
8. Nolle, A., Nemirovski, G.: ELITE: An Entailment-based Federated Query Engine for Complete and Transparent Semantic Data Integration. In: Proceedings of the 26th International Workshop on Description Logics (2013)

---

[4] http://www.dis.uniroma1.it/~mastro

9. Poggi, A., Lembo, D., Calvanese, D., De Giacomo, G., Lenzerini, M., Rosati, R.: Linking data to ontologies. Journal on Data Semantics X, 133–173 (2008)
10. Sirin, E., Parsia, B., Grau, B.C., Kalyanpur, A., Katz, Y.: Pellet: A practical owl-dl reasoner. Web Semantics: science, services and agents on the World Wide Web 5(2), 51–53 (2007)
11. Wache, H., Voegele, T., Visser, U., Stuckenschmidt, H., Schuster, G., Neumann, H., Hübner, S.: Ontology-based integration of information - a survey of existing approaches. In: IJCAI 2001 Workshop: Ontologies and Information Sharing, Citeseer, pp. 108–117 (2001)

Revisiting the Hardness of Query Answering
in Expressive Description Logics*

Magdalena Ortiz and Mantas Šimkus

Institute of Information Systems,
Vienna University of Technology, Austria
ortiz@kr.tuwien.ac.at, simkus@dbai.tuwien.ac.at

Abstract. Answering conjunctive queries over *Description Logic* (DL) knowledge bases is known to be 2ExpTime-hard for the DLs $\mathcal{ALCI}$, $\mathcal{SH}$, and their extensions. In this technical note, we revisit these results to identify other equally hard settings. In particular, we show that a simple adaptation of the proof for $\mathcal{SH}$ proves that query answering is 2ExpTime-hard already for $\mathcal{ALC}$ if we consider more expressive query languages such as positive existential queries and (restricted classes of) conjunctive regular path queries.

1 Introduction

Ontology-based data access, and the related setting of query answering over *Description Logic* (DL) *knowledge bases* (KBs), has received considerable attention in the DL community. Most work has been devoted to the so-called *lightweight DLs* of the DL-Lite and $\mathcal{EL}$ families, but *expressive DLs* like $\mathcal{ALC}$ and its extentions have also been considered, cf. [13] and its references. The first query answering algorithms for the latter kind of DLs had the common feature of requiring double exponential time [8,4,3], and the question of whether this was worst-case optimal remained open for a while. For all extensions of $\mathcal{ALC}$ that support inverse roles, this gap was closed by Lutz, who proved 2ExpTime-hardness of answering conjunctive queries (CQs) in $\mathcal{ALCI}$ [9]. An orthogonal 2ExpTime-hardness result was later shown for CQs over $\mathcal{SH}$ knowledge bases [5], closing the gap for all DLs that support transitive roles and role hierarchies. Recently, 2ExpTime-hardness was also proved for $DL\text{-}Lite_{bool}^{\mathcal{H}}$, a DL that does include full $\mathcal{ALC}$, but does support inverse roles and role hierarchies [2]. In contrast, answering unions of CQs (UCQs) is feasible in single exponential time for $\mathcal{ALCH}$ and $\mathcal{ALCHQ}$, and even for $\mathcal{SH}$ if suitable restrictions on the occurrences of transitive roles in the queries are imposed [12,9,6]. This shows that, for plain CQs and UCQs, the culprits for 2ExpTime-hardness are indeed inverse roles, and transitive roles in combination with role hierarchies.

The aforementioned lower bounds are for plain CQs. However, recent works have gone beyond CQs and unions thereof, by investigating more expressive query languages like *regular path queries* and their extensions [4,3,11,1], or restricted classes of Datalog

* This research has been partially supported by the Austrian Science Fund (FWF) projects T515 and P25518, and by the Vienna Science and Technology Fund (WWTF) project ICT12-015.

R. Kontchakov and M.-L. Mugnier (Eds.): RR 2014, LNCS 8741, pp. 216–223, 2014.
© Springer International Publishing Switzerland 2014

queries [10]. To our knowledge, it had not been investigated whether for these more expressive query languages, 2ExpTime-hardness holds already in the absence of inverse roles, transitive roles, and role hierarchies. To tackle the issue, in this this technical note we revisit the proof in [5] to show that 2ExpTime-hardness holds already for $\mathcal{ALC}$ in the following extensions of UCQs:

- positive existential queries (PQs),
- conjunctive regular path queries (CRPQs) without the Kleene star, and
- conjunctive 2-*way* regular path queries (C2RPQs) without the Kleene star and with only 2 variables.

2   Query Languages

We assume familiarity with DLs, and in particular with $\mathcal{ALC}$ KBs. Their semantics is given by interpretations $\mathcal{I} = \langle \Delta^\mathcal{I}, \cdot^\mathcal{I} \rangle$. We call $\mathcal{I}$ *tree-shaped* if the graph with nodes $\Delta^\mathcal{I}$ and edges (d, e) for all $(d, e) \in r^\mathcal{I}$ for a role name r, is a tree in the usual sense.

We focus here on Boolean queries of the form $\exists x.\varphi(x)$, where x is a tuple of variables and $\varphi(x)$ is a formula whose syntax depends on the considered query language. In *positive queries (PQs)*, $\varphi(x)$ is built using $\wedge$ and $\vee$ from atoms of the forms $A(x)$ and $r(x, y)$, where A is a concept name, r a role name, and x, y are variables from x. A *conjunctive query* (CQ) is a positive query built using only $\wedge$, and a *union of CQs* is a positive query that is in DNF, that is, it is a disjunction of conjunctions. *Conjunctive regular path queries (CRPQs)* are defined analogously to CQs, but atoms may additionally take the form $\mathcal{E}(x, y)$, where $\mathcal{E}$ is a regular expression over the alphabet of role names. *Conjunctive 2-way regular path queries (C2RPQs)* are similar, but regular expressions are over the alphabet of role names r and their inverses r^-. In this technical note we also consider a restricted class of C(2)RPQs that we call $(\circ, \cup)$-queries, which only allow for concatenation $\circ$ and union $\cup$ in complex roles, but disallows the Kleene star $*$. We note that $(\circ, \cup)$-queries are closely related to PQs. Indeed, every $(\circ, \cup)$-query can be rewritten as a PQ by using sequences of binary atoms in the place of $\circ$, and disjunction in the place of $\cup$. However, this requires the use of additional variables and may result in a larger query.

The semantics of queries is defined in terms of *matches*, which are mappings from the variables in x to objects in $\Delta^\mathcal{I}$ that make the query true; the latter notion is defined in the natural way, see e.g.,[4]. Here we consider the *query non-entailment problem*, which consists on deciding whether there exists a model of a given KB $\mathcal{K}$ that admits no match for a given query q, in symbols $\mathcal{K} \not\models q$. It is well known that every satisfiable $\mathcal{ALC}$ KB $\mathcal{K}$ with only one ABox individual has a tree-shaped model, and that this extends to query non-entailment: $\mathcal{K} \not\models q$ iff there is a tree-shaped model of $\mathcal{K}$ that admits no match for q (cf. [5,9]). All complexity bounds mentioned here are for *combined complexity*, i.e., the complexity measured in terms of the combined sizes of $\mathcal{K}$ and q.

3   2ExpTime-Hardness of CQs in $\mathcal{SH}$ Revisited

We recall the proof of 2ExpTime-hardness of answering CQs over $\mathcal{SH}$ KBs in [5]. It is done by a reduction from the word problem of an exponentially space bounded

Alternating Turing Machine $\mathcal{M}$, that is, the problem of deciding for each input word w to $\mathcal{M}$, whether there is an accepting computation of $\mathcal{M}$ on w that uses at most $2^{|w|}$ tape cells. The reduction builds a KB $\mathcal{K}_w$ and a query q_w such that $\mathcal{M}$ accepts w iff $\mathcal{K}_w \not\models q_w$. Since a computation of an ATM is naturally represented as a tree of configurations, the KB $\mathcal{K}_w$ is such that its tree-shaped models resemble accepting computations of $\mathcal{M}$ on w. We next recall the construction of $\mathcal{K}_w$ and q_w from [5]. The construction uses a very simple ABox of the form $A(a)$, for a an individual and A a concept name, hence we can restrict our attention to tree-shaped models. We relax slightly the construction of $\mathcal{K}_w$ to be an $\mathcal{ALC}$ KB, by omitting the (sole) use of a transitive role to connect a node and its child to a common successor, for some nodes of the tree-shaped models of $\mathcal{K}_w$. Hence our description of $\mathcal{K}_w$ uses a single role r.

Configuration nodes. Intuitively, the tree-shaped models $\mathcal{K}_w$ are trees of nodes representing configurations of $\mathcal{M}$. For a configuration K_h in which $\mathcal{M}$ is in state q and its head in position i reading a symbol a, there is a (q, a, i)-*configuration node* n that represents K_h. The node n stores K_h, and as a technical trick, it additionally stores another configuration K_p such that $\mathcal{M}$ may move from K_p to K_h; that is, n contains both the current configuration K_h and a possible previous configuration K_p. To properly store the current and previous configurations K_h and K_p, each (q, a, p)-configuration node n is in turn the root of a binary tree of depth $|w|$ whose $2^{|w|}$ leaves correspond to the tape cell positions, and store their $|w|$-bit address using a set of concept names $\mathbf{B} = \{B_1, \ldots B_{|w|}\}$. All arcs in this tree are r-arcs. Each leaf ℓ with address i, which is called an i-*cell* (or just a *cell* if i is unimportant) has a child ℓ_h and a child ℓ_p that store the symbol on tape position i in K_h and in K_p, respectively If i is the position of the head of $\mathcal{M}$ in K_h, then ℓ_h also stores the state q of $\mathcal{M}$, otherwise it stores a special marker nil which intuitively means 'the head is not on this position'. Similarly for K_p and ℓ_p. For these labels we use concept names for alphabet symbols, states of $\mathcal{M}$, and the special marker nil.

$\mathcal{K}_w$ contains axioms that ensure that the configurations K_h and K_p are described correctly: e.g., there is exactly one symbol on each tape position, the head is at exactly one position, and $\mathcal{M}$ is in exactly one state. There are also axioms in $\mathcal{K}_w$ to ensure that K_h is the result of correctly applying from K_p a transition of $\mathcal{M}$.

Computation trees. A tree-shaped model $\mathcal{I}_c$ of $\mathcal{K}_w$ is called a *computation tree* and it is a tree of configuration nodes connected via the role r. Its root n_0 has an r-successor that is a $(q_0, a_0, 0)$-configuration node describing (as current configuration) the initial configuration of $\mathcal{M}$. Each (q, a, p)-configuration node n with q an existential state has a 2-step r-successor n' that is a $(q', a', p + M)$-configuration node, for some transition (q, a, q', a', M) of $\mathcal{M}$ (here the transition is read as follows: $\mathcal{M}$ is in state q and reading a, it writes a, moves to state q', and the head moves in direction $M \in \{-1, 0, 1\}$). Similarly, each (q, a, p)-configuration node n with q a universal state, has a 2-step r-successor n' that is a $(q', a', p + M)$-configuration node for each transition (q, a, q', a', M) of $\mathcal{M}$. The axioms of $\mathcal{K}_w$ also ensure that every configuration node with no successors is in an accepting state. Figure 1 illustrates a fragment of a computation tree with four configuration trees. A configuration tree with a magnified i-cell is illustrated in Figure 2.

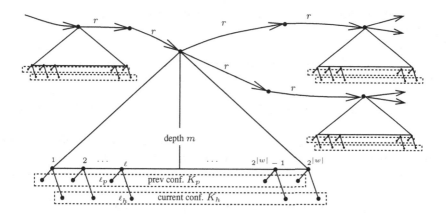

Fig. 1. A fragment of a computation tree

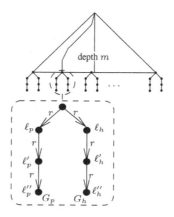

Fig. 2. A configuration tree with a magnified cell

Proper computation trees. To have a one to one correspondence between the tree-shaped models $\mathcal{I}_c$ of $\mathcal{K}_w$ and the accepting computations c of $\mathcal{M}$ on input w, it suffices to ensure that for each pair n, n' of successive configuration nodes, the current configuration of n coincides with the previous configuration of n'. This is captured by the notion of *properness*, which states that for every counter i value up to $2^{|w|}$, the node ℓ_h of the i-cell of n and the node ℓ_p of the i-cell of n' satisfy exactly the same concepts corresponding to head position, written symbol, and state of $\mathcal{M}$. Properness is not guaranteed by the axioms of $\mathcal{K}_w$ alone, and a tree-shaped model $\mathcal{I}_c$ of $\mathcal{K}_w$ (i.e., a computation tree) may be proper or not. Here where the query comes into play, by testing a computation tree is proper. More precisely, q_w should have a match in a computation tree $\mathcal{I}_c$ iff $\mathcal{I}_c$ is *not* proper. In this way each computation tree with no match corresponds to an accepting computation of $\mathcal{M}$ on w, and we obtain that there is a tree-shaped model $\mathcal{I}_c$ of $\mathcal{K}_w$ where there is no match for q_w iff there is an accepting computation of $\mathcal{M}$ on w. This (together with the tree-model property of $\mathcal{ALC}$) suffices to ensure that $\mathcal{K}_w \not\models q$ iff $\mathcal{M}$ accepts w as desired.

By using suitable auxiliary nodes and labels, properness can be characterized in such a way that it can easily tested by the query. We use a set $\mathbf{Z}$ of concept names $Z_{a,q}$ for a an alphabet symbol and q either a state $\mathcal{M}$ or the special marker nil. We have said that every i-cell ℓ has two children ℓ_p and ℓ_h, which respectively correspond to the i-th tape position in the previous and in the current configuration. By adding auxiliary r-children ℓ'_f to ℓ_f and ℓ''_f to ℓ'_f for $f \in \{h, p\}$, and labeling all ℓ_f and ℓ'_f with suitable values for the concepts $\mathbf{B} \cup \mathbf{Z}$ (exactly as done in [5]), one can obtain the following characterization. Two cells ℓ and m are called A-conspicuous, where A is a concept name, if (c1) A is true at the ℓ_h-node of n and the m_p-node of n', or (c2) A is true at the ℓ'_h-node of n and the m'_p-node of n'. In Proposition 4 of [5] it is proved that a computation tree is not proper iff the following property $(*)$ holds:

$(*)$ There exits a cell ℓ in a configuration tree, and a cell m in a successive configuration tree, such that ℓ and m are A-conspicuous for all $A \in \mathbf{B} \cup \mathbf{Z}$.

To test $(*)$ with the query, we ensure that each ℓ''_f satisfies a special concept G_f, but we relax condition (c) in Definition 1 of [5], which requires that ℓ''_f is also a child of ℓ_f, and that the arcs from ℓ_f to ℓ''_f and from ℓ'_f to ℓ''_f are t-arcs for a transitive role t. Instead, we only require ℓ''_f to be an r-child of ℓ'_f. The queries testing for $(*)$ are similar to those in [5], but differ slightly according to the query language being considered.

Positive Queries. We can define the query q_w that tests $(*)$ as a PQ as follows. As in [5], we obtain q_w by taking the conjunction of a CQ $q(A, u, v)$ for each $A \in \mathbf{B} \cup \mathbf{Z}$, where the variables u, v are shared by all $q(A, u, v)$, and the remaining variables are disjoint. That is, $q_w = \exists u, v. \bigwedge_{A \in \mathbf{B} \cup \mathbf{Z}} q(A, u, v)$, where $q(A, u, v)$ is as follows:

$$q(A, u, v) = \exists x_1^A, x_2^A, x_3^A, y_0^A, \ldots y_{|w|+1}^A, z_0^A, \ldots z_{|w|+3}^A \cdot$$
$$r(x_1^A, y_0^A) \wedge r(x_1^A, z_0^A) \wedge$$
$$r(y_0^A, y_1^A) \wedge \cdots \wedge r(y_{|w|}^A, y_{|w|+1}^A) \wedge A(y_{|w|+1}^A) \wedge$$
$$r(z_0^A, z_1^A) \wedge \cdots \wedge r(z_{|w|+2}^A, z_{|w|+3}^A) \wedge A(z_{|w|+3}^A) \wedge$$
$$\left(r(y_{|w|+1}^A, u) \vee \left(r(y_{|w|+1}^A, x_2^A) \wedge r(x_2^A, u)\right)\right) \wedge G_h(u) \wedge$$
$$\left(r(z_{|w|+3}^A, v) \vee \left(r(z_{|w|+3}^A, x_3^A) \wedge r(x_3^A, v)\right)\right) \wedge G_p(v)$$

The basic query $q(A, u, v)$ is illustrated in the left-hand-side of Figure 3, and the full query q_w on the right-hand-side. For readability, most labels have been omitted in the depiction of q_w. Each arc represents an atom of the form $r(x, y)$. The double dashed arcs between $y_{|w|+1}^A$ and u, and between $z_{|w|+3}^A$ and v, represent a disjunction. The only difference between this $q(A, u, v)$ and the one in [5] is that the disjunction of atoms $r(y_{|w|+1}^A, u) \vee \left(r(y_{|w|+1}^A, x_2^A) \wedge r(x_2^A, u)\right)$ replaces the atom $t(y_{|w|+1}^A, u)$, and similarly $r(z_{|w|+3}^A, v) \vee \left(r(z_{|w|+3}^A, x_3^A) \wedge r(x_3^A, v)\right)$ replaces $t(z_{|w|+3}^A, v)$.

Intuitively, $q(A, u, v)$ deals with A-conspicuousness, and q_w tests $(*)$ by taking the conjunction for all $A \in \mathbf{B} \cup \mathbf{Z}$. Note that the shared variables u, v are needed to ensure that all the components $q(A, u, v)$ speak about the same pair of cells ℓ, m.

To see that q_w has a match iff $(*)$ holds, let ℓ, m be cells of two successive configurations that are A-conspicuous for all $A \in \mathbf{B} \cup \mathbf{Z}$. We can find a match for q_w as follows.

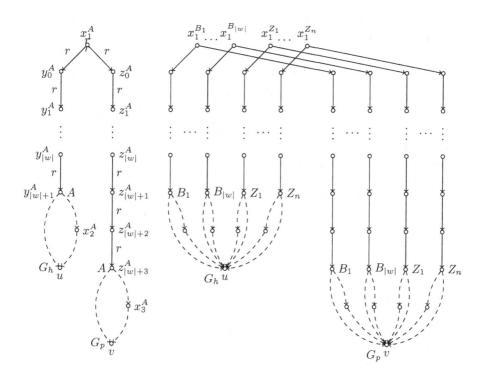

Fig. 3. The basic query $q(A, u, v)$ and the full query q_w

First we match u on the ℓ_h'' node of ℓ, which satisfies G_h, and v on the m_p'' node of m, which satisfies G_p. Consider an arbitrary $A \in \mathbf{B} \cup \mathbf{Z}$. We distinguish two cases:

- If (c1) applies, we match $y_{|w|+1}^A$ on the ℓ_h node of ℓ, x_2^A on the ℓ_h' node of ℓ, $z_{|w|+3}^A$ on m_p node of m, and x_3^A on m_p' node of m.
- Otherwise, if (c2) applies, we match $y_{|w|+1}^A$ on the ℓ_h' node of ℓ and $z_{|w|+3}^A$ on m_p' node of m. In this case, the matches for x_2^A and x_3^A become irrelevant.

The matches of all other variables are then uniquely determined by conjunctions of atoms $r(z, z')$, in such a way that x_1^A will be matched to the root of the configuration node of ℓ in the latter case, and to its parent in the former. As the z_i^A chains are exactly two r-arcs longer than the y_i^A chains, m must be a leaf in a configuration node that follows that of ℓ. We can argue analogously that every match for q_w, the variables u and v are respectively matched to the nodes ℓ_h'' and m_p'' of a pair ℓ, m of cells of two successive configurations that are A-conspicuous for all $A \in \mathbf{B} \cup \mathbf{Z}$.

In this way we obtain that, for every computation tree $\mathcal{I}_c$, we have $\mathcal{I}_c$ is proper iff $\mathcal{I}_c \models q_w$. This ends the reduction to PQ entailment in $\mathcal{ALC}$.

Note that converting q_w into a union of CQs (i.e., into DNF) results in an exponentially larger formula. This blow-up may be unavoidable. In fact, for CQs, it has been shown that query entailment is feasible in EXPTIME for $\mathcal{ALCH}$ and $\mathcal{ALCHQ}$, and even for $\mathcal{SH}$ if suitable restrictions on the occurrences of transitive roles in the queries are imposed [12,9,6]. It follows from these results that there is no CQ whose size is polynomial in w and $\mathcal{M}$ that can test for properness of computation trees.

$(\circ, \cup)$**-Queries.** Defining the query q_w as a $(\circ, \cup)$-query is straightforward. We only need to replace in $q(A, u, v)$ the disjunction $r(y_{|w|+1}^A, u) \vee \left(r(y_{|w|+1}^A, x_2^A) \wedge r(x_2^A, u) \right)$ by the atom $\left(r \cup (r \circ r) \right)(y_{|w|+1}^A, u)$, and the disjunction $r(z_{|w|+3}^A, v) \vee \left(r(z_{|w|+3}^A, x_3^A) \wedge r(x_3^A, v) \right)$ by the atom $\left(r \cup (r \circ r) \right)(z_{|w|+3}^A, v)$ (and we can drop the variables x_2^A and x_3^A). Then we can define q_w as above, as the conjunction of the (modified) $q(A, u, v)$ for all $A \in \mathbf{B} \cup \mathbf{Z}$. A match for this modified q_w in a computation tree is a match for the positive query above, and vice versa.

An alternative $(\circ, \cup)$-query for testing properness is obtained by replacing in each $q(A, u, v)$ the sequence of atoms $r(x_1^A, y_0^A) \wedge r(y_0^A, y_1^A) \wedge \cdots \wedge r(y_{|w|}^A, y_{|w|+1}^A)$ by a single atom $r \circ \cdots \circ r(x_1^A, y_{|w|+1}^A)$ for a chain $r \circ \cdots \circ r$ of length $|w| + 2$, and the sequence $r(x_1^A, z_0^A) \wedge r(z_0^A, z_1^A) \wedge \cdots \wedge r(z_{|w|}^A, z_{|w|+3}^A)$ by $r \circ \cdots \circ r(x_1^A, z_{|w|+3}^A)$ for a chain of length $|w| + 4$; note that we can get rid of all but one variable y_i^A, and all but one z_i^A. Using the *test* constructor A? for a name A sometimes allowed in CRPQs (with semantics $A?^{\mathcal{I}} = \{e, e \mid e \in A^{\mathcal{I}}\}$) we can even replace in $q(A, u, v)$ the whole sequence of atoms from x_1^A to u by a single atom, and the whole sequence from x_1^A to v by another atom, using only one variable x^A additionally to u and v. However, the number of variables in the resulting q_w still depends linearly on $|w|$ and $\mathcal{M}$.

Using the inverse roles allowed in C2RPQs, we can even go one step further and write the whole query $q'(A, u, v)$ as one single atom with variables u and v:

$$q'(A, u, v) = G_h? \circ (r^- \cup r^- \circ r^-) \circ A? \circ \underbrace{r^- \circ \cdots \circ r^-}_{|w|+2 \text{ times}} \circ \underbrace{r \circ \cdots \circ r}_{|w|+4 \text{ times}} \circ A? \circ \left(r \cup (r \circ r) \right)(u, v)$$

The conjunction of these queries also gives a query q_w that correctly tests properness, but using only two variables. We note that in C2RPQs the tests A? add no expressive power, as they can be simulated by adding a axiom $A \sqsubseteq \exists r_A$ to the KB for a fresh role r_A, and replacing A? by $r_A \circ r_A^-$. Summing up, we obtain:

Theorem 1. *Query entailment in $\mathcal{ALC}$ is 2ExpTime-hard for:*

1. *positive queries,*
2. *any extension of CQs that allows for atoms of the form $\left(r \cup (r \circ r) \right)(z, z')$ for a role name r and variables z, z',*
3. *the class of $*$-free CRPQs, and*
4. *the class of $*$-free C2RPQs with only two variables.*

We note that for PQs and CRPQs (with no inverses), it is not clear whether the reduction can be done using a bounded number of variables. The same holds for the lower bounds for $\mathcal{ALCI}$ and $\mathcal{SH}$ [9,5]. In contrast, CQ entailment in $\mathcal{SHI}$ is 2ExpTime-hard already for queries with only two variables [7], similarly to C2RPQs in $\mathcal{ALC}$.

4   Conclusions

We have seen that query answering in $\mathcal{ALC}$ and its extensions becomes 2ExpTime-hard even for rather restricted settings. However, once this 2ExpTime-hard boundary has been crossed, one can significantly extend both the DL and the query language without an additional increase in complexity. Query entailment remains in 2EXPTIME even for *positive* 2-way regular path queries, which extend all the query languages mentioned above, and for $\mathcal{ZIQ}$, $\mathcal{ZOQ}$ ad $\mathcal{ZOI}$, which respectively extend the well known $\mathcal{SHIQ}$, $\mathcal{SHOQ}$ and $\mathcal{SHOI}$ [4,3].

In this paper we have focused on identifying query languages for which query entailment in $\mathcal{ALC}$ is 2EXPTIME-hard. It would also be interesting to study which are the minimal DL constructs needed to show 2EXPTIME-hardness, similarly as done in [2], but trying to avoid the combined use of role hierarchies and inverse roles. In line with aforementioned paper, it is worth remarking that the presence of disjunction is crucial. Indeed, even C2RPQs in (disjunction-free) Horn-$\mathcal{SHOIQ}$ can be answered in single exponential time [11]. For a more detailed discussion of the topic, and references to other related results, the reader may refer to [13].

References

1. Bienvenu, M., Calvanese, D., Ortiz, M., Šimkus, M.: Nested regular path queries in description logics. In: Proc. of KR 2014. AAAI Press (2014)
2. Bourhis, P., Morak, M., Pieris, A.: The impact of disjunction on query answering under guarded-based existential rules. In: Proc. of IJCAI 2013, pp. 796–802. IJCAI/AAAI (2013)
3. Calvanese, D., Eiter, T., Ortiz, M.: Regular path queries in expressive description logics with nominals. In: Proc. of IJCAI 2009, pp. 714–720 (2009)
4. Calvanese, D., Eiter, T., Ortiz, M.: Answering regular path queries in expressive description logics via alternating tree-automata. Information and Computation 237, 12–55 (2014)
5. Eiter, T., Lutz, C., Ortiz, M., Šimkus, M.: Query answering in description logics with transitive roles. In: Proc. of IJCAI 2009, pp. 759–764 (2009)
6. Eiter, T., Ortiz, M., Šimkus, M.: Conjunctive query answering in the Description Logic $\mathcal{SH}$ using knots. Journal of Computer and System Sciences 78(1), 47–85 (2012)
7. Glimm, B., Kazakov, Y.: Role conjunctions in expressive description logics. In: Cervesato, I., Veith, H., Voronkov, A. (eds.) LPAR 2008. LNCS (LNAI), vol. 5330, pp. 391–405. Springer, Heidelberg (2008)
8. Glimm, B., Lutz, C., Horrocks, I., Sattler, U.: Conjunctive query answering for the description logic SHIQ. J. Artif. Intell. Res. (JAIR) 31, 157–204 (2008)
9. Lutz, C.: The complexity of conjunctive query answering in expressive description logics. In: Armando, A., Baumgartner, P., Dowek, G. (eds.) IJCAR 2008. LNCS (LNAI), vol. 5195, pp. 179–193. Springer, Heidelberg (2008)
10. Motik, B., Sattler, U., Studer, R.: Query answering for OWL-DL with rules. J. Web Sem. 3(1), 41–60 (2005)
11. Ortiz, M., Rudolph, S., Šimkus, M.: Query answering in the Horn fragments of the description logics $\mathcal{SHOIQ}$ and $\mathcal{SROIQ}$. In: Proc. of IJCAI 2011, pp. 1039–1044 (2011)
12. Ortiz, M., Šimkus, M., Eiter, T.: Worst-case optimal conjunctive query answering for an expressive description logic without inverses. In: Proc. of AAAI 2008, pp. 504–510 (2008)
13. Ortiz, M., Šimkus, M.: Reasoning and query answering in Description Logics. In: Eiter, T., Krennwallner, T. (eds.) Reasoning Web 2012. LNCS, vol. 7487, pp. 1–53. Springer, Heidelberg (2012)

An Ontology for Container Terminal Operations

Luca Pulina

POLCOMING, Università degli Studi di Sassari
Viale Mancini 5 – 07100 Sassari – Italy
lpulina@uniss.it

Abstract. In this paper we describe the DESCTOP ontology for container terminal operations. It is a semantic-based representation of the main functions of a container terminal, in order to formally represent the operational flows which the terminal operators should be supported in.

1 Context and Motivation

A Container Terminal (CT) is a place where containers – also known as, Intermodal Transport Units (ITUs) – are transshipped between diverse transport vehicles, e.g., vessels and trucks. As reported in [7] (among others), a CT is mainly composed of three interacting systems: a land-side system, aimed to manage ITUs arrive and depart by trucks; a yard storage and handling system, to govern the (temporary) ITU storage and transfer between quayside and land-side; and a quayside system, to deal with the management of berths and quay cranes when a vessel arrives.

A container terminal is a very complex environment: terminal operators must take several key decisions to cope with different scheduling and planning issues, involving different equipments and their interaction. The main target of such decisions is to improve both the efficiency and the effectiveness of the container terminal with respect to some well-established performance indicators. In order to deal with such critical decisions, terminal operators can leverage on Decision Support Systems (DSSs, see, e.g, [14]), computer-based information systems aimed to support decision-making activities (see, e.g., [11] for DSSs in the context of container terminal).

This paper presents an ontology for the semantic-based representation of the operations in a container terminal, focusing on the operational flows related to the yard storage and the quayside system. This ontology here described aims to be the conceptual layer of the knowledge-based DSS designed in the context of the DESCTOP (DEcision Support for Container Terminal OPerations) project[1], which is finalized to design a DSS to improve the efficiency of the CT operations at the Cagliari International Container Terminal (CICT, http://www.cict.it). We can report several success stories about the usage of ontologies to support decision-making in the context of intermodal logistics – see, e.g., [5] –, as well

---

[1] http://visionlab.uniss.it/desctop

R. Kontchakov and M.-L. Mugnier (Eds.): RR 2014, LNCS 8741, pp. 224–229, 2014.

Fig. 1. Container terminal equipment: (a) a Quay Crane (serving a vessel); (b) the container Storage Yard; and (c) a Rubber-Tyred Gantry crane

as in other application domains, e.g., in the clinical one [2,12] and in systems diagnostics [6].

The work here presented builds on and extend part of the work presented in [3]. We describe the DESCTOP ontology in Section 2 and we conclude the paper in Section 3 with some final remarks.

2   The DESCTOP Ontology

In this section we describe the DESCTOP ontology, which represents the knowledge layer of the DSS investigated in the project. In particular, the usage of an ontology allows us to design an Ontology-Based Data Access (OBDA) information system, in order to effectively accomplish tasks such as:

- Monitoring the terminal informations by means of the computation of critical Key Performance Indicators (KPIs).
- Information search and retrieval for reporting purpose.
- Discover new relevant data that could be used for the optimization of terminal operations.

By means of the DESCTOP ontology we can represent all the necessary knowledge to monitor the activities related to both yard storage and quayside systems. In a few words, ITUs are stored in the storage yard, in rectangular regions called blocks, and they can be stacked one on top of the other. There are two main typologies of operations in a container terminal, namely unloading and loading. Considering the first, ITUs arrive at the container terminal via vessels. They are unloaded from the vessel to an internal truck whereby a Quay Crane (QC). Such truck will shuttle the ITU to a Rubber-Tyred Gantry (RTG) crane that will pick up the ITU from the truck and place it into the assigned position in the storage yard. In the case of a loading operation the flow is in the other way round. In Figure 1 we show some container terminal equipment.

The information system leverages the conceptual layer represented by the DESCTOP ontology allowing for an operator to monitor several kinds of KPIs, involving different components of the terminal, e.g., trucks, ITUs, vessels, yard cranes. Examples of monitored KPIs are quay crane rate – i.e., the relationship between the total amount of ITUs moved from a vessel with respect to the quay crane working time – and the average waiting time of a truck in order to be loaded in a given period. These KPIs are implemented as SPARQL 1.1 queries.

About the ontology language, our choice fall to OWL2 QL – see [10] – because it guarantees that (*i*) conjunctive query answering can be implemented using efficient algorithms with respect to the size of data, and (*ii*) the consistency of the ontology can be evaluated using efficient algorithms with respect to the size of the ontology itself. Furthermore, because of OWL2 QL falls in a Description Logics (DLs) of the DL-Lite family, it enables to implement the OBDA approach called, in the words of [9], *OBDA with databases*, that allows a reduction of conjunctive queries over ontologies to first-order queries over standard relational databases [4,13,1,8].

In Figure 2 we present a graphical outline of the TBox of the DESCTOP ontology. Looking at the figure, we pinpoint the following classes:

Driver is the class representing the equipment drivers.

ITU denotes the goods packaged in containers. It has relationships with classes representing type, size, and weight of a container.

ManagedVessel is the vessel subject of ITUs loading/unloading. This class is also in relationship with classes representing line, sail direction, and services related to a vessel.

OperationType represents the kind of operation involving ITUs (load, unload, move).

QC is devoted to Quay Cranes representation.

QCMove aims at monitoring Quay Crane activities (and compute related KPIs).

RTG represents Rubber-Tyred Gantry cranes.

RTGMove is devoted to monitoring RTG activities (and compute related KPIs).

SYBlock models the position of a ITU in the Storage Yard. SYBlock is the domain of data properties related to columns, lane, block, and height in the storage yard.

Truck is the internal truck representation.

TruckTransport is devoted to monitor internal trucks activities (and compute related KPIs).

WorkingDay denotes a working day, represented by both a day and a shift.

Concerning the object properties, we describe in the following the ones related to QCMove as domain. The object properties related to the remaining equipment are organized in a similar way and the can be found in detail in the full documentation of the DESCTOP ontology.

QCInteractsWithTruck allows a relation with the class Truck, in order to keep trace which truck has been used to load/unload during a move.

QCInteractsWithVessel allows a relation with the class ManagedVessel, in order to record which QC was working during a load/unload operation.

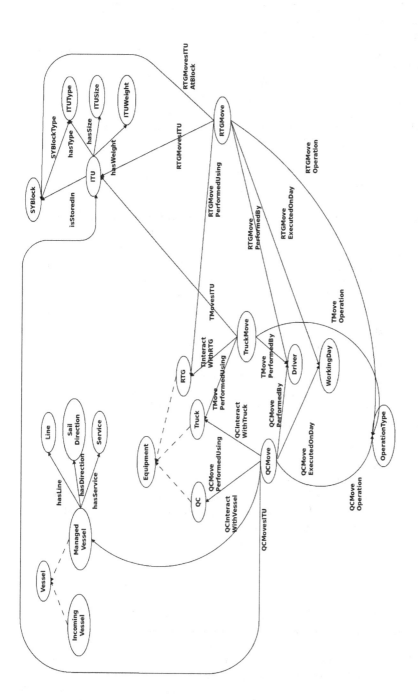

Fig. 2. The DESCTOP ontology. Ellipses denote concepts (classes) and pointed edges denote class attributes. Directed edges represent roles, and they are labeled with their names. Dotted edges represent concept inclusion.

`QCMoveExecutedOnDay` relates to `WorkingDay`, to take into account day and shift of a given QC move.

`QCMoveOperation` relates to `OperationType`.

`QCMovePerformedBy` relates to `Driver`, in order to allow the computation of KPIs related to the drivers productivity.

`QCMovePerformedUsing` specifies the QC used for the considered move.

`QCMovesITU` relates a QC move with a specific ITU.

The OWL file related to the DESCTOP ontology is available at `http://visionlab.uniss.it/desctop/rr14`.

3   Conclusions

This paper described an ontology for the representation of container terminal operations. The full documentation of the DESCTOP ontology is available at the website: `http://visionlab.uniss.it/desctop/ontodesctop`.

In order to implement the OBDA with databases approach on the DESCTOP information system, it has been also developed the related Relational Database schema. It is available for download at: `http://visionlab.uniss.it/desctop/rr14`, while the full documentation of the PostgreSQL Database schema implementation is available at: `http://visionlab.uniss.it/desctop/desctopdb.html`.

To conclude, we are currently working on a simulator based on Operational Research techniques aimed to produce realistic raw data to populate the ABox. The resulting data will be made available as a benchmark for OBDA systems. Finally, we are designing a more expressive version of the DESCTOP ontology, including, e.g., cardinality constraints.

Acknowledgments. The author wish to thank the anonymous reviewers for their comments and suggestions to improve the quality of the paper. This work is partially supported by Regione Autonoma della Sardegna e Autorità Portuale di Cagliari con L.R. 7/2007, Tender 16 2011, CRP-49656 "Metodi innovativi per il supporto alle decisioni riguardanti l'ottimizzazione delle attività in un terminal container", the DESCTOP project (`http://visionlab.uniss.it/desctop`).

References

1. Artale, A., Calvanese, D., Kontchakov, R., Zakharyaschev, M.: The DL-Lite family and relations. Journal of Artificial Intelligence Research 36, 1–69 (2009)
2. Bouamrane, M.-M., Rector, A., Hurrell, M.: Using OWL ontologies for adaptive patient information modelling and preoperative clinical decision support. Knowledge and Information Systems 29(2), 405–418 (2011)
3. Bourguet, J.-R., Cicala, G., Pulina, L., Tacchella, A.: OBDA and Intermodal Logistics: Active Projects and Applications. In: Faber, W., Lembo, D. (eds.) RR 2013. LNCS, vol. 7994, pp. 210–215. Springer, Heidelberg (2013)

4. Calvanese, D., De Giacomo, G., Lembo, D., Lenzerini, M., Rosati, R.: Tractable Reasoning and Efficient Query Answering in Description Logics: The *DL-Lite* family. Journal of Automated Reasoning 39(3), 385–429 (2007)
5. Casu, M., Cicala, G., Tacchella, A.: Ontology-based data access: An application to intermodal logistics. Information Systems Frontiers (2012)
6. Cicala, G., Oreggia, M., Tacchella, A.: Towards an ontology-based framework to generate diagnostic decision support systems. In: Baldoni, M., Baroglio, C., Boella, G., Micalizio, R. (eds.) AI*IA 2013. LNCS, vol. 8249, pp. 25–36. Springer, Heidelberg (2013)
7. Huynh, N., Walton, M., Davis, J.: Finding the number of yard cranes needed to achieve desired truck turn time at marine container terminals. Transportation Research Record: Journal of the Transportation Research Board 1873(1), 99–108 (2004)
8. Kontchakov, R., Lutz, C., Toman, D., Wolter, F., Zakharyaschev, M.: The Combined Approach to Query Answering in DL-Lite. In: Principles of Knowledge Representation and Reasoning: Proceedings of the Twelfth International Conference, KR 2010, Toronto, Ontario, Canada, May 9-13, AAAI Press (2010)
9. Kontchakov, R., Rodríguez-Muro, M., Zakharyaschev, M.: Ontology-based data access with databases: A short course. In: Rudolph, S., Gottlob, G., Horrocks, I., van Harmelen, F. (eds.) Reasoning Weg 2013. LNCS, vol. 8067, pp. 194–229. Springer, Heidelberg (2013)
10. Motik, B., Patel-Schneider, P.F., Parsia, B., Bock, C., Fokoue, A., Haase, P., Hoekstra, R., Horrocks, I., Ruttenberg, A., Sattler, U., et al.: OWL 2 Web Ontology Language: Structural Specification and Functional-Style Syntax. W3C Recommendation 27 (2009)
11. Murty, K.G., Liu, J.: Y.-W. Wan, R. Linn. A decision support system for operations in a container terminal. Decision Support Systems 39(3), 309–332 (2005)
12. Musen, M.A., Shahar, Y., Shortliffe, E.H.: Clinical decision-support systems. In: Biomedical Informatics, Health Informatics, pp. 698–736. Springer, New York (2006)
13. Poggi, A., Lembo, D., Calvanese, D., De Giacomo, G., Lenzerini, M., Rosati, R.: Linking data to ontologies. Journal on Data Semantics X 10, 133–173 (2008)
14. Turban, E., Aronson, J., Liang, T.-P.: Decision Support Systems and Intelligent Systems. Pearson Prentice Hall (2005)

Hydrowl: A Hybrid Query Answering System for OWL 2 DL Ontologies

Giorgos Stoilos

School of Electrical and Computer Engineering
National Technical University of Athens, Greece

Abstract. This system description paper introduces the OWL 2 query answering system Hydrowl. Hydrowl is based on novel *hybrid* techniques which in order to compute the query answers combine at run-time a reasoner ans_1 supporting a (tractable) fragment of OWL 2 (e.g., OWL 2 QL and OWL 2 RL) with a fully-fledged OWL 2 DL reasoner ans_2. The motivation is that if most of the (query answering) work is delegated to the (usually) very scalable system ans_1 while the interaction with ans_2 is kept to a bare minimum, then we can possibly provide with scalable query answering even over expressive fragments of OWL 2 DL. We discuss the system's architecture and we present an overview of the techniques used. Finally, we present some first encouraging experimental results.

1 Introduction

Conjunctive query (CQ) answering over ontological knowledge expressed in the OWL 2 DL language has attracted the interest of many researchers as well as application developers the last decade [1, 2]. Unfortunately, query answering over OWL 2 DL ontologies is of very high computational complexity [3, 4] and even after modern optimisations and intense implementation efforts [5] OWL 2 DL systems are still not able to cope with datasets containing billions of data.

The need for efficient query answering has motivated the development of several fragments of OWL 2 DL [6], like OWL 2 EL, OWL 2 QL, and OWL 2 RL for which query answering can be implemented (at-most) in polynomial time with respect to the size of the data. Consequently, for many of these languages there already exist highly scalable systems which have been applied successfully to industrial-strength applications, like OWLim [1] and Oracle's RDF Semantic Graph [7]. The attractive properties of these systems have led application developers to use them even in cases where the input ontology is expressed in the far more expressive OWL 2 DL language. Clearly, in such cases these systems would most likely be *incomplete*—that is, for some user query and dataset they will fail to compute all certain answers. However, techniques that attempt to deliver complete query answering even when using scalable systems that are not complete for OWL 2 DL have also been proposed [8, 9].

In this system description paper we present the architecture and main characteristics of the Hydrowl[1] query answering system. Hydrowl supports

---

[1] http://www.image.ece.ntua.gr/~gstoil/hydrowl/

R. Kontchakov and M.-L. Mugnier (Eds.): RR 2014, LNCS 8741, pp. 230–238, 2014.
© Springer International Publishing Switzerland 2014

expressive ontology languages and like the latter mentioned approaches it is based on novel techniques which attempt to use scalable but possibly incomplete systems as much as possible in order to achieve favourable performance. All inferences involving "problematic" constructors of OWL 2 DL that these systems do not support (e.g., existential restrictions and disjunctions) are either explicated ("materialised") at a pre-processing step as additional axioms or are restricted as much as possible during on-line query evaluation. Our first experimental evaluation using Hydrowl has provided with many encouraging results several of which we report in Section 4.

We have so far tried to keep the design of Hydrowl as modular as possible. Hence, we feel that one of its interesting features is that any application developer can plug his/her system into Hydrowl with minimum implementation effort and, moreover, many different combinations of systems are possible. This is also supported by the fact that we have already provided implementations in Hydrowl that combine the OWL 2 RL reasoner OWLim [1] with HermiT [10], with HermiT-BGP [5], and with Rapid [11].

2   Techniques Used in Hydrowl

In the current section we briefly outline the techniques used in Hydrowl.

As mentioned above, for a given OWL 2 DL ontology $\mathcal{T}$, dataset $\mathcal{A}$, and query $\mathcal{Q}$, Hydrowl still tries to use as much as possible an incomplete but scalable system ans to compute the certain answers of $\mathcal{Q}$ over $\mathcal{T} \cup \mathcal{A}$. Clearly, in that case, there can be entailments of $\mathcal{T} \cup \mathcal{A}$ related to $\mathcal{Q}$ that ans will miss. To recover such missing inferences Hydrowl follows two different but not necessarily incompatible approaches.

In the first approach, inferences involving unsupported constructors are explicated ("materialised") in a form that ans can eventually "recognise". For example, let ans be an OWL 2 RL system and let $\mathcal{T} = \{A \sqsubseteq \exists R, \exists R \sqsubseteq B\}$. Since ans is an OWL 2 RL system it cannot handle axioms with existential restrictions in the right hand side. Hence, Hydrowl will compute for ans a new set of axioms $\mathcal{R}$ that will contain the axiom $A \sqsubseteq B$, i.e., it will materialise the entailment $\mathcal{T} \models A \sqsubseteq B$. It can be verified that when ans is applied over $\mathcal{T} \cup \mathcal{R}$ it is able to return all answers to *every* ground query and *every* datasets over $\mathcal{T}$—that is, for every ground CQ $\mathcal{Q}$ and $\mathcal{A}$ we have $\mathsf{cert}(\mathcal{Q}, \mathcal{T} \cup \mathcal{A}) \subseteq \mathsf{ans}(\mathcal{Q}, \mathcal{T} \cup \mathcal{R} \cup \mathcal{A})$.[2] Such a set of axioms $\mathcal{R}$ is called the *repair* of $\mathcal{T}$ for ans [8, 12] and the process of computing it *repairing*.

However, repairing captures only ground entailments hence even after repairing ans is still incomplete for queries containing existential variables (which in the following we call non-SPARQL queries). To also support such queries we need to further materialise the non-ground inferences that are related to the existential variables of the query. Hydrowl accomplishes this by combining ans with a second system ans′ which can explicate such type of information. One such

---

[2] Note that this allows ans to be unsound, i.e., return wrong answers. However, to the best of our knowledge, the vast majority of OWL 2 RL systems are sound.

family of systems are query rewriting systems which take as input a (possibly non-SPARQL) query Q and a TBox T and compute a so-called query rewriting Rew [13–15]. Roughly speaking, a rewriting Rew for Q, T consists of two parts, a set of datalog rules Rew_D which captures ground entailments of T and a union of conjunctive queries Rew_Q which captures all inferences related to non-ground entailments. Consequently, for $\text{Rew}_Q \uplus \text{Rew}_D$ a rewriting for a non-SPARQL CQ Q over a TBox T we have that $\text{cert}(Q, T \cup A) \subseteq \text{ans}(\text{Rew}_Q, T \cup R \cup A)$ for every dataset A. Summarising, this approach of Hydrowl to query answering follows the following three steps:

1. Compute a repair R of T for ans.
2. Load the dataset A, the input TBox T, and the repair R to ans.
3. For a CQ Q, if Q is SPARQL then directly evaluate it over ans; otherwise compute a rewriting $\text{Rew}_D \uplus \text{Rew}_Q$ for Q, T and evaluate Rew_Q over ans.

Note that steps 1 and 2 are usually required to be performed only once as a pre-processing (changes in A can be handled incrementally). Moreover, note that the important component both in computing R as well as in computing Rew_Q is a query rewriting system [8, 12]. Hence, we call this approach rewriting-based query answering.

Interestingly, by recent theoretical results [16–18], it follows that for systems complete for OWL 2 RL repairs always exists for ontologies expressed in Horn-$\mathcal{SHIQ}$ (a fairly expressive fragment of OWL 2) and they might also exist even for arbitrary OWL 2 DL ontologies. Unfortunately, there can be ontologies where a repair does not exist. The second approach followed by Hydrowl can work even if no repair has been pre-computed at all (although some best effort "partial" repair can be assumed). Instead, for an ontology T and a system ans, Hydrowl first constructs the set U of atomic (concept and role) queries over T for which ans is complete, called *query base* of ans for T. Then, for an arbitrary query Q, the query base can be used to efficiently determine if ans is complete for Q (and any dataset A) or not [19]; in the former case, Hydrowl uses ans to evaluate Q, while in the latter it resorts to a fully-fledged OWL 2 DL reasoner ans'. There are three benefits here. First, it is trivial to see that in theory query bases always exist.[3] Second, ans is expected to be mostly complete (T usually includes few "problematic" constructors) leaving few queries that need to be evaluated using the fully-fledged OWL 2 DL system. Third, it has been shown [19] that even in cases that Hydrowl needs to use ans', the scalable system can still be exploited in order to speed up the evaluation by ans'. We call this approach of Hydrowl hybrid query answering and is summarised by the following three steps:

1. Load the dataset A and the input TBox T to ans.
2. Compute a query base U of ans for T.
3. For a (ground) CQ Q, if it can be determined using U that ans is complete for Q then directly evaluate Q using ans; otherwise evaluate Q using a fully-fledged OWL 2 DL reasoner ans' together with ans.

---

[3] Note, however, that there are limitation in automatically extracting them.

Note that indeed the two previous approaches are not incompatible. For example, before computing a query base one can pre-compute some partial repair $\mathcal{R}$ and also load it to ans. Then clearly, the more the partial $\mathcal{R}$ approximates *the* (full) repair the smaller the query base of ans for $\mathcal{T}$ is. Moreover, a query rewriting system can also be used in Step 3 of the hybrid query answering approach in order to capture some non-ground entailments and hence provide some support for non-SPARQL CQs, however, this has not been forulated and implemented yet.

3   Architecture of Hydrowl

Hydrowl is implemented in JAVA and is available under the AGPL license. Figure 1 presents its main components, where JAVA classes are marked as rectangles, procedures by ovals and data-flow (TBox) with light-blue arrows. The two approaches outlined in the previous section are implemented by the two main classes HybridEvaluator and RewritingBasedEvaluator. As illustrated, to implement these approaches the methods use internally an incomplete and a complete reasoner and this communication is performed through interfaces, namely IncompleteReasoner and CompleteReasoner in the case of HybridEvaluator and IncompleteReasoner and QueryRewritingSystem in the case of RewritingBasedEvaluator. HybridEvaluator is using an additional componenent called (Q,T)-CompletenessChecker with which it decides whether the user query can be evaluated using the incomplete reasoner or the complete one needs to be employed.

The use of interfaces makes the architecture highly modular as by implementing them one can readily use the query answering techniques of Hydrowl

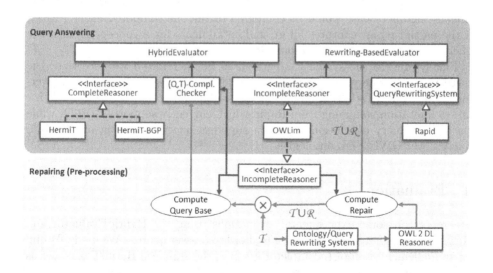

Fig. 1. Main components of Hydrowl

with their reasoner of choice. So far we have provided implementations of CompleteReasoner using the standard HermiT reasoner [10] and HermiT-BGP [5], an implementation of QueryRewritingSystem using Rapid [11], and an implementation of IncompleteReasoner using OWLim, however, we envision that OWL 2 DL systems such as Pellet, query rewriting systems such as Ontop and Clipper, as well as OWL 2 RL systems and triple stores such as Apache Jena, RDFox, and Stardog can be easily integrated.

For both approaches to work, a repair $\mathcal{R}$ or a query base $\mathcal{U}$ for the incomplete system ans used in query answering should have been computed previously at a pre-processing step. This can be done using the repairing package of Hydrowl. Internally, this package uses a rewriting system, an OWL 2 DL reasoner and the system ans with which it again communicates through an interface. The rewriting system is used to produce an initial repair (i.e., some first materialisation of the ground inferences of $\mathcal{T}$) while both the OWL 2 DL reasoner and ans are used as minimisation steps to produce the final repair $\mathcal{R}$. The existing implementation uses again Rapid and HermiT but the choice of these systems compared to the ones used as implementations of CompleteReasoner and QueryRewritingSystem is irrelevant (i.e., one can use completely different systems during query answering). However, clearly, the incomplete reasoner used in this step to produce $\mathcal{R}$ or $\mathcal{U}$ must be the same (or at least equivalent in expressivity) as the one used during query answering.

Subsequently, the query base is used by (Q,T)-CompletenessChecker to determine if the incomplete reasoner is (in)complete for the given user query while the repair needs to be loaded to QueryRewritingSystem in order for the incomplete system to be complete for all ground entailments over the input ontology and data. As mentioned in the previous section, in the hybrid query answering approach one can also compute and load some partial (or even full) repair to the incomplete reasoner. In that case, the "more complete" the partial repair the smaller the query base would be compared to the one computed using $\mathcal{T}$ (e.g., if the partial repair captures all ground entailments for a concept A then the query base won't contain the atomic query :-$A(x)$).

Finally, due to the systems integrated so far in Hydrowl we note that, the rewriting-based query answering approach supports repairing and query answering (of arbitrary CQs) over ontologies expressed in the $\mathcal{ELHI}$ fragment of OWL 2 DL (a limitation stemming from Rapid which currently supports $\mathcal{ELHI}$), while the hybrid query evaluation approach supports query answering of SPARQL queries over OWL 2 DL ontologies.

4 Evaluation

We report on some experimental evaluations using the HybridEvaluator and RewritingBasedEvaluator classes of Hydrowl to answer queries. We used OWLim as an implementation of IncompleteReasoner, the standard HermiT reasoner as an implementation of CompleteReasoner and Rapid as an implementation of QueryRewritingSystem.

Table 1. Query Answering Times

Query	LUBM				UOBM					
	1	3	8	9	3	4	9	11	12	14
HermiT-BGP	2.5	1.4	1.4	105	204	5.8	21.6	1.7	1.2	48.7
Hydrowl	.07	.07	.24	.13	.02	.01	.01	.07	.04	35.3

4.1 Hybrid Query Answering

First, using the repairing package of Hydrowl we computed query bases for
OWLim for ontologies LUBM and UOBM. For LUBM we required 14.5 seconds while for UOBM we required 48.7 seconds. Some additional manual editing
was required for UOBM in order to remove the atomic queries :-Woman(x) and
:-PeopleWithManyHobbies(x) from the query base as OWLim is incomplete for
them but this is not recognised automatically by the Compute QueryBase procedure of Hydrowl. This is because this procedure is based on a computation of a
repair using Rapid which currenlty only supports $\mathcal{ELHI}$ while these queries require reasoning over disjunctions and functional number restrictions. Second, we
used HybridEvaluator in order to evaluate all the test queries of LUBM (we used
5 universities) and of UOBM (we used 1 department) and we compared against
the HermiT-BGP system [5]. Table 1 presents the results (in seconds) for all the
interesting queries (for the rest both systems have similar response times). In
grey colour we have marked those queries where Hydrowl uses both OWLim and
HermiT. As can be seen, in all queries the hybrid query answering approach of
Hydrowl is faster than HermiT-BGP. In some cases the difference is quite significant and this is even in the cases where Hydrowl uses both HermiT and OWLim.
It is worth noting query 3 over UOBM which requires non-deterministic reasoning. HermiT-BGP non-deterministicaly checks whether many individuals are
instances of the class Student while Hydrowl using both HermiT and OWLim
manages to restrict this search space to only a few individuals. Similarly, evaluating query 14 requires non-deterministic reasoning over the class Woman.

4.2 Repairing-Based Query Answering

First, we wanted to evaluate whether repairs for large and complex ontologies
can be computed efficiently in practice. Using the repairing package of Hydrowl
we managed to compute repairs for 151 out of the 152 ontologies of our dataset.
In the vast majority of cases a repair could be computed in less than a few
minutes (usually within seconds) and only for the very large ones we required
several minutes; Table 2 presents results for the latter. Despite their size and
complexity we see that we can compute repairs for them in less than 1 hour
which, given that this usually occurs once, we feel is a reasonable amount of
time. Actually, if we discard a very expensive minimisation step of repairing,

Table 2. $|\mathcal{T}|$ ($|\mathcal{R}|$): number of axioms of the input TBox (repair), t: time in seconds

$\mathcal{T}$	$|\mathcal{T}|$	$|\mathcal{R}|$	t	$\mathcal{T}$	$|\mathcal{T}|$	$|\mathcal{R}|$	t
Not-Galen	5471	3015(4153)	298(42)	Galen-doc	4229	6051(6176)	1152(28)
Fly	19845	10361(12368)	2884(178)	Galen	4229	3012(3062)	257(24)

Table 3. Loading times for Fly and UOBM for the various ABoxes

	Universities				
	1	2	5	10	20
UOBM	4.1	6.8	16.2	31.9	73.2
UOBM∪$\mathcal{R}$	4.4	8.3	24.3	44.9	108.1

(a) UOBM

	$\mathcal{A}$	$2 \times \mathcal{A}$	$3 \times \mathcal{A}$	$4 \times \mathcal{A}$	$5 \times \mathcal{A}$
Fly	14.0	21.9	22.7	27.9	31.5
Fly∪$\mathcal{R}$	31.9	55.1	68.5	93.0	119.3
Fly∪$\mathcal{R}^-$	33.2	62.1	70.1	100.6	118.2

(b) Fly

then we can compute some (non-minimal) repair very efficiently while its size is not considerably larger than the minimal one (see Table 2 numbers in brackets).

Next, we loaded the repairs we computed for UOBM and Fly into OWLim; Table 3 presents loading times of the original ontology with (i) data of various sizes (for UOBM we used 1 to 20 universities and for Fly we multiplied the original ABox up to 5 times) and (ii) with and without the computed repairs. As can be seen, the overhead introduced by additionally loading the repair ($\mathcal{R}$) is significant only in the Fly ontology, mostly due to its size, however, note that loading is also usually performed only once. In Fly we have also loaded the non-minimal repair ($\mathcal{R}^-$) and as it turns out there is no significant difference compared to the minimal one (recall that computing it is much more efficient).

Table 4. Results for Answering the Fly Queries

$\mathcal{Q}_1$		$\mathcal{Q}_2$		$\mathcal{Q}_4$		$\mathcal{Q}_5$	
t_{Rapid}	t_{OWLim}	t_{Rapid}	t_{OWLim}	t_{Rapid}	t_{OWLim}	t_{Rapid}	t_{OWLim}
0.31	0.31	0.90	1.28	0.07	0.04	0.05	0.02

Finally, we have used RewritingBasedEvaluator to answer all 4 non-SPARQL queries of Fly (using the original ABox); Table 4 presents the results where t_{Rapid} is the time required by Rapid and t_{OWLim} the time required by OWLim (total time is their sum). As we can see in most cases we were able to compute and evaluate a rewriting almost instantaneously. The good behaviour of Hydrowl can be attributed to the fact that most hard work is pushed to a pre-processing step that is materialising all ground entailments into the repair and explicating them by loading the ontology, the repair, and the data into OWLim.

Acknowledgements. The work was funded by a Marie Curie Career Reintegration Grant within European Union's 7th Framework Programme (FP7/2007-2013) under REA grant agreement 303914.

References

1. Kiryakov, A., Bishoa, B., Ognyanoff, D., Peikov, I., Tashev, Z., Velkov, R.: The Features of BigOWLIM that Enabled the BBC's World Cup Website. In: Workshop on Semantic Data Management (SemData) (2010)
2. Motik, B., Horrocks, I., Kim, S.M.: Delta-Reasoner: A Semantic Web Reasoner for an Intelligent Mobile Platform. In: Proceedings of the 21st International World Wide Web Conference (WWW 2012), pp. 63–72 (2012)
3. Ortiz, M., Calvanese, D., Eiter, T.: Data complexity of query answering in expressive description logics via tableaux. Journal of Automated Reasoning 41(1), 61–98 (2008)
4. Glimm, B., Lutz, C., Horrocks, I., Sattler, U.: Conjunctive query answering for the description logic $\mathcal{SHIQ}$. Journal of Artificial Intelligence Research (JAIR) 31, 157–204 (2008)
5. Kollia, I., Glimm, B.: Optimizing sparql query answering over owl ontologies. J. Artif. Intell. Res. (JAIR) 48, 253–303 (2013)
6. Motik, B., Cuenca Grau, B., Horrocks, I., Wu, Z., Fokoue, A., Lutz, C. (eds.): OWL 2 Web Ontology Language Profiles. W3C Recommendation (2009)
7. Wu, Z., Eadon, G., Das, S., Chong, E.I., Kolovski, V., Annamalai, M., Srinivasan, J.: Implementing an inference engine for RDFS/OWL constructs and user-defined rules in oracle. In: Proc. of ICDE, pp. 1239–1248. IEEE (2008)
8. Stoilos, G., Cuenca Grau, B., Motik, B., Horrocks, I.: Repairing ontologies for incomplete reasoners. In: Aroyo, L., Welty, C., Alani, H., Taylor, J., Bernstein, A., Kagal, L., Noy, N., Blomqvist, E. (eds.) ISWC 2011, Part I. LNCS, vol. 7031, pp. 681–696. Springer, Heidelberg (2011)
9. Zhou, Y., Nenov, Y., Cuenca Grau, B., Horrocks, I.: Complete query answering over horn ontologies using a triple store. In: Alani, H., et al. (eds.) ISWC 2013, Part I. LNCS, vol. 8218, pp. 720–736. Springer, Heidelberg (2013)
10. Glimm, B., Horrocks, I., Motik, B., Stoilos, G., Wang, Z.: Hermit: An owl 2 reasoner. Journal of Automated Reasoning (JAR) (in Press, 2014)
11. Trivela, D., Stoilos, G., Chortaras, A., Stamou, G.: Optimising resolution-based rewriting algorithms for dl ontologies. In: Proceedings of the 26th Workshop on Description Logics (DL 2013), Ulm, Germany (2013)
12. Stoilos, G.: Ontology-based data access using rewriting, OWL 2 RL systems and repairing. In: Presutti, V., d'Amato, C., Gandon, F., d'Aquin, M., Staab, S., Tordai, A. (eds.) ESWC 2014. LNCS, vol. 8465, pp. 317–332. Springer, Heidelberg (2014)
13. Calvanese, D., De Giacomo, G., Lembo, D., Lenzerini, M., Rosati, R.: Tractable reasoning and efficient query answering in description logics: The DL-Lite family. Journal of Automated Reasoning 39(3), 385–429 (2007)
14. Artale, A., Calvanese, D., Kontchakov, R., Zakharyaschev, M.: The dl-lite family and relations. Journal of Artificial Intelligence Research (JAIR) 36, 1–69 (2009)
15. Pérez-Urbina, H., Motik, B., Horrocks, I.: Tractable query answering and rewriting under description logic constraints. Journal of Applied Logic 8(2), 186–209 (2010)

16. Eiter, T., Ortiz, M., Simkus, M., Tran, T.K., Xiao, G.: Query rewriting for Horn-$\mathcal{SHIQ}$ plus rules. In: Proc. of AAAI (2012)
17. Cuenca Grau, B., Motik, B., Stoilos, G., Horrocks, I.: Computing datalog rewritings beyond horn ontologies. In: Proceedings of the Twenty-Third International Joint Conference on Artificial Intelligence (IJCAI 2013) (2013)
18. Kaminski, M., Nenov, Y., Grau, B.C.: Datalog rewritability of disjunctive datalog programs and its applications to ontology reasoning. In: Proceedings of AAAI 2014 (2014)
19. Stoilos, G., Stamou, G.: Hybrid query answering for owl ontologies. In: Proceedings of the 21st European Conference on Artificial Intelligence (ECAI 2014) (2014)

Ontology-Based Answer Extraction Method

Hajer Baazaoui-Zghal and Ghada Besbes

Riadi, ENSI Campus Universitaire de la Manouba, Tunis, Tunisia
hajer.baazaouizghal@riadi.rnu.tn, ghada.besbes@gmail.com

Abstract. Question Answering (QA) systems aim at providing answers to Natural Language questions in an open domain context and can provide a solution to the problem of response accuracy. This paper describes an answer extraction method based on ontologies. Our goal consists on performing an efficient question answering by extracting answers based on the question graph, lexico-syntactic patterns and score computation.

Keywords: Answer extraction, ontology, lexico-syntactic patterns.

1 Ontology-Based Answer Extraction Method

Question Answering (QA) systems are considered as advanced information retrieval systems, allowing the user to ask a question in Natural Language (NL) and returning the precise answer instead of a set of documents. The aim of this paper is to design, implement and experiment a new answer extraction method based on ontologies. The proposed system relies on two main components: question analysis component and answer extraction component. The goal is to perform an efficient similar question search and a detailed question analysis [1] in order to obtain useful information for the answer extraction. First, the user submits a question in NL to obtain the reformulated question, submitted to a search engine (on the web or on a document collection) and relevant phrases are extracted from the retrieved documents. Using the questions typed attributed graph, the system extracts relevant passages and then relevant answers containing the questions subject. If exists an answer that satisfies the pattern reformulated question, the answer will be directly returned to the user (6) and stored in the question base (7), otherwise, a score computation will take place for each answer (8) and the ordered list of potential answers will be displayed to the user along with their scores (9). This feedback gives the opportunity for answer validation (10) and the update of the patterns stored in the question base (12).

The score measure relies on two criteria: the rank of the document from which the candidate answer was extracted and the similarity between the answer and the reformulated question. The similarity used in this score computation is the overall similarity used in the similar question search component. In fact, it is the average between semantic similarity (calculated using WordNet distance) and statistic similarity. It is calculated using Vector Space Model [2] and dynamically defined vectors representing both candidate answer and reformulated question. A stop words elimination and a lemmatization processes are previously applied, the vectors are built and a cosine similarity gives the final results.

R. Kontchakov and M.-L. Mugnier (Eds.): RR 2014, LNCS 8741, pp. 239–240, 2014.

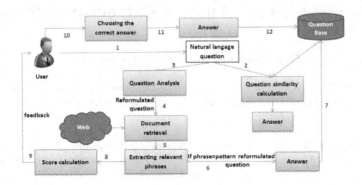

Fig. 1. General system's architecture

2 Similar Question Search and Answer Extraction Evaluation

During the experimentations, we calculate the similarities between the user's question and each question extracted from the question base after filtering. We extract the most similar questions to the user's question and we return an ordered set of answers. For performance evaluation, we use the measures: (1) Success at n (S@n), which means the percentage of queries for which we return the correct similar question in the top n (1, 2, 5, and 10) returned results, (2) Mean Reciprocal Rank (MRR) calculated over all tested questions and (3) Precision which is the proportion of the number of correct answers to the number of returned answers.

As example experiments for s@1=64% shows that the correct answer is extracted from the first rank. The precision results show a good performance with 0,53.

3 Conclusion and Future Work

This paper presents a new answer extraction method based on ontologies. Our contribution can be summarized in extracting answers from documents automatically. Experiments were conducted and showed an improvement of the precision, success and MRR of the extracted results. As perspectives, we plan to integrate modular ontologies and Case-based Reasonning to the QA proposed system.

References

1. Besbes, G., Baazaoui-Zghal, H., Moreno, A.: Ontology-based question analysis method. In: Larsen, H.L., Martin-Bautista, M.J., Vila, M.A., Andreasen, T., Christiansen, H. (eds.) FQAS 2013. LNCS, vol. 8132, pp. 100–111. Springer, Heidelberg (2013)
2. Salton, G., Wong, A., Yang, C.S.: A vector space model for automatic indexing. Commun. ACM 18(11), 613–620 (1975)

Collective, Incremental Ontology Alignment Through Query Translation

Thomas Kowark and Hasso Plattner

Hasso Plattner Institute
August-Bebel-Str. 88, 14482 Potsdam, Germany
{firstname.lastname}@hpi.de

1 Introduction

In software repository analysis, researchers are confronted with the problem of having to perform similar analyses on heterogeneous data sources, e.g., in order to verify their findings on data from different projects. Complete data translation is often not feasible due to the size of the underlying repositories and the complexity of the translation process. From an effort-benefit-ratio point of view, query translation is a more efficient choice.

To be carried out automatically, a translation requires an alignment between the source and target data sources, i.e., the ontologies describing them. Though the necessary matching tasks can be partially automated, progress in this field is currently slowing down [3]. One of the proposed additions that could further improve the completeness of alignments is user involvment. Current matching tools limit the involvement mainly to providing yes/no answers on suggested correspondences. On the contrary, Ellis et al. presented an approach that integrates the matching into an information retrieval task that provides immediate benefits for the users [1]. We follow their core idea and integrate ontology matching into the task of query translation. The direct output of this task is a translated query that users can readily execute on the target data source. Simultaneously, the steps taken during translation allow the extraction of an alignment between the underlying ontologies. These alignments are reused in future translation tasks so users only have to match elements without existing correspondence. To collectivize the matching efforts, we implemented our approach as a platform called *RepMine*, where users can benefit from and share the resulting alignments.

2 Query Translation and Alignment Extraction

As shown in Figure 1, a query parser first creates a conceptual, graph-based representation of an input query issued on repository R1 [2]. This graph contains the entities touched by a query, their relations amongst each other, as well as attribute constraints. Supported by automated matching tools, users transform this query graph (QG_1) into an output graph (QG_1') that reflects an equivalent query on the output repository. By transforming the graph, users simultaneously share their knowledge about correspondences between ontology elements.

R. Kontchakov and M.-L. Mugnier (Eds.): RR 2014, LNCS 8741, pp. 241–242, 2014.

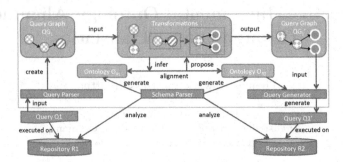

Fig. 1. Concept of using query translations for simultaneous ontology matching

Two cases have to be distinguished: Re-labeling and re-structuring of the query graph. If labels are exchanged (e.g., 'Developer' ⇔ 'Programmer'), equivalence of the concepts is assumed. Complex label substitutions express subsumptions (e.g., 'ProjectMember' ⇔ 'Developer ∪ Manager'). If re-labeling does not suffice, restructuring has to take place. It provides instructions for instance construction, e.g., a 'hasEdited' relation between a user and a file can be assumed, if a file revision that belongs to the file has been created by the user. Alternatively, it can also indicate subsumptions, e.g., if a 'Bug' translates to an 'Issue' with a 'Label' named 'bug', we can infer that Bug ⊂ Issue.

Our system uses existing alignments to support subsequent query translations. Firstly, only elements without existing correspondences have to be translated. Secondly, OWL inference is used to determine transitive alignments. Given a third repository R3 and alignments between O_{R1} and O_{R2}, as well as O_{R2} and O_{R3}, translation of queries between O_{R1} and O_{R3} is possible without requiring further manual transformation input.

A challenge in this setup are conflicting alignments resulting from query graph transformations. We aim to reduce these issues by alerting users if their query transformations would invalidate existing correspondences and requiring them to resolve the conflicts before accepting the changes, e.g., by deleting one alignment. Our system is currently based on a simple rule set. In future work, we will formalize the alignment extraction process and test it by recreating previous software repository mining studies on different repositories by means of query translation and using the resulting alignment for data transformation.

References

1. Ellis, J.B., Hassanzadeh, O., Srinivas, K., Ward, M.J.: Collective ontology alignment. In: Proceedings of the Ontology Matching Workshop (2013)
2. Kowark, T., Uflacker, M., Zeier, A.: Towards a Shared Platform for Virtual Collaboration Monitoring in Design Research. In: Plattner, H., Meinel, C., Leifer, L. (eds.) Design Thinking - Envisioning Co-Creation. Springer (2011)
3. Shvaiko, P., Euzenat, J.: Ontology matching: State of the art and future challenges. IEEE Trans. on Knowl. and Data Eng. 25(1), 158–176 (2013), http://dx.doi.org/10.1109/TKDE.2011.253

Disjunctive Constraints in RDF and Their Application to Context Schemas

Emanuele Rabosio[1], Álvaro Cortés-Calabuig[2], and Jan Paredaens[2]

[1] Politecnico di Milano, Italy
emanuele.rabosio@polimi.it
[2] University of Antwerp, Belgium
{alvaro.cortes,jan.paredaens}@ua.ac.be

1 Introduction

RDF is a data model whose relevance is growing in the last years. Recently, some proposals have enriched the model with integrity constraints, well known within relational databases. In this work we extend an existing framework [1,3] with two new types of integrity constraints of disjunctive nature, inspired by similar kinds of dependencies studied for the relational model. The problem of the logical implication for the two novel categories is also analyzed. Moreover, as an application scenario, we propose a complete and independent set of constraints to model the context in RDF, where the context is a notion employed in databases to perform information filtering on the basis of the user's current situation.

2 Disjunctive Constraints for RDF

In the following we consider two pairwise disjoint sets U and V. The former is a recursive enumerable infinite set of *URIs*, and the latter a recursive enumerable infinite set of *variables*; variables are denoted prefixing their names with \$. A *term* is either a URI or a variable. An *RDF-graph* $\mathcal{G}$ is a finite set of triples (s, p, o), subject, property, object, $s, p, o \in U$. An embedding of a finite set S of triples of terms in an RDF-graph $\mathcal{G}$ is a total function $e : V_S \cup U \to U$, such that: (a) $e(u) = u$ for each $u \in U$, (b) if $(t_1, t_2, t_3) \in S$ then $(e(t_1), e(t_2), e(t_3)) \in \mathcal{G}$. Given an RDF graph $\mathcal{G}$ and a constraint $\mathcal{C}$, the notation $\mathcal{G} \models \mathcal{C}$ indicates that $\mathcal{G}$ satisfies $\mathcal{C}$. The definitions of the two novel categories of constraints follow:

Definition 1 (DEGC). *A disjunctive equality generating constraint (DEGC) is a pair* $(S, \{E_1, \ldots, E_n\})$, *where: S is a finite set of triples of terms, and $E_1 \ldots E_n$ are finite sets of equalities in the form $(t_1 = t_2)$, with $t_1, t_2 \in V_S \cup U$.*

Definition 2 (DEGC Satisfaction). *A graph $\mathcal{G}$ satisfies the DEGC $(S, \{E_1, \ldots, E_n\})$ iff for every embedding e of S in $\mathcal{G}$, there exists $E_i \in \{E_1, \ldots, E_n\}$ such that for every $(t_1 = t_2) \in E_i$ holds that $e(t_1) = e(t_2)$.*

Example 1. Consider $\mathcal{C}_1 = DEGC(\{(\$x, a, \$y)\}, \{\{(\$x = b), (\$y = c)\}, \{(\$x = d), (\$y = e)\}\})$. $\{(f, g, h)\} \models \mathcal{C}_1$, $\{(b, a, c)\} \models \mathcal{C}_1$, $\{(d, a, e)\} \models \mathcal{C}_1$, $\{(b, a, e)\} \not\models \mathcal{C}_1$.

R. Kontchakov and M.-L. Mugnier (Eds.): RR 2014, LNCS 8741, pp. 243–244, 2014.
© Springer International Publishing Switzerland 2014

Definition 3 (DTGC). *A disjunctive triple generating constraint (DTGC) is a pair $(S, \{S_1, \ldots, S_n\})$ where S, $S_1, \ldots, S_n$ are finite sets of triples of terms.*

Definition 4 (DTGC Satisfaction). *An RDF-graph $\mathcal{G}$ satisfies the DTGC $(S, \{S_1, \ldots, S_n\})$ if for every embedding e of S in $\mathcal{G}$ there exist $S_i \in \{S_1, \ldots, S_n\}$ and an embedding e' of $S \cup S_i$ in $\mathcal{G}$ such that for all $(t_1, t_2, t_3) \in S$ holds $(e(t_1), e(t_2), e(t_3)) = (e'(t_1), e'(t_2), e'(t_3))$.*

Example 2. Consider $\mathcal{C}_2 = DTGC(\{(\$x, a, b)\}, \{\{(\$x, a, c)\}, \{(\$x, d, \$y)\}\})$. $\{(e, a, f)\} \vDash \mathcal{C}_2$, $\{(e, a, b), (e, d, g)\} \vDash \mathcal{C}_2$, $\{(e, a, b), (e, a, c)\} \vDash \mathcal{C}_2$, $\{(e, a, b)\} \nvDash \mathcal{C}_2$.

SPARQL ASK queries can be employed to check whether an RDF graph satisfies a given DEGC or DTGC.

One of the most important questions studied within relational constraints is *logical implication*. The problem is now analyzed in the RDF framework for the new classes of constraints introduced above. Let $\mathcal{SC}$ be a finite set of constraints and $\mathcal{C}$ be a constraint. As in [3], $\mathcal{C}$ is a logical consequence of $\mathcal{SC}$ iff for all graphs $\mathcal{G}$ holds $(\forall \mathcal{C}' \in \mathcal{SC}(\mathcal{G} \vDash \mathcal{C}')) \Rightarrow \mathcal{G} \vDash \mathcal{C}$. We study whether a single constraint of one of the two new types is a logical consequence of a set of constraints of the same type. Our results are summarized by the following theorems, proven in [4]:

Theorem 1. *The implication problem for DEGCs is decidable.*

Theorem 2. *The implication problem for DTGCs is undecidable.*

As an application, DEGCs and DTGCs are employed with other types of constraints to define a complete and independent set of constraints for an RDFS representation of context schemas. Context schemas [2] are trees with nodes of two kinds (black and white), used to describe the available contexts in a given scenario. Here we just provide two sample constraints on our representation: *1) the only allowed classes are Node and its subclasses WhiteNode, BlackNode, Root, Leaf: DEGC(\{(\$x, type, \$y)\}, \{\{(\$y=Node)\}, \{(\$y=WhiteNode)\}, \{(\$y=BlackNode)\}, \{(\$y=Root)\}, \{(\$y=Leaf)\}\}); 2) each non-root node has at least one parent: DTGC(\{(\$x, type, Node)\}, \{\{(\$x, type, Root)\}, \{(\$x, childOf, \$y)\}\}).*

Acknowledgments. E. Rabosio is partially supported by the Politecnico di Milano Polisocial Award 2013 project ObiGame.

References

1. Akhtar, W., Cortés-Calabuig, Á., Paredaens, J.: Constraints in RDF. In: Schewe, K.-D. (ed.) SDKB 2010. LNCS, vol. 6834, pp. 23–39. Springer, Heidelberg (2011)
2. Bolchini, C., Quintarelli, E., Tanca, L.: Carve: Context-aware automatic view definition over relational databases. Information Systems 38(1), 45–67 (2013)
3. Cortés-Calabuig, A., Paredaens, J.: Semantics of constraints in RDFS. In: Proc. of AMW. CEUR-WS.org (2012)
4. Rabosio, E., Cortés-Calabuig, A., Paredaens, J.: Appendix to "Disjunctive constraints in RDF and their application to context schemas",
http://home.deib.polimi.it/rabosio/Papers/appendixRCP2014.pdf

Linked Open Data in the Earth Observation Domain: The Vision of Project LEO

Manolis Koubarakis[1], Charalampos Nikolaou[1], George Garbis[1],
Konstantina Bereta[1], Panayiotis Smeros[1], Stella Gianakopoulou[1],
Kallirroi Dogani[1], Maria Karpathiotaki[1], Ioannis Vlachopoulos[1],
Dimitrianos Savva[1], Kostis Kyzirakos[2], Stefan Manegold[2], Bernard Valentin[3],
Nicolas James[3], Heike Bach[4], Fabian Niggemann[4], Philipp Klug[4],
Wolfgang Angermair[5], and Stefan Burgstaller[5]

[1] National and Kapodistrian University of Athens, Greece
{koubarak,charnik,ggarbis,konstantina.bereta,psmeros,sgian,
kallirroi,mkarpat,johnvl,dimis}@di.uoa.gr
[2] Centrum Wiskunde & Informatica, The Netherlands
{kostis.kyzirakos,stefan.manegold}@cwi.nl
[3] Space Applications Services, Belgium
{bernard.valentin,nicolas.james}@spaceapplications.com
[4] VISTA Geowissenschaftliche Fernerkundung, Germany
{bach,niggemann, klug}@vista-geo.de
[5] PC-Agrar Informations und Beratungsdienst, Germany
{Angermair,Burgstaller}@eurosoft.de

Lots of Earth Observation (EO) data has become available at no charge in Europe and the US recently and there is a strong push for *more open EO data*. *Linked data* is a new data paradigm which studies how one can make RDF data available on the Web, and interconnect it with other data with the aim of increasing its value. In the last few years, linked *geospatial* data has received attention as researchers and practitioners have started tapping the wealth of geospatial information available on the Web. As a result, the *linked open data (LOD) cloud* has been rapidly populated with geospatial data some of it describing EO products (e.g., CORINE Land Cover and Urban Atlas published by project TELEIOS). The abundance of this data can prove useful to the new satellite missions (e.g., Sentinels) as a means to increase the usability of the millions of images and EO products that are expected to be produced by these missions.

However, open EO data that are currently made available by space agencies such as ESA and NASA are *not* following the linked data paradigm. Therefore, from the perspective of a user, the EO data and other kinds of geospatial data necessary to satisfy his or her information need can only be found in different data silos, where each silo may contain only part of the needed data. *Opening up these silos* by publishing their contents as RDF and interlinking them with semantic connections will allow the development of data analytics applications with great environmental and financial value.

Our earlier project TELEIOS (http://www.earthobservatory.eu/) concentrated on developing data models, query languages, scalable query evaluation techniques, and efficient data management systems that can be used to

R. Kontchakov and M.-L. Mugnier (Eds.): RR 2014, LNCS 8741, pp. 245–246, 2014.

prototype applications of linked EO data. However, developing a methodology and related software tools that support the whole life-cycle of linked open EO data (e.g., publishing, interlinking etc.) has *not* been tackled by this project. The main objective of the European project "Linked Open Earth Observation Data for Precision Farming" (LEO) presented in this paper is to go beyond TELEIOS by designing and implementing software supporting *the life-cycle of linked open EO data* and its combination with linked geospatial data, and by developing a precision farming application that heavily utilizes such data.

The scientific and technical objectives of LEO can be briefly described as follows:

1. To specify the whole life-cycle of linked open EO data and auxiliary geospatial data (e.g., maps, meteorological data) that are typically made available by public bodies and utilized in EO applications (e.g., precision farming) and publish the developed tools as an infrastructure that can be easily used by data publishers and application developers.
2. To design and implement an extraction and transformation tool that takes as input vector or raster EO data and open geospatial data and their metadata available in some well-known format (e.g., a shapefile), transforms it into RDF and makes it available on the LOD cloud.
3. To develop concepts, techniques and tools that will allow data publishers to discover geospatial, temporal and similarity relations among open EO data and other open geospatial data and metadata. The developed linking tool will be an extension of the well-known tool Silk (`http://silk.wbsg.de/`) which currently does not support the discovery of such kind of relations.
4. To develop tools for (i) cross-platform searching over linked EO metadata using keywords expressing a user information need, time predicates and spatial predicates, and (ii) a tool for browsing and visualizing time evolving linked geospatial data and the creation, sharing, and collaborative editing of 'temporally-enriched' thematic maps which are produced by combining different sources of such data and other geospatial information available in standard OGC file formats (e.g., KML). The latter tool will be an extension of the tool Sextant (`http://sextant.di.uoa.gr/`) developed in TELEIOS which will be re-developed for mobile platforms (tablets and smartphones).
5. To demonstrate the value of the developed tools by (i) performing large-scale publication and linking of open EO data from the GMES Space Component Data Access warehouse managed by ESA, and geospatial datasets made available by other public bodies in Europe, and (ii) developing a precision farming application that shows how geo-information services based on linked open EO data, linked geospatial data and specialized algorithms can contribute to an environmentally friendly increase in the efficiency of agricultural production.

Acknowledgements. This work has been funded by the FP7 project LEO (611141) (`http://linkedeodata.eu/`).

Combining Fuzzy and Probabilistic Reasoning for Crowd-Sourced Categorization and Tagging

Tanel Tammet[1] and Ago Luberg[2]

[1] Tallinn University of Technology, Ehitajate tee 5, Tallinn 19086 Estonia
[2] Eliko Competence Center, Mäealuse 2/1, Tallinn 12618 Estonia

Abstract. The paper presents a novel method for extending automated reasoners for first order logic with both fuzzy and probabilistic reasoning. We encode probabilities in object logic and avoid meta-logic altogether, thus simplifying the task of extending existing provers. The paper focuses on achieving high efficiency while tackling probabilistic reasoning tasks. The motivation and examples for the presented method stem from the practical experience of calculating object categories and tags from the large amounts of crowdsourced tourism data harvested, extracted and aggregated from the web.

Keywords: crowdsourced categorization, tagging, web harvesting.

1 Introduction

The context of the proposed method is building a world-wide database of the sightseeing popularity of concrete places (POI-s) and wider areas in the world, using purely crowd-sourced data. By sightseeing popularity we mean the estimate of a number of people visiting the place and considering it as an interesting place for sightseeing, as opposed to very popular places with no or little potential for sightseeing, like hospitals, schools, gas stations, bus stops and airports. The sightseeing popularity database we build is used by our Sightsmap system [6] (see also http://sightsmap.com) for showing a zoomable and pannable touristic popularity heatmap for any area in the world as an overlay on the standard Google maps.

The problem targeted by the current paper is writing rules for combining different kinds of crowd-sourced category information into a set of properly weighted tags usable for beforementioned purposes. Experience has shown that attempts to combine a fuzzy measure with a probability-based measure into a single "confidence" score is satisfiable only in very simple cases. When we start to consider cumulating evidence, the principal differences between fuzzy logic and probabilities create inconsistencies.

It is worth noting that a significant number of different theories and methods for probabilistic reasoning have been developed, see [2], [5], but due to widely different requirements in actual tasks and real-world domains, no single leading approach has emerged. Several systems use probabilities or a fuzzy criteria to recommend the items. In article by [4] a similarity coefficient is offered to find

R. Kontchakov and M.-L. Mugnier (Eds.): RR 2014, LNCS 8741, pp. 247–248, 2014.

the best suitable point of interests for the given user. The similarity is calculated by the profile and the item vectors. The similarity is larger if the angle between the vectors is smaller. [1] have presented an uncertainty of situations based on the contextual conditions. [3] are creating rules by the recommendation process.

2   Proposed Rule System

The purpose of the current paper is to propose a rule system combining fuzzy and probabilistic logic in object logic. In other words, we do not propose a new logic but show a practical way to encode both fuzzy and probabilistic logic in classical first order reasoning systems. The encoding is suitable for presenting object categorization rules in the context of uncertain information and does not significantly increase the complexity of the derivation algorithms, i.e. does not slow down the provers.

The first principle of the approach is to avoid metalogic and encode as much as possible in classical first order logic.

The second principle is to avoid combining fuzzy and probabilistic measures into a common "confidence" measure, keeping them separated at all stages during the inference process.

Acknowledgments. This research has been supported by the European Regional Development Fund.

References

1. Ciaramella, A., Cimino, M.G., Lazzerini, B., Marcelloni, F.: Situation-aware mobile service recommendation with fuzzy logic and semantic web. In: International Conference on Intelligent Systems Design and Applications, pp. 1037–1042 (2009)
2. Fuhr, N.: Probabilistic datalog: Implementing logical information retrieval for advanced applications. Journal of the American Society for Information Science 51, 2000 (1999)
3. Golovin, N., Rahm, E.: Automatic optimization of web recommendations using feedback and ontology graphs. In: Lowe, D.G., Gaedke, M. (eds.) ICWE 2005. LNCS, vol. 3579, pp. 375–386. Springer, Heidelberg (2005)
4. Kang, E.-y., Kim, H., Cho, J.: Personalization method for tourist point of interest (poi) recommendation. In: Gabrys, B., Howlett, R.J., Jain, L.C. (eds.) KES 2006. LNCS (LNAI), vol. 4251, pp. 392–400. Springer, Heidelberg (2006)
5. Pearl, J.: Probabilistic Reasoning in Intelligent Systems: Networks of Plausible Inference. Morgan Kaufmann Publishers Inc., San Francisco (1988)
6. Tammet, T., Luberg, A., Järv, P.: Sightsmap: crowd-sourced popularity of the world places. In: ENTER 2013: Information and Communication Technologies in Tourism 2013. Springer (2013)

Visual Editor for Answer Set Programming: Preliminary Report

Barbara Nardi

Dipartimento di Matematica, Università della Calabria, 87030 Rende, Italy
b.nardi@mat.unical.it

1 Introduction

Answer Set Programming (ASP) [1, 2] is a declarative programming paradigm which has been proposed in the area of non-monotonic reasoning and logic programming. The applications of ASP range from classical scientific applications in the field of Artificial Intelligence [1] to real-world applications [3]. Although modeling in ASP is often not particularly difficult, the task of writing ASP programs is still mostly performed by expert programmers and is uncomfortable for novices. In order to facilitate the design of ASP applications, some editors and programming environment were proposed in the last few years [4–7]. However, the task of designing an ASP program consists of writing text files (more or less computer-assisted), and could be daunting to the beginners due to its "cryptic" syntax. A similar problem was solved by Query By Example (QBE) graphical interfaces [8] in the field of databases. QBE tools are nowadays well-recognized solutions to this issue (e.g., a QBE tool is the default in the user-oriented Microsoft Access). Following this idea, we have developed a new visual editor prototype conceived for easing the design of ASP programs that is preliminarily described in this paper. It is worth noting that existing programming tools for ASP do not provide satisfactory solutions in this respect. Indeed, existing graphical tools are either limited to modeling domain entities [7], or feature extensions of QBE interfaces, as in [6], that do not capture in an intuitive graphical form way the richer language of ASP. Moreover old visual logic languages [9] can serve only as a source of inspiration, since the main language constructs of ASP were not considered.

2 Visual Editor

The ASP programmer does not need to provide an algorithm for solving a problem; rather, he specifies the properties of the desired solution for its computation by means of a collection of logic rules called logic program. Consider as an example the well-known NP-complete problem 3-COLORING: given an undirected graph $G = (V, E)$, assign each vertex one of three colors – say, red, green, or blue – such that adjacent vertices always have distinct colors. 3-COLORING can be encoded in ASP as follows:

```
node(v). edge(i,j). ∀v ∈ V  ∀(i,j) ∈ E
col(X,red) v col(X,blue) v col(X,yellow) <- node(X).
<- edge(X,Y), col(X,C), col(Y,C).
```

R. Kontchakov and M.-L. Mugnier (Eds.): RR 2014, LNCS 8741, pp. 249–250, 2014.

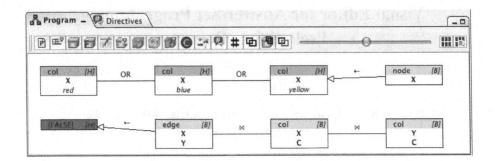

Fig. 1. 3-Colorability encoding in the new visual editor

The first line asserts facts representing the input graph G, the second line states that each vertex needs to have some color. The last line contains a constraint that disallows situations in which two connected vertices are associated with the same color. This logic program can be also represented in a graphical way and composed by a programmer by exploiting our new graphical environment.

In Figure 1 it is reported a snapshot of our tool in which the ASP encoding for the well-known 3-colorability problem is depicted in a new visual syntax. In particular, rules are represented by means of a graph in the new graphic language, where atoms are boxes and are connected by edges symbolizing logical connectives. Predicate arguments can be specified in apposite boxes, and are immediately visible. Joins between variables can be created by dragging and dropping variables.

The visual editor prototype is already integrated in the ASPIDE [6] environment, and it currently supports disjunctive ASP programs with constraints. As future work we will develop a visual syntax for advanced constructs, such as aggregates, and functions.

References

1. Baral, C.: Knowledge Representation, Reasoning and Declarative Problem Solving. Cambridge University Press (2003)
2. Gelfond, M., Lifschitz, V.: Classical Negation in Logic Programs and Disjunctive Databases. New Generation Computing 9, 365–385 (1991)
3. Grasso, G., Leone, N., Manna, M., Ricca, F.: Asp at work: Spin-off and applications of the dlv system. In: Balduccini, M., Son, T.C. (eds.) Logic Programming, Knowledge Representation, and Nonmonotonic Reasoning. LNCS, vol. 6565, pp. 432–451. Springer, Heidelberg (2011)
4. Perri, S., Ricca, F., Terracina, G., Cianni, D., Veltri, P.: An integrated graphic tool for developing and testing DLV programs. In: Proceedings of SEA 2007, pp. 86–100 (2007)
5. Sureshkumar, A., Vos, M.D., Brain, M., Fitch, J.: Ape: An ansprolog* environment. In: SEA 2007. Workshop Proceedings, vol. 281, pp. 101–115 (2007)
6. Febbraro, O., Reale, K., Ricca, F.: Aspide: Integrated development environment for answer set programming. In: Delgrande, J.P., Faber, W. (eds.) LPNMR 2011. LNCS, vol. 6645, pp. 317–330. Springer, Heidelberg (2011)
7. Busoniu, P.A., Oetsch, J., Pührer, J., Skocovsky, P., Tompits, H.: Sealion: An eclipse-based ide for asp with advanced debugging support. TPLP 13(4-5), 657–673 (2013)
8. Catarci, T., Costabile, M.F.: Visual query systems. J. Vis. Lang. Comput. 7(3), 243–245 (1996)
9. Agustí-Cullell, J., Puigsegur, J., Robertson, D.S.: A visual syntax for logic and logic programming. J. Vis. Lang. Comput. 9(4), 399–427 (1998)

Adaptive Stream Query Processing Approach for Linked Stream Data: (Extended Abstract)⋆

Zia Ush Shamszaman

Insight Centre for Data Analytics, National University of Ireland, Galway, Ireland
zia.shamszaman@insight-centre.org

1 Introduction and Motivation

Over the last few years, numerous efforts [1–4, 7] have been proposed based on SPARQL-like query languages on harvesting Linked Stream Data (LSD) processing in RDF and related formats. While each existing processor has advantages, neither of them wins in diverse settings. They differ on a wide range of aspects including the execution method, operational semantics, streaming operators and more. Considering state-of-the-art solutions, recent evaluations by [5, 6, 8] show that C-SPARQL [2] suffers from duplicate results for simple queries and misses some certain output in complex queries but provides more correct results than others. On the otherhand CQELS [7] performs better than others in terms of throughput and functionalities [6]. This diversity in output result is true for other processors including EP-SPARQL [1] and StreamingSPARQL [3].

2 Bridging the Gap between LSD Processing and Real Life Applications

In this research we investigate existing LSD processors to selectively combine the strength based on the applications requirements and data properties. The main goal is to bridge the gap between LSD query processing and real world applications, creating an adaptive layer which allows to react to changing requirements for better performance and quality of results. Our approach to achieve this goal is to consider differences and similarities of existing engines at a granular level and see how data properties and application requirements affects those dimensions. The initial set of dimensions I plan to consider for adaptability is originated by my initial analysis of state-of-the-art engines and their fine-grained characteristics. Such list of dimensions include query execution strategy, time model, abstract operators for the processing model, quality of service/quality of information, log management and privacy requirements. I also intend to contribute to RDF Stream Processing (RSP) in W3C community designing a standard model for LSD and make sure that it can support adaptive stream query processing.

⋆ This research has been partially supported by Science Foundation Ireland (SFI) under grant No. SFI/12/RC/2289 and EU FP7 CityPulse Project under grant No.603095. http://www.ict-citypulse.eu.

R. Kontchakov and M.-L. Mugnier (Eds.): RR 2014, LNCS 8741, pp. 251–252, 2014.

References

1. Anicic, D., Fodor, P., Rudolph, S., Stojanovic, N.: EP-SPARQL: a unified language for event processing and stream reasoning. In: Proc. of the 20th WWW Conference, pp. 635–644. ACM (2011)
2. Barbieri, D.F., Braga, D., Ceri, S., Della Valle, E., Grossniklaus, M.: C-sparql: Sparql for continuous querying. In: Proceedings of the 18th International Conference on World Wide Web, WWW 2009, pp. 1061–1062. ACM, New York (2009)
3. Bolles, A., Grawunder, M., Jacobi, J.: Streaming SPARQL - extending SPARQL to process data streams. In: Bechhofer, S., Hauswirth, M., Hoffmann, J., Koubarakis, M. (eds.) ESWC 2008. LNCS, vol. 5021, pp. 448–462. Springer, Heidelberg (2008)
4. Calbimonte, J.-P., Jeung, H., Corcho, O., Aberer, K.: Enabling query technologies for the semantic sensor web. International Journal On Semantic Web and Information Systems (IJSWIS) 8(1), 43–63 (2012)
5. Dell'Aglio, D., Calbimonte, J.-P., Balduini, M., Corcho, O., Della Valle, E.: On correctness in rdf stream processor benchmarking. In: Alani, H., Kagal, L., Fokoue, A., Groth, P., Biemann, C., Parreira, J.X., Aroyo, L., Noy, N., Welty, C., Janowicz, K. (eds.) ISWC 2013, Part II. LNCS, vol. 8219, pp. 326–342. Springer, Heidelberg (2013)
6. Le-Phuoc, D., Dao-Tran, M., Pham, M.-D., Boncz, P., Eiter, T., Fink, M.: Linked stream data processing engines: Facts and figures. In: Cudré-Mauroux, P., et al. (eds.) ISWC 2012, Part II. LNCS, vol. 7650, pp. 300–312. Springer, Heidelberg (2012)
7. Le-Phuoc, D., Dao-Tran, M., Xavier Parreira, J., Hauswirth, M.: A native and adaptive approach for unified processing of linked streams and linked data. In: Aroyo, L., Welty, C., Alani, H., Taylor, J., Bernstein, A., Kagal, L., Noy, N., Blomqvist, E. (eds.) ISWC 2011, Part I. LNCS, vol. 7031, pp. 370–388. Springer, Heidelberg (2011)
8. Zhang, Y., Duc, P.M., Corcho, O., Calbimonte, J.-P.: SRBench: A streaming RDF/SPARQL benchmark. In: Cudré-Mauroux, P., et al. (eds.) ISWC 2012, Part I. LNCS, vol. 7649, pp. 641–657. Springer, Heidelberg (2012)

Combining Logic and Business Rule Systems

Pieter Van Hertum

Department of Computer Science, KU Leuven
pieter.vanhertum@cs.kuleuven.be

1 Introduction and Motivation

To model complex knowledge, for solving knowledge intensive problems, one needs formalisms and systems that can represent this knowledge in a natural, compact and modular way. Business Rule Systems (BRS) are a widespread way of modelling knowledge intensive problems, with many applications in, among others, planning, supply chain management and expert systems. These systems model knowledge in the form of "if-then" rules "**if** body **then** head". This formalism is used in academic as well as industrial context, mostly because of its readability and the ease with which one can modify the behaviour of rules. Rich interface tools are available, like for example IBM's JRules and Drools. A second advantage of this formalism is the ease to reason with the rules in an automated way for which efficient algorithms exist.

From our point of view, the rigid rule based formulation of the knowledge has some major disadvantages. The knowledge that is modelled in a system, is strictly packed in these procedural style rules. These rules are equipped with a procedural style semantics. As a result, the only method of inference on these rules is execution (given information that occurs in the body of some of the rules, infer new information occurring in the heads of these rules). These limitations are not just an issue of the specific Business Rule System, but of the underlying rules and the ambiguity in their semantics. Take these two rules:

- **if** *Raining* **then** *Wet(Car)*
- **if** *Age(Person1) > Age(Person2)* **then** *Older(Person1, Person2)*

These two rules behave the same semantically, as long as forward reasoning (execution) is used. If it is raining, the car will be wet and if the age of the first person is larger than the age of the second one, we will call the first one older. We notice however that the intended semantics can be different if the knowledge is used in a different direction. Say that *Ann* is older than *Bob*, it can be derived that the age of *Ann* is a larger number than the age of *Bob*. If the car is wet on the other hand, it doesn't have to rain.

Alternatively, formalisms based on first order logic could be used to model business logic. They result in an increase in expressive power, but one might be concerned about the impact on the efficiency since an increase in expressivity can imply an increase in complexity. However, a lot of improvements on these computational tools have been made since these formalisms were developed.

R. Kontchakov and M.-L. Mugnier (Eds.): RR 2014, LNCS 8741, pp. 253–254, 2014.
© Springer International Publishing Switzerland 2014

Where once it was thought it was impossible to solve real life problems with first order logic, research now shows otherwise [1].

In what follows we propose two new ways of reasoning in the context of BRS that are going to be investigated in this research. The first new way of reasoning, is called **Exploration of specification**. When developing a rule set for a Business Rule System, an important part is testing the system for correctness. To do this properly, a user should be able to ask questions like "Given partial information X, is result Y ever possible?".

Another inference we want to study is **Generation of explanations**. A rule system is in many contexts used for decision problems. In many cases, it is very interesting for a user to know why a certain decision is made. This is not only useful for debugging purposes, this can also be a rich feature in the day to day use of such a system.

Another aspect of this research is solving the problems a rule system has when reasoning with incomplete information. In many real life applications, large amounts of data can be relevant to the problem at hand. However, sometimes, a small fraction of that data can be enough to take a decision. In the current state of the art, a business rule system is not capable of reasoning with incomplete information in a general way.

2   Completed and Future Work

In our research group, the reasoning tool IDP is developed, based on FO($\cdot$), a set of conservative extensions of first order logic [2]. We believe this tool is perfectly suited for addressing the issues mentioned above since (1) the FO($\cdot$) logic is specifically developed for handling incomplete knowledge ; (2) the IDP system can be easily extended with new inferences and language constructs; (3) multiple concepts and constructs that are needed are already incorporated in the IDP system.

A feasibility study of modelling Business Rule knowledge in FO($\cdot$) has been done. In [3] it is also studied what features the IDP system has that extend the current BR systems and what features are still missing in the IDP system to correctly formalize BR knowledge. The next stage of the research, the development of the new language constructs from the ICLP publication is followed by the development of the new inferences, discussed above.

References

1. Bruynooghe, M., Blockeel, H., Bart, B., De Cat, B.: Predicate logic as a modeling language: Modeling and solving some machine learning and data mining problems with idp3. CoRR abs/1309.6883 (2013)
2. De Cat, B., Bogaerts, B., Bruynooghe, M., Denecker, M.: Predicate logic as a modelling language: The IDP system. CoRR abs/1401.6312 (2014)
3. Van Hertum, P., Vennekens, J., Bogaerts, B., Devriendt, J., Denecker, M.: The effects of buying a new car: an extension of the IDP knowledge base system. TPLP 13(4-5-Online-Supplement) (2013)

Author Index